Data Driven Strategies: Theory and Applications

Wang Jianhong

School of Engineering and Sciences
Tecnologico de Monterrey, Monterrey, Mexico

Ricardo A. Ramirez-Mendoza

School of Engineering and Sciences
Tecnologico de Monterrey, Monterrey, Mexico

Ruben Morales-Menendez

School of Engineering and Sciences
Tecnologico de Monterrey, Monterrey, Mexico

CRC Press
Taylor & Francis Group
Boca Raton London New York

CRC Press is an imprint of the
Taylor & Francis Group, an **informa** business

First edition published 2023
by CRC Press
6000 Broken Sound Parkway NW, Suite 300, Boca Raton, FL 33487-2742

and by CRC Press
4 Park Square, Milton Park, Abingdon, Oxon, OX14 4RN

CRC Press is an imprint of Taylor & Francis Group, LLC

Library of Congress Cataloging-in-Publication Data (applied for)

ISBN: 978-0-367-74659-9 (hbk)
ISBN: 978-0-367-75008-4 (pbk)
ISBN: 978-1-003-16070-0 (ebk)

DOI: 10.1201/9781003160700

Typeset in Times New Roman
by Innovative Processors

Acknowledgments

We are grateful to a number of outstanding people who we have the opportunity to meet, learn from and work with during our PhD, postdoctoral and whose support and encouragement have made this exciting and challenging journey possible.

Tecnologico de Monterrey is one top university all over the world. There we can calm down to our teaching and research about advanced control theory and engineering. Every day we always have coffee and talk about our new ideas about the same topic. Due to the nice scenes and mountains, they often give us some inspirations. Without these natural sceneries and support from Tecnologico de Monterrey, we could not do any new directions for later research.

Jiangxi University of Science and Technology is one of the best places the first author could have hoped for doing his teaching and research, and he acknowledges the support from Jiangxi University of Science and Technology, which is an exceptional lab and research environment with amazing and interesting people.

Finally, our special thanks go to our family and friends, without them we would not have made it here. Thanks for understanding that time was often scarce and for always encouraging us. We want to thank our parents and families for their love and endless support, thanks for always being there for us.

After all, finishing this book is not the end of a journey, but the beginning of a much bigger one. In the next years, we will try our best to combine the functional analysis and geometry into system identification and advanced control theory. Maybe we can open some new directions in these research fields.

Wang Jianhong
Ricardo A. Ramirez-Mendoza
Ruben Morales-Menendez

Preface

One of the motivations behind this book was to collect together the many results of the direct data driven strategy from two different points, i.e. data driven identification and data driven control. For this reason, we have, rather unashamedly, included a number of ideas that were developed at Tecnologico de Monterrey and in this sense some of the discussions in the book are included as background material that some readers may wish to skip on an initial reading.

This book includes all of our recent contributions about direct data driven strategy for control and identification respectively. These recent contributions are all published in some international journals. On the other hand, we have clearly attempted to incorporate all the major developments in this field, some of which are rather recent and as yet may not be widely known.

Over the past few decades the state of this direct data driven strategy has come close to our living information age. But it is still nowhere near closed enough for many control applications. In this respect the field is wide open for researchers to come up with fresh ideas that will bridge the gap between ideal performance and achievable practice.

Wang Jianhong
Ricardo A. Ramirez-Mendoza
Ruben Morales-Menendez

Contents

Introduction of Data Driven Strategy

1.1 Introduction

In designing the feedback controller for every control system structure, we always assume that the mathematical model corresponding to the plant is known in prior, i.e. the process of designing the controller is based on the priori knowledge of the plant. But this assumption does not hold in reality, i.e. the considered plant is unknown, in which case the mathematical model of the plant may be identified by many system identification methods. Generally, to design the feedback controller in the closed loop system, there exist two kinds of approaches based on the priori knowledge of the plant. These two approaches are divided into model based control and direct data driven control. The main difference between these two approaches is whether the mathematical model for the plant is needed to be identified prior. The model based control approach needs to construct the mathematical model of the plant by using a mechanical modeling technique or system identification idea, then this mathematical model is used to design the later controller. So the controller performance of this model based control approach depends on the accuracy of the identified plant. To alleviate this dependency, some scholars propose to apply the input-output measured data to design the controller directly and avoid the tedious identification process of the plant. From a system identification idea point of view, the second approach, as stated above, is called the direct data driven control approach. The system identification idea plays an important role in theory and engineering, because it not only provides a method to construct an accurate mathematical model for the plant, but is a source of ideas for designing controllers directly for the closed loop system. Furthermore, as the direct data driven control approach avoids the identification process concerning the plant, and simplifies the whole designing process of the controller, in theory or in the engineering field it is more widely considered to achieve the control performance than the classical model based control method. Many common direct data driven control approaches include virtual reference feedback tuning control, subspace predictive control, iterative feedback tuning control, etc. The essence of direct data driven control is to extract all the useful and learnable information from the input-output measured data that can be extracted with a model class suggested. The concept of learnable information is formally defined, which in many cases but not always creates a special data dependent criterion. The feasible for direct data driven control is that in model building we are up against the fundamental problem in science, which is to learn from nature, and a successful application of direct data driven control requires a measure of thinking.

The main advantage of direct data driven control is that the controller is designed directly by using only input-output measured data without identifying the plant, i.e. the mathematical model for the plant is unknown. Because of some safety and production restrictions, the open loop system is not used widely in many industrial production processes. So in such a situation, it is very urgent to design a controller in a closed loop system. But the difficulty with the controller design in the closed loop system is that the correlation is considered between the input and external disturbance induced by the feedback loop. When considering the problem of designing a controller in a closed loop system, the controller is always designed based on a known mathematical model for the plant. But the mathematical model for the plant cannot be easily determined in the industrial process, as the cost of developing the mathematical model for the plant is very high. The original idea of data driven control arises from system identification, which applies the input-output measured data to obtain the mathematical equation for the plant, so some existing theoretical results from system identification can be extended to data driven control. As the goal of system identification is to identify the plant and data driven control concerns in designing the controller, some similarities exist between them. For example, after the mathematical equation, corresponding to the plant is obtained through the least squares method or other statistical methods, after which the model structure validation is used to testify whether the identified plant is accepted or refused. If the identified plant is refused, then all the four steps in system identification are restarted to identify a new mathematical equation for the plant.

Direct data driven control and model predictive control, two separable subjects in academic research, are gaining more popularity in recent years. Direct data driven control and model based control are two main strategies for designing controllers whether in open loop or closed loop structures. More specifically, model based control is used to identify or construct a mathematical equation for the considered system firstly, then this constructed mathematical equation is used to design the feedback controller or feed-forward controller. So for a model based control strategy, the performance of the latter controller is closely dependent on the identified model corresponding to that considered system. When constructing a mathematical equation or identifying an unknown system, physical modeling and system identification are applied widely as the most important step, i.e. no model then no controller. To avoid the identification step during the whole controller design, the idea of system identification is extended to get a new control strategy, i.e. direct data driven control. As the system uses the measured data to identify the unknown system, then can we think that the measured data could be applied to design the unknown controller, not the unknown system? This is the main essence of direct data driven control, and the common element between system identification and direct data driven control is to let the measured data speak something about the unknown system and unknown controller. Because there is no need for the mathematical equation for the unknown system, direct data driven control is widely studied in theory research and practice, such as paper machines, chemical processes, and water networks. Generally, direct data driven control includes some kind of forms, for example, virtual reference feedback tuning and iterative correlation tuning, etc. As system identification is the original idea for direct data driven control, lots of new concepts, coming from system identification, can also be applied to direct data driven control, for example, persistent excitation, model validation and optimal input signal, etc. Consider a special case where constrain conditions are imposed in unknown parameters in system identification, then lots of existing numerical optimization methods are introduced to identify the unknown parameters, and Lyapunov theory is used to analyze the parameter convergence and algorithm stability. Similarly, if constrain conditions are also considered in controller design, then one novel control strategy is yielded to be model predictive control (MPC).

Model predictive control is a powerful methodology that has been widely considered and used in a variety of industrial applications, such as building energy management. More specifically, model predictive control is a special form of suboptimal control problem, whose control objective is to keep the state of a system near some desired points. MPC combines elements of several ideas that we have put forth, for example, certainty equivalent control, multistage look ahead, and rollout algorithms. MPC tends to have some applied properties for classical linear quadratic control, i.e. there are two main reasons for replacing the classical linear quadratic control with MPC: (1) The considered system may be nonlinear, and using a model that is linearized around the desired point for control purposes may be inappropriate. (2) There may be control and state constraints, which are not handled adequately through a quadratic penalty on state and control. The solution obtained from a linear quadratic model is not suitable for this because the quadratic penalty on state and control tends to blur the boundaries of the constraints. Generally, MPC converts one optimal control problem into one numerical optimization problem with equality or inequality constraints, which correspond to the control and state constraints. Moreover, when the considered system is either deterministic, or else it is stochastic, it is replaced with a deterministic version by using typical values in place of all uncertain quantities, such as the certainty control approach in implementing MPC. Roughly speaking, at each stage, an optimal control problem is solved over a fixed length horizon, starting from the current state. The first component of the corresponding optimal policy is then used as the control of the current stage, while the remaining components are discarded. The optimization process is then repeated at the next stage, once the next state is revealed or the optimization algorithm is terminated iteratively.

From the above detailed description of MPC, MPC corresponds to one numerical optimization problem, whose cost function or loss function is one error value between the actual output and its desired output reference. In reality, the desired output is given, but the actual output is unknown in priori, so firstly we need to model the considered system and collect its actual output through persistently exciting the system with one appropriate input signal. It means the considered system is identified and used to calculate the actual output. There are two modeling approaches used to identify the considered system, the first principle and system identification. The first principle needs lots of priori information about the considered system, such as Newton's law, mathematical or physical laws, etc. The main essence of system identification is to excite the considered system, then use these collected input-output data to identify or estimate the unknown parameters, as the parameters are estimated online and used to describe the considered system, whether in open loop or closed loop conditions. The advantage of the second system identification approach is that no other priori information is needed, but only input-output data. In this big data period, this requirement for input-output data is tolerable. Roughly speaking, when obtaining the actual output in the cost function for MPC, the input-output data, corresponding to the considered system in an open loop or closed loop, are collected to identify the system through some statistical methods, for example, the least squares method, maximum likelihood method, etc. Then the identified system is applied to express or describe the actual output, so the actual output depends on the accuracy of the identified system. In practice, system identification is a well-developed technique for estimating system parameters from operational data typically taken during dedicated system testing or excitation, so system identification is also named as data driven.

Data driven control methods for system analysis and control such as virtual reference feedback tuning, subspace prediction control, model free control based on neural networks, and our considered iterative correlation tuning control have become increasingly popular over the recent years. The common property among them is that the controller is designed directly by the input-output measured data without any prior knowledge about the considered plant i.e. system identification and some physical principles are not needed to construct the mathematical equation for the considered plant. More specifically, in the case of the parameterized controller, the problem of designing the unknown controller is transformed into identifying those unknown controller parameters. As data driven control methods not only alleviate the dependency on the plant, but also simplify the controller design, so now they are widely studied in engineering and theoretical research.

1.2   Outline

The entire structure of this book is plotted in Fig. 1.1, where the closed relations among all chapters are explicitly seen. Chapter 1 introduces some prior knowledge about data driven control, model predictive control, and the whole structure of this book. Then from Chapter 2, our contributions are given one by one. More specifically, in Chapter 2, the combination of data driven and model predictive control is proposed to be one novel advanced control strategy-data driven model predictive control, where the prediction output in the cost function for model predictive control is constructed based on the interval predictor and bounded error prediction. In Chapter 3, some new results about closed loop system identification are described from the point of stealth identification, then to complete the theory analysis for system identification, model structure validation is considered for the studied closed loop system identification in Chapter 4. Based on the results of model structure validation for closed loop system identification, its inverse thinking is proposed to be one data driven control strategy-iterative correlation tuning control, whose interact property is also considered in Chapter 6. Furthermore, our new research on nonlinear system identification is put forth in Chapter 5, which corresponds to our previous study on linear system identification. To be best of our knowledge, new theories must be used to solve some problems in engineering, so in Chapter 7, our proposed data driven control and date driven identification are applied to solve two cases of engineering problems. For example, state of charge estimation and optimal input design for aircraft flutter model parameters identification. Concluding remarks are given at the end of each chapter, and in Chapter 8, we then provide a brief summary on the results presented in this book and an outlook to possible directions for future research on these topics.

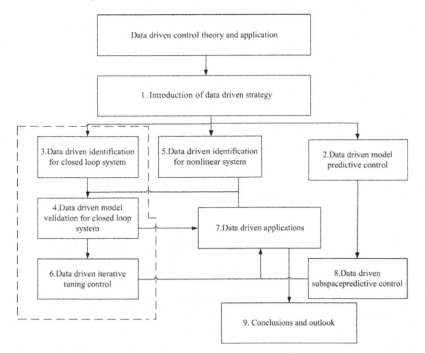

Figure 1.1: Structure of this book

1.3   Contributions

The main theme of this book is the development of system identification and advanced control theory into one recent novel data driven control strategy. In this book, the contributions of the proposed theories and their applications in engineering proposed by the authors over the recent years, are given in detail. As research on system identification and some control strategies, such as model predictive control and adaptive control are very mature, therefore, especially in this information period, it is time for the identification of control. System identification is used to construct the mathematical model for the considered plant, which is unknown. Based on this mission, we propose that we can extend system identification to construct the prediction output directly, or harvest the expected closed loop controller from the measured input-output data. Then the identification of the unknown plant is neglected, i.e. the unknown controller is obtained from the measured data. This thought was a stroke of luck. Firstly, based on our previous research on system identification theory, we found that many existing results from the system identification theory can be applied to serve or solve our thinking. In this book, all contents are divided into two categories, the first category corresponds to our recent research on the system identification theory, and the second category is our applied system identification into some advanced control theory, resulting in the yield of two data driven control strategies. For the sake of convenience, whatever the system identification and advanced control, we call them data driven estimation and data driven control, as they have one common property, i.e. lots of measured data are used to construct the mathematical model or closed loop controller.

The goal of system identification strategy is to build a mathematical model of a dynamic system based on some initial information about the considered system and the measured data collected from the experiment. The detailed processes of the system identification strategy consist of designing and conducting one identification experiment in order to collect the measurement data, selecting the structure of the dynamic system and specifying the unknown parameters to be identified, and eventually fitting the model parameters to the obtained data. As a consequence, the quality of the obtained model is evaluated through the model validation procedure. Generally, the system identification strategy is an iterative process and if the quality of the obtained model is not satisfactory, some or all of the listed phases can be repeated to yield one satisfied model. As the system identification theory is sometimes considered a mature field, with a wide and solid literature, many tools or methods for system identification, as well as for control theory, and stability analysis, have emerged in recent years.

For the first case-data driven estimation or identification, our contributions are concerned with closed loop system identification and nonlinear system identification, which correspond to the more complex systems. Through the identification process, new identification strategies and model structure validation are considered our main subjects, other subjects are also covered, such as the optimization algorithm, convergence analysis, consistent analysis, etc. The most important point is that stealth is introduced to achieve the closed loop system identification, i.e. stealth identification. For the nonlinear system identification, the Polynomial nonlinear state space model and tailor made parameterization are proposed to denote the considered nonlinear system. Based on this priori research on system identification or data driven estimation, we then start to think of the principle of data driven control.

For the second case-data driven control, our contributions are concerned with iterative correlation tuning control and data driven model predictive control. Our proposed iterative correlation tuning control originates from the last system identification process, i.e. model structure validation. It means that the inverse idea for model structure validation is used to get iterative correlation tuning control, and in the implementation of this iterative correlation tuning control, lots of other topics are applied, such as stochastic algorithm, convex optimization, game theory, and hybrid theory, etc. As direct data driven control and model predictive control are two separate subjects in academic research, our next mission is to combine them, which means that we continue research on applying system identification to model predictive control, i.e. data driven model predictive control equals to system identification for model predictive control.

Generally, the main advantage of direct data driven control is that the controller is designed directly by using only input-output measured data without identifying the plant, i.e. the mathematical model for the plant is unknown. Because of some safety and production restrictions, the open loop system is not used widely in many industrial production processes. So in such situations, it is very urgent to design controllers in the closed loop system. But the difficulty with the controller design in a closed loop system is that the correlation is considered between the input and external disturbance, induced by the feedback loop. If considering the problem of designing a controller in the closed loop system, the controller is always designed based on a known mathematical model for the plant. To avoid the identification process for an unknown model for the plant and to design the controller directly in the closed loop system, the direct data driven control scheme has been studied profoundly, where the feedback controller and forward controller exist in the same closed loop system simultaneously.

The book is intended as a text book for graduate students as well as a basic reference for practicing engineers facing the problem of designing control systems. Control researchers from other areas will find a comprehensive presentation filed with bridges to various other control design techniques.

CHAPTER 2

Data Driven Model Predictive Control

2.1 Introduction

In this chapter, some newly derived results about system identification are applied in the model predictive control. To emphasize that the model predictive control is based on system identification, the name of the data driven model predictive control is used here. More specifically, two concepts—bounded error and interval predictor model, coming from the system identification theory, are introduced into model predictive control. Further, the stability analysis and the robustness of the data driven model predictive control are also studied.

(1) Bounded error identification is applied to model predictive control. After introducing the family of models and some basic assumptions, we present bounded error identification to construct the interval predictor, using the neighborhood of a given data point. To guarantee the obtained interval predictor to be a minimum interval predictor, two optimal vectors used to adjust the width of the obtained interval predictor are suggested to be piecewise affine forms, using the Karush-Kuhn-Tucker (KKT) optimality conditions. When the interval predictor is applied to model predictive control, the midpoint of that interval is chosen in an optimization problem to obtain the optimal control input for model predictive control. The property that controls input exists in its own cost function inspires us to use parallel distributed algorithms to solve the unconstrained or constrained optimization problem, and the detailed computational process of the Newton algorithm and Augmented Lagrangian algorithm are given for the sake of completeness.

(2) Interval prediction model is studied for model predictive control strategy with unknown but bounded noise. After introducing the family of models and some basic information, some computational results are presented to construct the interval predictor model, using a linear regression structure whose regression parameters arc included in a sphere parameter set. A size measure is used to scale the average amplitude of the predictor interval, then one optimal model that minimizes this size measure is efficiently computed by solving a linear programming problem. The active set approach is applied to solve

the linear programming problem, and based on these optimization variables, the predictor interval of the considered model with a sphere parameter set can be directly constructed. To apply the interval prediction model to model predictive control, the midpoint of that interval is substituted in a quadratic optimization problem with an inequality constrained condition to obtain the optimal control input. After formulating it as a standard quadratic optimization and deriving its dual form, the Gauss-Seidel algorithm is applied to solve the dual problem and the convergence of the Gauss-Seidel algorithm is provided too.

(3) As the cooperative distributed model predictive control scheme is widely applied in large scale networks of systems, asymptotic stability is a very important index for measuring the performance of cooperative distributed model predictive control. Based on the obvious inequality from the classical Lyapunov stability condition, we derive a set of linear matrix inequalities to replace the common inequalities by using the Schur complement and S-procedure. Furthermore, when combining local state and input constraint sets, a set of more complex linear matrix inequalities is used to guarantee the asymptotical stability for cooperative distributed model predictive control.

2.2 Application of bounded error identification into model predictive control

Model predictive control has developed considerably over the past two decades, both within the control field and in the industry. This success can be attributed to the fact that model predictive control is perhaps the most general way of posing the constraint control problem in the time domain. Model predictive control formulation integrates optimal control, stochastic control, control of processes with dead time, multivariable control, and future references when available. One important advantage of model predictive control is that because of the finite control horizon used, constraints and, in general, nonlinear processes which are frequently found in industry, can be handled. The rationale behind model predictive control is the following: at each time step, an L2 or alternative variation of the cost function is locally optimized over time to obtain the open loop control as a function of time instant, only a small portion of which is actually applied to the system. The time horizon is then shifted, and the process is repeated at a later time step based on state feedback [1]. Although model predictive control is a robust type of control in most reported applications, some new and very promising results allow one to think that this control technique will experience greater expansion within this community in the near years [2]. However, although lots of applications have been reported in both industries and research institutions, model predictive control has not yet reached its popularity in the industry.

The most important element in model predictive control is the prediction of output value. After deriving the prediction of output value by the prediction error method and then substituting it with the one considered cost function, we take the derivative of the cost function concerning input value to obtain one optimal input [3]. But the problem of deriving the prediction of output value is dependent on external noise, which is always assumed to be independent and identically distributed white noise. Because white noise is an ideal case, it does not exist in engineering, additionally, deriving statistical properties of noise is often very difficult in practice as it is usually not possible to measure noise directly [4]. To relax this strict probabilistic description of noise, an assumption on the noise bound is less restrictive, as noise is bounded and the bound can be roughly calculated from the specification of the used sensor. Here we investigate the model predictive control in presence of bounded noise, which is similar to set membership predictive control [5]. The idea of set membership predictive control is from set membership identification in the classical system identification theory. As this chapter is different from set membership identification, we apply the interval predictor model, coming from classical system identification theory to model predictive control. To have a better understanding of how to apply the interval predictor model to model predictive control, we first give a short review of the interval predictor model.

In the classical system identification theory, one parametric model structure corresponding to an identified system is selected firstly, and then the parameters in the parametric model structure are estimated using the measured input-output data. During the system identification process, many identification methods are proposed to identify these unknown parameters, for example, the classical least squares method, instrumental variable method, maximum likelihood estimation method, prediction error method, Bayesian method, etc. [6]. One common property of these identification methods is that the prior information about noise is known. Based on some probabilistic assumptions on noise, the unknown parameters are identified as the specific numerical values. But the probabilistic assumptions on noise are not realistic and these probabilistic assumptions are not realized easily in reality. So to relax the probabilistic assumptions on noise, we always assume that the noises are unknown but bounded. This unknown but bounded assumption is weaker than the original probabilistic assumption, as it needs no prior distribution on noise. The commonly used method applied to solve the unknown but bounded case is called set membership identification [7]. In the set membership identification, the obtained result is not a detailed numerical value, but a guaranteed interval with respect to each parameter. This guaranteed interval means that each parameter can be included in this interval with one guaranteed accuracy which is assessed by some probabilistic inequalities. From the idea of set membership identification, after the unknown parameters are identified, the identified parametric model may be applied to determine one prediction for the output value of the system, together with probabilistic confidence intervals around the prediction [8]. The confidence interval can accurately describe the actual probability that the future predictor will fall into the obtained interval. The future predictor is important for the next controller design and state estimation, so in [9], a novel approach for the construction of prediction models is proposed. The advantage of this novel approach is that instead of using a standard identification way, where one constructs a parametric model by minimizing an identification cost, the identified model is used to design the prediction interval [10]. This novel approach directly considers the interval model and applies measured data to ascertain the reliability of such an interval predictor model [11]. It means that we directly obtain the interval predictor model from measured data and avoid the identification process for the parametric model structure.

In this section, the framework is to propose new system output predictors based on the bounded error identification technique and stored data and apply them to model predictive control. When constructing new system output predictors, a local approximation of the considered system is generated in a bounding sense. For every data point given, data from the neighborhood of this given point is gathered and a prediction for this given point is formed. As the bounded error identification technique returns a guaranteed interval predictor to bind the considered system output, i.e. a lower and upper bound of the system output is obtained, so the midpoint of the interval prediction can be used as a central estimate, which is a new system output predictor at every given point. The contribution of the paper is two folds. Firstly, after introducing the family of models under study and some basic information or assumptions about the bounded error identification technique, a guaranteed interval predictor is obtained to bind the system output. To assure the obtained guaranteed interval predictor is a minimum interval prediction under our assumptions, we find the minimum interval predictor depends on two vectors. Then if these two vectors are chosen appropriately, the minimum interval predictor will be constructed easily. In this case, two linear optimization problems are solved to obtain these two optimal vectors. As absolute values exist in these two linear optimization problems, we apply the Karush-Kuhn-Tucker optimality conditions to see that the optimal vectors are all piecewise affine forms. Secondly, based on this constructed interval predictor, the midpoint of this interval predictor can be used as a central estimate, then the prediction of the system output can be obtained by that central estimate. The goal of model predictive control is to control the system to track the desired output reference and reject disturbances. Moreover, the considered controller may enforce input and output constraints. So when applying that central estimate to the prediction of the system output, such a control objective can be formulated as a quadratic programming problem with an inequality constrained condition. Through observing a cost function in this quadratic programming problem, we find the control input at each time instant exists in each cost function separable, then it would be possible to decompose the original quadratic programming problem into many independent sub-problems. This motivates us to apply the parallel distributed algorithm to solve each separable and independent sub-problem. For the sake of completeness, the approximate Newton method is proposed to solve the quadratic programming problem without constrained conditions, and further to solve the quadratic programming problem with inequality constrained condition, the augmented Lagrangian strategy is carried out to complete the goal. Whatever approximate Newton method or augmented Lagrangian strategy is used here, the parallel distributed idea spans the whole computation process, where the detailed gradient vector and Hessian matrix corresponding to each cost function in the sub-problem are also given.

2.1.1   Problem formulation

To introduce a problem for prediction based on nonparametric the local estimation and bounded error identification technique, consider an unknown discrete time or possibly a nonlinear system. At time instant k, a measured output of a considered system is denoted as $y_k \in R$, $u_k \in R^{n_u}$ is a vector of noise corrupted measurements of the system or the control input. The exact functional form that relates y_k and u_k is unknown, but a bounding inequality is assumed as follows. As any discrete time instant k, the measured output y_k and vector x_k are related by inequality.

$$\left| y_k - r(u_k)\theta(u_k) \right| \le \sigma \tag{2.1}$$

where $\sigma \in R$ is a known positive constant, $r(u_k) \in R^{n_\theta}$ is a known function and $\theta(u_k) \in R^{n_\theta}$ is an unknown function. From (2.1), it means that at each sample time k, the measured system output can be approximated locally by an affine expression of a known regressor vector $r(u_k)$. The error of above local approximation is bounded by the constant σ. Function $r(u_k): R^{n_u} \to R^{n_\theta}$ defines the components of the regressor vector and the dimension of $\theta(u_k)$. All the above descriptions on system and some options for regressor vector $r(u_k)$ can be seen in [17]. As in this paper, the values of the parameters are not estimated, our goal is to require a prediction for a given point, and only then a bounding assumption of function $\theta(u_k)$ is needed.

Assumption 2.1: There exist known positive constant L_1 and such that $\left\| r(u) \right\| \le C$, and

$$\left\| \theta(u_i) - \theta(u_j) \right\| \le \frac{L_1}{C} s(u_i, u_j), \forall i, j \tag{2.2}$$

where u_i, u_j denote two measurements at two different sample time instants i, j, $\| \; \|$ is a Euclidean norm and $s(u_k) \in \left\{ 0, R^+ \right\}$ is a known nonnegative function. Function $s(u_k)$ defines the locality of (2.2), constant σ and L_1 define the level of uncertainty present in the approximation. Note that σ and L_1 are a priori information, constant L_1 bounds the normalized rate of change corresponding to function $\theta(u_k)$, constant σ bounds the error present in the local and affine approximation $r(u_k)\theta(u_k)$.

As the idea of a bounded error identification strategy is to provide an interval predictor which gives an estimate of the maximum and minimum bounds on system output y_k. Based on this interval predictor, a central estimate y_k can be obtained as the midpoint of the interval predictor. Then this system output predictor will be used in model predictive control to track the desired output reference and reject disturbance.

2.2.2   Predictor based on bounded error identification

Assume a set of noise corrupted measurements y_i and $u_i, i = 1, 2 \cdots N$ be available and denote the data set as $D = \{(u_i, y_i) : i = 1, 2 \cdots N\}$. A family of interval predictors, used to bind the system output, uses a neighborhood of u_k to infer the prediction. Let the neighborhood of a given sample value u_k be a nonempty subset $D_\mu(u_k) = \{(u_{s_l}, y_{s_l}) : s_l \in \{1, 2 \cdots N\}, l - 1, 2 \cdots N_\mu\} \subseteq D$ of measured data, where

$N_\mu = \mu N$ and $\mu \in (0,1]$ is a constant. Constant μ determines the cardinality of set $D_\mu(u_k)$. Therefore the interval predictors use every element of set D when $\mu = 1$ or a subset $D_\mu(u_k)$ if $\mu < 1$ to obtain the interval bounds. Then using the bounded error identification strategy, the interval predictor is formulated as

Theorem 2.1 (Interval predictor): Given a vector $u_k \in R^{n_u}$, a nonempty set of

measurements $D_\mu(u_k) = \{(u_{s_l}, y_{s_l}) : s_l \in \{1, 2 \cdots N\}, l - 1, 2 \cdots N_\mu\} \subseteq D$ and two

vectors $\underline{\lambda}, \overline{\lambda} \in R^{N_\mu}$, an interval predictor

$$f_\lambda(u_k) = \left[\underline{f_\lambda}(u_k), \overline{f_{\overline{\lambda}}}(u_k) \right]$$

with $\underline{f_\lambda}(u_k) \leq \overline{f_{\overline{\lambda}}}(u_k)$, can be defined as

$$\begin{cases} \underline{f_\lambda}(u_k) = -\left|\underline{\lambda}\right|^T b_1 + \underline{\lambda}^T b_2 - \sigma \\ \overline{f_{\overline{\lambda}}}(u_k) = \left|\overline{\lambda}\right|^T b_1 + \overline{\lambda}^T b_2 + \sigma \end{cases} \tag{2.3}$$

where $|\lambda|$ denotes a vector whose elements are the solute values of the elements of vector λ. Vectors b_1 and b_2 are defined as

$$b_1 = \begin{bmatrix} \sigma + L_1 s(u_k, u_{s_1}) \\ \sigma + L_1 s(u_k, u_{s_2}) \\ \vdots \\ \sigma + L_1 s(u_k, u_{s_{N_\mu}}) \end{bmatrix}, \quad b_2 = \begin{bmatrix} y_{s_1} & y_{s_2} & \cdots & y_{s_{N_\mu}} \end{bmatrix}^T \tag{2.4}$$

Equation (2.3) is one interval predictor model, it contains the system output y_k with some guaranteed probability. When measured data are collected in data set

$$D = \{(u_i, y_i) : i = 1, 2 \cdots N\}$$

or

$$D_\mu(u_k) = \{(u_{s_l}, y_{s_l}) : s_l \in \{1, 2 \cdots N\}, l = 1, 2 \cdots N_\mu\} \subseteq D$$

whatever open loop or closed loop is used the following relation holds:

$$y_k \in f_\lambda(u_k) = \left[\underline{f_\lambda}(u_k), \overline{f_{\overline{\lambda}}}(u_k) \right], \forall k = 1, 2 \cdots N \tag{2.5}$$

It means the system output predictor will be included in this guaranteed interval with one guaranteed accuracy. To do deep analysis of the guaranteed interval (2.5), define the feasible solution set $\Theta(u_k)$ as

$$\Theta(u_k) = \left\{ \theta \in R^{n_\theta} : \left| y_{s_l} - r\left(u_{s_l}\right)\theta \right| \le \sigma + L_1 s\left(u_k, u_{s_l}\right), \left(u_{s_l}, y_{s_l}\right) \in D_\mu\left(u_k\right) \right\} \quad (2.6)$$

Set $\Theta(u_k)$ can be rewritten in matrix form as

$$\begin{cases} \Theta(u_k) = \left\{ \theta \in R^{n_\theta} : \left| A_1 \theta - b_2 \right| \le b_1 \right\} \\ A_1 = \left[r\left(u_{s_1}\right) \quad r\left(u_{s_2}\right) \quad \cdots \quad r\left(u_{s_{N_\mu}}\right) \right]^T \end{cases} \quad (2.7)$$

Using above notations, we obtain the following Theorem 2.2.

Theorem 2.2: Given data sequence (u_k, y_k) about the unknown system, and an interval predictor $f_\lambda(u_k)$ is given by (5), if $A_1 \underline{\lambda} = A_1 \overline{\lambda} = r(u_k)$, then it holds that

$$\underline{f_\lambda}(u_k) \le y_k \le \overline{f_{\overline{\lambda}}}(u_k)$$

Proof: The second inequality was proven in [12], so here we only prove the first inequality. As $\theta_k \in \Theta(u_k)$ and from inequality $\left| y_k - r(u_k)\theta_k \right| \le \sigma$, we obtain

$$y_k \ge r(u_k)\theta_k - \sigma \ge \min\left\{ r(u_k)\theta, \theta \in \Theta(u_k) \right\} - \sigma = \min\left\{ A_1 \underline{\lambda}^T \theta, \theta \in \Theta(u_k) \right\} - \sigma$$

$$= \min\left\{ \underline{\lambda}^T \left(A_1 \theta - b_2 \right) + \underline{\lambda}^T b_2, \theta \in \Theta(u_k) \right\} - \sigma$$

$$= \min\left\{ \left| \underline{\lambda} \right|^T \left(-b_1 \right) + \underline{\lambda}^T b_2, \theta \in \Theta(u_k) \right\} - \sigma$$

$$= -\left| \underline{\lambda} \right|^T b_1 + \underline{\lambda}^T b_2 - \sigma \quad (2.8)$$

Thus concluding the proof.

By observing the interval predictor (2.3) and (2.5) again, from the set membership identification theoretical perspective, $\{b_1, b_2, \sigma\}$ are prior known, so in order to construct this interval predictor, two vectors $\underline{\lambda}$ and $\overline{\lambda}$ must be chosen appropriately. A good option about choosing two variables $\underline{\lambda}$ and $\overline{\lambda}$ is to provide the minimum interval prediction and present optimal properties in the worst case prediction error sense. The minimum interval predictor is achieved by two optimal vectors such that

$$\overline{\lambda} = \arg\min_{\lambda} \left| \lambda \right|^T b_1 + \lambda^T b_2$$

$$\text{subject to } A_1 \lambda = r(u_k) \quad (2.9)$$

$$\underline{\lambda} = \arg\min_{\lambda} - \left| \lambda \right|^T b_1 + \lambda^T b_2$$

$$\text{subject to } A_1 \lambda = r(u_k) \quad (2.10)$$

In this worst case prediction error sense, these two vectors $\underline{\lambda}$ and $\overline{\lambda}$ are obtained by solving the above two linear optimization problems respectively.

Now we apply KKT optimality conditions to solve linear optimization problem (2.9). Notice that there exist absolute operations in linear optimization problem (2.9), one slack variable s is introduced to eliminate the absolute operations, $|\lambda| \leq s$.

Using the slack variables in (2.9), the linear optimization problem (2.9) can be reformulated as that

$$\min_{s,\lambda} s^T b_1 + \lambda^T b_2$$

$$\text{subject to } A_1 \lambda = r(u_k), \ \lambda \leq s, \text{ and } -s \leq \lambda \tag{2.11}$$

Since KKT optimality conditions are necessary and sufficient conditions for optimal solution, then the Lagrangian function of (2.11) can be written as

$$L(\lambda, s, \mu_1, \mu_2, \mu_3) = s^T b_1 + \lambda^T b_2 - \mu_1 \left(A_1 \lambda - r(u_k) \right) - \mu_2 (s - \lambda) - \mu_3 (s + \lambda) \tag{2.12}$$

where $\mu_i, i = 1, 2, 3$ are the Lagrangian multipliers, since $s = |\lambda|$ for an optimal solution (λ, s), the KKT conditions are equivalent to the following relations

$$\frac{\partial L}{\partial \lambda} = b_2 - \mu_1 A_1 + \mu_2 - \mu_3 = 0 \tag{a}$$

$$\frac{\partial L}{\partial s} = b_1 - \mu_2 - \mu_3 = 0 \tag{b}$$

$$\frac{\partial L}{\partial \mu_1} = A_1 \lambda - r(u_k) = 0 \tag{c}$$

$$\mu_2 (s - \lambda) = 0, \mu_2 \geq 0 \tag{d}$$

$$\mu_3 (s + \lambda) = 0, \mu_3 \geq 0 \tag{e}$$

From (d) and (e), we can see that $\lambda > 0$ implies $\mu_3 = 0$, and that $\lambda < 0$ implies $\mu_2 = 0$, so next we continue to discuss the two cases.

(i) $\lambda > 0 \Rightarrow \mu_3 = 0$

Adding (a) and (b) to get that

$$\mu_1 A_1 = b_1 + b_2$$

Multiplying μ_1 on both sides of (c), we get that

$$\mu_1 A_1 \lambda = \mu_1 r(u_k)$$

Substituting $\mu_1 A_1 = b_1 + b_2$ into $\mu_1 A_1 \lambda = \mu_1 r(u_k)$, we obtain

$$(b_1 + b_2)\lambda = \mu_1 r(u_k)$$

As we are not assured that $(b_1 + b_2)$ is invertible, so in order to obtain the optimal vector λ, we use the pseudo inverse matrix to remedy the rank deficiency, i.e.

$$\lambda = (b_1 + b_2)^T \left[(b_1 + b_2)(b_1 + b_2)^T \right]^{-1} \mu_1 r(u_k) \tag{2.13}$$

Furthermore we see that $\lambda > 0$ implies $\mu_3 = 0$, but $\mu_2 > 0$, from (a) we get

$$b_2 - \mu_1 A_1 = -\mu_2 < 0$$

Also from (b), we have that

$$b_1 = \mu_2 > 0$$

(ii) $\lambda < 0 \Rightarrow \mu_2 = 0$

This case is similar to case (i), then we also obtain

$$\lambda = (b_2 - b_1)^T \left[(b_2 - b_1)(b_2 - b_1)^T \right]^{-1} \mu_1 r(u_k) \tag{2.14}$$

Furthermore if $\lambda < 0$, then $\mu_2 = 0$ and $\mu_3 > 0$, then from (a) and (b), we have that

$$b_2 - \mu_1 A_1 = \mu_3 > 0, b_1 = \mu_3 > 0$$

Combining the above analysis on linear optimization problem (2.9), the appropriate choice on vector $\bar{\lambda}$ is formulated as Theorem 2.3.

Theorem 2.3: Suppose that linear optimization problem (2.9) is feasible, then there exists $\mu_1 \in R^{N_\mu}$, such that for an optimal solution $\bar{\lambda}^*$, it holds that

$$\bar{\lambda}^* = \begin{cases} (b_1 + b_2)^T \left[(b_1 + b_2)(b_1 + b_2)^T \right]^{-1} \mu_1 r(u_k), & b_2 - \mu_1 A_1 < 0 \cap b_1 > 0 \\ (b_2 - b_1)^T \left[(b_2 - b_1)(b_2 - b_1)^T \right]^{-1} \mu_1 r(u_k), & b_2 - \mu_1 A_1 > 0 \cap b_1 > 0 \end{cases} \tag{2.15}$$

Remark: If $\bar{\lambda} = 0$, then $\bar{f}_{\bar{\lambda}}(u_k) = \sigma$, which is a constant.
Similarly a good choice on vector $\underline{\lambda}$ can be seen.

Theorem 2.4: Suppose that linear optimization problem (2.10) is feasible, then there exists $\mu_2 \in R^{N_\mu}$, such that for an optimal solution $\underline{\lambda}^*$, it holds that

$$\underline{\lambda}^* = \begin{cases} (b_1 - b_2)^T \left[(b_1 - b_2)(b_1 - b_2)^T \right]^{-1} \mu_2 r(u_k), & b_2 + \mu_2 A_1 > 0 \cap b_1 > 0 \\ (-b_2 - b_1)^T \left[(-b_2 - b_1)(-b_2 - b_1)^T \right]^{-1} \mu_2 r(u_k) & b_2 - \mu_1 A_1 < 0 \cap b_1 > 0 \end{cases} \tag{2.16}$$

From the optimal choices on $\underline{\lambda}^*$ and $\overline{\lambda}^*$, they are all piecewise affine forms, and they automatically adapt to how the actual data samples are spread and can easily handle sparse data sets or data lying asymmetrically.

2.2.3 Application bounded error identification into model predictive control

We start to apply the above interval predictor, derived from bounded error identification, into model predictive control strategy. A central estimate $\hat{y}_k$ can be obtained by the expression

$$\hat{y}_k = \frac{1}{2}\left(\left|\overline{\lambda}^*\right|^T - \left|\underline{\lambda}^*\right|^T\right)b_1 + \frac{1}{2}\left(\overline{\lambda}^* + \underline{\lambda}^*\right)b_2 = \lambda_1 b_1 + \lambda_2 b_2$$

$$\lambda_1 = \frac{1}{2}\left(\left|\overline{\lambda}^*\right|^T - \left|\underline{\lambda}^*\right|^T\right), \lambda_2 = \frac{1}{2}\left(\overline{\lambda}^* + \underline{\lambda}^*\right) \tag{2.17}$$

where, optimal vectors $\underline{\lambda}^*$ and $\overline{\lambda}^*$ are derived from (2.15) and (2.16), the midpoint of the interval predictor is used to regard the central estimate. Then $\hat{y}_k$ is deemed as the system output predictor at sample instant k.

2.2.3.1 Model predictive control problem

The goal of model predictive control is to control the system in order to track a desired output reference and reject disturbances from $k = 1$ up to some finite time step N, where this time step N can be very large. Assuming that the control input $u(l), l = 0, -1, -2, \cdots$ are known, then a control objective can be formalized by one optimization problem.

$$\min_{u_1 \cdots u_N} \sum_{k=1}^{N} \left[\hat{y}_k - y_{des}(k)\right]^T Q_1 \left[\hat{y}_k - y_{des}(k)\right] + u_k^T S_1 u_k \tag{2.18}$$

where $y_{des}(k)$ is the desired output reference, Q_1, S_1 are positive semi-definite weighting matrices selected by the designer. Observing (2.17) and (2.18), the output predictor $\hat{y}_k, k = 1, 2 \cdots N$ exists in cost function (2.18).

Replacing $\hat{y}_k$ as $\lambda_1 b_1 + \lambda_2 b_2$ into cost function (2.18), we obtain an explicit form corresponding to optimization problem.

$$\min_{u_1 \cdots u_N} \sum_{k=1}^{N} \left[\lambda_1 b_1 + \lambda_2 b_2 - y_{des}(k)\right]^T Q_1 \left[\lambda_1 b_1 + \lambda_2 b_2 - y_{des}(k)\right] + u_k^T S_1 u_k \tag{2.19}$$

As in cost function (2.19), $\{\lambda_1, \lambda_2, y_{des}(k), Q_1, S_1\}$ are known, equation (2.4) implies that vector b_2 is consisted of measurement output around the neighborhood of u_k, and set $D_\mu(u_k) = \left\{\left(u_{s_l}, y_{s_l}\right): s_l \in \{1, 2 \cdots N\}, l = 1, 2 \cdots N_\mu\right\}$ is available. So vector b_2 is also known, and control input u_k only exists in vector b_1. To express this dependency on control input u_k, we rewrite b_1 as $b_1(u_k)$. Then control input u_k, $k = 1, 2 \cdots N$ can be derived by solving one optimization problem.

$$\min_{u_1 \cdots u_N} \sum_{k=1}^{N} \left[\lambda_1 b_1 (u_k) + \lambda_2 b_2 - y_{des}(k) \right]^T Q_1 \left[\lambda_1 b_1 (u_k) + \lambda_2 b_2 - y_{des}(k) \right]$$

$$+ u_k^T S_1 u_k = \min_{u_1 \cdots u_N} f_k (u_k) \quad f_k (u_k) = \left[\lambda_1 b_1 (u_k) + \lambda_2 b_2 - y_{des}(k) \right]^T$$

$$Q_1 \left[\lambda_1 b_1 (u_k) + \lambda_2 b_2 - y_{des}(k) \right] + u_k^T S_1 u_k \tag{2.20}$$

From (2.20), we see that control input u_k only exists in cost function $f_k(u_k)$, not other $f_i(u_i), i \neq k$. This property tells us that parallel distributed algorithm can be applied to solve optimization problem (2.20).

2.2.3.2 Unconstrained parallel and distributed algorithm

The original optimization problem (2.20) can be divided into the N sub-optimization problem, i.e. control input u_k at time instant k can be gained by the following sub-problem

$$\min_{u_k} f_k (u_k) \tag{2.21}$$

As $f_k(u_k)$ is twice continuously differentiable, Newton's algorithm is described by the equation

$$u_k (t+1) = u_k (t) - \gamma \left[\nabla^2 f_k (u_k(t)) \right]^{-1} \nabla f_k (u_k(t)) \tag{2.22}$$

where $u_k(t)$ denotes the control input at iteration step t, γ is a step size, $\nabla f_k (u_k(t))$ is the gradient vector, $\nabla^2 f_k (u_k(t))$ denotes the Hessian matrix. Taking derivatives of cost function $f_k(u_k)$ with respect to u_k, we obtain the gradient vector.

$$\nabla f_k (u_k) = \frac{\partial f_k (u_k)}{\partial u_k} = 2 \left[\lambda_1 b_1 (u_k) + \lambda_2 b_2 - y_{des}(k) \right]^T Q \left[\lambda_1 \frac{\partial b_1 (u_k)}{\partial u_k} \right] + 2 S u_k$$

$$\frac{\partial b_1 (u_k)}{\partial u_k} = L_1 \begin{bmatrix} \dfrac{\partial s \left(u_k, u_{s_1} \right)}{\partial u_k} \\[2mm] \dfrac{\partial s \left(u_k, u_{s_2} \right)}{\partial u_k} \\[2mm] \vdots \\[2mm] \dfrac{\partial s \left(u_k, u_{s_{N_\mu}} \right)}{\partial u_k} \end{bmatrix} \tag{2.23}$$

Taking derivatives of $\nabla f_k (u_k(t))$ with respect to u_k again, the Hessian matrix is given.

$$\nabla^2 f_k\left(u_k(t)\right) = \frac{\partial^2 f_k\left(u_k\right)}{\partial\left(u_k\right)^2} = 2\left[\lambda_1 b_1\left(u_k\right) + \lambda_2 b_2 - y_{des}\left(k\right)\right]^T$$

$$Q_1\left[\lambda_1 \frac{\partial^2 b_1\left(u_k\right)}{\partial\left(u_k\right)^2}\right] + 2\lambda_1\left[\frac{\partial b_1\left(u_k\right)}{\partial u_k}\right] Q_1 \left[\frac{\partial b_1\left(u_k\right)}{\partial u_k}\right]^T \lambda_1 + 2S_1$$

$$(2.24)$$

As Newton's algorithm converges much faster than other classical algorithms, and $f_k(u_k)$ has a local minimum u_k^* for which $\nabla^2 f_k\left(u_k^*\right)$ is positive definite. Let $u_k\left(t+1\right)$ be generated by Newton iteration with step size $\gamma \equiv 1$, we have

$$u_k\left(t+1\right) - u_k^* = \left[\nabla^2 f_k\left(u_k(t)\right)\right]^{-1}\left[\nabla^2 f_k\left(u_k(t)\right)\left(u_k\left(t+1\right) - u_k^*\right) - \nabla f_k\left(u_k(t)\right)\right]$$

$$= \left[\nabla^2 f_k\left(u_k(t)\right)\right]^{-1} \int_0^1\left[\nabla^2 f_k\left(u_k(t)\right) - \nabla^2 f_k\left(u_k^* + \xi\left(u_k(t) - u_k^*\right)\right)\right]$$

$$d\xi\left(u_k(t) - u_k^*\right)$$

$$(2.25)$$

We obtain for Euclidean norm

$$\left\|u_k\left(t+1\right) - u_k^*\right\| = \left\|\left[\nabla^2 f_k\left(u_k(t)\right)\right]^{-1}\right\|\left(\int_0^1\left[\nabla^2 f_k\left(u_k(t)\right) - \nabla^2 f_k\right.\right.$$

$$\left.\left.\left(u_k^* + \xi\left(u_k(t) - u_k^*\right)\right)\right]d\xi\right)\left\|\left(u_k(t) - u_k^*\right)\right\|$$

$$(2.26)$$

Using the continuity of $\nabla^2 f_k\left(u_k(t)\right)$, it follows that given any $\alpha \in (0,1)$, there exists some $\varepsilon > 0$ such that

$$\left\|u_k\left(t+1\right) - u_k^*\right\| = \alpha\left\|\left(u_k(t) - u_k^*\right)\right\|$$

$$(2.27)$$

For all $u_k(t)$ with

$$\left\|\left(u_k(t) - u_k^*\right)\right\| \le \varepsilon$$

After simple computations on (2.27), we have

$$\left\|u_k\left(t+1\right) - u_k^*\right\| = \alpha^{t+1}\left\|\left(u_k(0) - u_k^*\right)\right\| \le \alpha^{t+1}\varepsilon$$

$$(2.28)$$

Taking limit operation as $t \to +\infty$, then the fast convergence of Newton's algorithm is established.

$$\lim_{t \to +\infty}\left\|u_k\left(t+1\right) - u_k^*\right\| = \lim_{t \to +\infty} \alpha^{t+1}\varepsilon = 0$$

It means when iteration step $t \to +\infty$, the iteration value $u_k(t+1)$ will approach to local minimum u_k^* closely.

2.2.3.3 Constrained parallel and distributed algorithm

In optimization problem (2.20) or (2.21), no constraints are considered, it is an ideal case. Moreover, the controller may enforce input and output constraints.

$$\begin{cases} u_{min} \le u_k \le u_{max} \\ y_{min} \le \hat{y}_k \le y_{max} \end{cases} \quad k = 1, 2 \cdots N \tag{2.29}$$

where u_{min}, u_{max} and y_{min}, y_{max} denote the lower and upper bounds on input and output respectively. Combining (2.20) and (2.29), the model prediction control problem is formalized by the following optimization problem with inequality constrained condition:

$$\min_{u_1 \cdots u_N} \sum_{k=1}^{N} f_k(u_k); f_k(u_k) = \left[\underbrace{\lambda_1 b_1(u_k) + \lambda_2 b_2}_{\hat{y}_k} - y_{des}(k) \right]^T$$

$$Q_1 \left[\underbrace{\lambda_1 b_1(u_k) + \lambda_2 b_2}_{\hat{y}_k} - y_{des}(k) \right] + u_k^T S_1 u_k$$

$$\text{subject to} \begin{cases} u_{min} \le u_k \le u_{max} \\ y_{min} \le \hat{y}_k \le y_{max} \end{cases} \quad k = 1, 2 \cdots N$$

Before solving above optimization problem with constrained condition, all input and output constraints can be combined as follows:

$$\begin{cases} e_j u = m_j, j = 1 \cdots r \\ u_k \in P_k, k = 1, 2 \cdots N \end{cases} \tag{2.30}$$

where $u = (u_1 \ u_2 \ \cdots \ u_N)$, P_k is a bounded polyhedral set.

Let e_{jk} denote the subvector of e_j that corresponding to u_k, and for a given j, let $I(j)$ be the set of indices k of subvectors u_k that appear in the jth constraint $e_j u = m_j$, that is

$$I(j) = \left\{ k \setminus e_{jk} \ne 0 \right\}, j = 1, 2 \cdots r$$

We transform one equivalent problem by introducing additional variables $z_{jk}, k \in I(j)$, as follows:

$$\min_{u_1 \cdots u_N} \sum_{k=1}^{N} f_k(u_k)$$

$$\text{subject to } e_{jk} u_k = z_{jk}, j = 1, 2 \cdots r, k \in I(j)$$

$$u_k \in P_k, k = 1, 2 \cdots N$$

$$\sum_{k \in I(j)} z_{jk} = m_j, j = 1, 2 \cdots r \tag{2.31}$$

For each $j = 1, 2 \cdots r$, consider Lagrange multipliers p_{jk} for the equality constraints $e_{jk}u_k = z_{jk}, k \in I(j)$. The recursive relation of the multipliers is

$$p_{jk}(t+1) = p_{jk}(t) + c(t)\left(e_{jk}u_k(t+1) - z_{jk}(t+1)\right), j = 1,2\cdots r, k \in I(j) \quad (2.32)$$

where $u_k(t+1)$ and $z_{jk}(t+1)$ minimize the Augmented Lagrangian function

$$\sum_{k=1}^{N} f_k(u_k) + \sum_{j=1}^{r}\sum_{k\in I(j)} p_{jk}(t)\left(e_{jk}u_k - z_{jk}\right) + \frac{c(t)}{2}\sum_{j=1}^{r}\sum_{k\in I(j)}\left(e_{jk}u_k - z_{jk}\right)^2$$

$$\text{subject to } \sum_{k\in I(j)} z_{jk} = m_j, j = 1,2\cdots r, u_k \in P_k, k = 1,2\cdots N \quad (2.33)$$

The minimization can be achieved iteratively by alternate minimizations with respect to u_k and z_{jk}, then the iteration has the form

$$u_k = \underset{\varsigma_k \in P_k}{\arg\min}\left\{f_k(u_k) + \sum_{\{k/k\in I(j)\}}\left\{p_{jk}(t)e_{jk}\varsigma_k + \frac{c(t)}{2}\left(e_{jk}\varsigma_k - z_{jk}\right)^2\right\}\right\},$$

$$\forall k = 1, 2\cdots N$$

$$\{z_{jk}/k \in I(j)\} = \underset{\left\{\varsigma_{jk}\backslash k\in I(j), \sum_{k\in I(j)}\varsigma_{jk}=m_j\right\}}{\arg\min}\left\{-\sum_{k\in I(j)} p_{jk}(t)\varsigma_{jk} + \frac{c(t)}{2},\right.$$

$$\left.\sum_{k\in I(j)}\left(e_{jk}\varsigma_k - \varsigma_{jk}\right)\right\}, \forall j = 1, 2\cdots r \quad (2.34)$$

where the above minimization with respect to $\{\varsigma_{jk}/k\in I(j)\}$ involves a separable quadratic cost and a equality constant, the minimum is attained for

$$\varsigma_{jk} = e_{jk}u_k + \frac{p_{jk}(t) - \lambda_j}{c(t)}, j = 1,2\cdots r, k \in I(j) \quad (2.35)$$

where λ_j is a scalar Lagrangian multiplier, the choice of λ_j must satisfy that

$$\sum_{k\in I(j)} \varsigma_{jk} - m_j$$

It is equivalent to

$$\lambda_j = \frac{1}{n_j}\sum_{k\in I(j)} p_{jk}(t) + \frac{c(t)}{n_j}\sum_{k\in I(j)}\left(e_{jk}u_k - m_j\right), \forall j = 1,2\cdots r \quad (2.36)$$

where n_j is the number of elements of set $I(j)$, that is

$$n_j = |I(j)|$$

From (2.35), the optimal values $z_{jk}(t+1)$ are given by

$$z_{jk}(t+1) = e_{jk}u_k(t+1) + \frac{p_{jk}(t) - \lambda_j(t+1)}{c(t)}, j = 1, 2 \cdots r, k \in I(j) \qquad (2.37)$$

Comparing (2.37) and (2.32), we see that

$$p_{jk}(t+1) = {}_j(t+1), j = 1, 2 \quad r, k \in I(j)$$

Rewriting the multiplier update formula (2.32) and adding then, we obtain

$$\lambda_j(t+1) = \lambda_j(t) + \frac{c(t)}{n_j}\left[\sum_{k \in I(j)}\left(e_{jk}u_k(t+1) - z_{jk}(t+1)\right)\right], j = 1, 2 \cdots r \qquad (2.38)$$

or equivalently

$$\lambda_j(t+1) = \lambda_j(t) + \frac{c(t)}{n_j}\left(e_j u(t+1) - m_j\right), j = 1, 2 \cdots r \qquad (2.39)$$

By replacing $p_{jk}(t)$ by $\lambda_j(t)$ in (2.35) and (2.36), the update formula for z_{jk} is obtained.

$$z_{jk} = e_{jk}u_k + \frac{\lambda_j(t) - \lambda_j}{c(t)}, j = 1, 2 \cdots r, k \in I(j) \qquad (2.40)$$

where λ_j is given by

$$\lambda_j = \lambda_j(t) + \frac{c(t)}{n_j}\sum_{k \in I(j)}\left(e_{jk}u_i - m_j\right), j = 1, 2 \cdots r \qquad (2.41)$$

Combining these two equations, the iteration for z_{jk} becomes

$$z_{jk} = e_{jk}u_k - \frac{1}{n_j}\left(e_j u - m_j\right), j = 1, 2 \cdots r, k \in I(j) \qquad (2.42)$$

By substituting (2.42) into (2.34) to eliminate z_{jk}, the parallel distributed iteration for minimizing the Augmented Lagrangian is given by

$$u_k = \arg\min_{\varsigma_k \in P_k}\left\{f_k(\varsigma_k) + \sum_{\{j \backslash k \in I(j)\}}\left\{\lambda_j(t)e_{jk}\varsigma_k + \frac{c(t)}{2}\right.\right.$$

$$\left.\left.\left(e_{jk}(\varsigma_k - u_k) + \frac{1}{n_j}\left(e_j u - m_j\right)\right)^2\right\}\right\}, k \in I(j) \qquad (2.43)$$

The above parallel distributed algorithm for constrained condition can be modified as the classical alternating direction method of multipliers.

2.2.4 Simulation example

Now we propose a simulation example to illustrate the nature of the above results. In this simulation example, the unknown regressive structure is assumed as follows

$$y(k) = 0.1u(k-1) + 0.2u(k-2) + e(k) = (u(k-1) \quad u(k-2)) \begin{pmatrix} 0.1 \\ 0.2 \end{pmatrix} + e(k)$$

Setting regression vector as $\varphi^T(k) = (u(k-1) \quad u(k-2))$, and seeking an explanatory interval predictor model of the form

$$y(k) = \varphi^T(k)\theta(k) + e(k), |e(k)| \le \gamma$$

In order to fit this above interval predictor model to the measured data, we choose $u(k) = \cos(k)$ and collect $N = 500$ observations as the data sequence $D_N = \{\varphi(k), y(k)\}_{k=1}^N$. Choosing that size measurement as $\mu = \gamma + 0.8r$, and solving the linear programming problem by the Newton method on the basis of our measured data $D_N = \{\varphi(k), y(k)\}_{k=1}^N$, we obtain one optimal center $\theta = \begin{pmatrix} 0.1 \\ 0.18 \end{pmatrix}$

with bounded radius $r = 0.2$, and level of magnitude bound $\gamma = 0.1$. The resulting interval predictor model is shown in Fig. 2.1, with the measured data clustered around at the point $\theta = \begin{pmatrix} 0.1 \\ 0.18 \end{pmatrix}$.

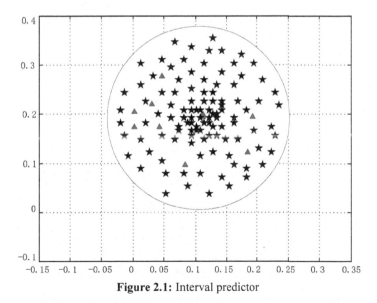

Figure 2.1: Interval predictor

2.2.5   Conclusion

This section aims to reduce the gap among system identification, numerical optimization and model predictive control. Model predictive control strategy is considered from the system identification theoretical perspective, and the numerical optimization theory is applied to solve the optimization problem, then the optimal control input for model predictive control is obtained. An interval predictor is presented to obtain new predictors to be used in the bounded error identification case, then the midpoint of the interval predictor is used as system output estimate. To obtain the minimum interval predictor, two optimal vectors are derived to be piecewise affine forms. After introducing this system output estimate into model predictive control, the property that controls input only exists in its own cost function inspires us to use parallel distributed algorithms to solve the unconstrained or constrained optimization problem.

2.3   Application of interval predictor model into model predictive control

In this section, we continue to do deep research for the construction of the interval prediction model and then apply it to the model predictive control strategy. The contribution of the paper is two folds. First, after introducing the family of models under study and some basic information about the interval predictor model, we present the computational results for constructing the interval predictor model using a linear regression structure, whose regression parameters are included in a sphere. Given a size measure to scale the average amplitude of the predictor interval, one optimal model that minimizes a size measure is efficiently computed by solving a linear programming problem, our first contribution is to apply the active set approach to solve this linear programming problem and then propose a Newton iterative form for the optimization variables. Based on these optimization variables, the predictor interval of the considered model with a sphere parameter set can be directly constructed. Furthermore, as for a fixed non-negative number coming from the size measure, we propose a good choice by using the Karush-Kuhn-Tucker optimality conditions. Based on this constructed interval predictor model, the midpoint of this interval predictor can be used as a central estimate, then the prediction of the output value can be obtained by that central estimate. The goal of model predictive control is to control the system to track a desired output reference and reject disturbances. Moreover, the considered controller may enforce input and output constraints. So after introducing that central estimate corresponding to the prediction of the output value, such a control objective can be formulated as a quadratic programming problem with an inequality constrained condition. It is well known that the dual of the quadratic programming problem is an unconstrained optimization problem [13]. After simple but tedious calculations, we formulate those input and output constraints into a standard inequality form and give a detailed process about how to derive the dual of the quadratic programming problem. As the dual problem has a simple constraint set, so it is amenable to the use of the Gauss-Seidel algorithm, whose convergence can be shown, if the step-size parameter is chosen appropriately.

2.3.1 Interval predictor model

Interval predictor model returns an interval as output. The following concepts can be seen in [14] and [15]. Define $\Phi \subseteq R^n$ and $Y \subseteq R$ be given sets, and they are denoted as the instance set and outcome set. The interval predictor model is a rule that assigns to each instance vector $\varphi \in \Phi$ a corresponding output interval. An interval predictor model is a set valued map

$$I : \varphi \rightarrow I(\varphi) \subseteq Y \tag{2.44}$$

where φ is a regressor vector, $I(\varphi)$ is the predictor interval, also $I(\varphi)$ is called an informative interval.

Consider the parametric model family M, the output of a system is expressed as $y = M(\varphi, q)$, for some parameters $q \in Q \subseteq R^{n_q}$. Through selecting a feasible set Q, an interval predictor model is obtained as the following relation:

$$M = \left\{ y = M(\varphi, q), q \in Q \subseteq R^{n_q} \right\} \tag{2.45}$$

In a dynamic setting, at each time instant the instance vector φ may contain past values of input and output measurements, then behave as a linear regression function. From standard auto-regressive structures, a parametric interval predictor model is derived.

$$y(k) = \varphi^T(k)\theta(k) + e(k), |e(k)| \leq \gamma \tag{2.46}$$

where $y(k)$ and $\varphi(k)$ denote the output measurement and regressor vector at time instant k, $\theta(k) \in R^n$ is the time varying unknown parameter, $e(k)$ is the external noise. But here any prior probability information of noise $e(k)$ is unknown. We only assume that noise $e(k)$ is unknown but bounded and γ is its magnitude bound.

Assume time varying unknown parameter $\theta(k) \in R^n$ satisfies

$$\theta(k) \in \Delta \subseteq R^n \tag{2.47}$$

where Δ is one assigned bounded set. Here we assume Δ is a sphere with center θ and radius r.

$$\Delta = \left\{ \theta + \delta : \theta, \delta \in R^n, \|\delta\| \leq r \right\} \tag{2.48}$$

Combining equations (2.45), (2.46) and (2.48), the parameters indicating the feasible set Q are the center θ and radius of sphere Δ, and the magnitude bound γ on noise $e(k)$.

Substituting (2.48) into (2.46), we obtain the output of the system.

$$y(k) = \varphi^T(k)(\theta + \delta) + e(k) = \varphi^T(k)\theta + \varphi^T(k)\delta + e(k) \tag{2.49}$$

Using the bounded radius r and magnitude bound γ, regression vector $\varphi(k)$, the output of the parametric model is one interval.

$$I\left(\varphi(k)\right) = \left[\varphi^T(k)\theta - \left(r\|\varphi(k)\| + \gamma\right), \ \varphi^T(k)\theta + \left(r\|\varphi(k)\| + \gamma\right)\right] \qquad (2.50)$$

Equation (2.50) is one interval model, it contains the output of the parametric model $y(k)$ with some guaranteed probability. When the observations are collected in the data sequence $D_N = \left\{\varphi(k), y(k)\right\}_{k=1}^{N}$, whether open or closed loop, the following relation holds:

$$y(k) \in I\left(\varphi(k)\right), for \ k = 1, 2 \cdots N \qquad (2.51)$$

where equation (2.51) means that the interval $I(\varphi(k))$ is consistent with a given data sequence D_N.

Observing the interval (2.50) again, we see that this interval is dependent on three parameters – (θ, r, γ). So if these three parameters are identified, the interval can be constructed based on equation (2.50). In order to obtain these three parameters, one linear programming problem is constructed.

Introducing a size measure $\mu = \gamma + ar$, where a is a fixed non-negative number, the optimal model that minimizes μ can be derived by solving a linear programming problem.

Theorem 2.5: Given an observed data sequence $D_N = \left\{\varphi(k), y(k)\right\}_{k=1}^{N}$, a model order n, and a size objective $\mu = \gamma + ar$, three parameters used to construct the optimal interval predictor model are computed by solving the following linear programming problem with respect to three variables

$$\theta \in R^n, \gamma, r$$

$$\begin{cases} \min\limits_{\theta,\gamma,r} \quad \gamma + ar \\ \text{subject to } \varphi^T(k)\theta - r\|\varphi(k)\| - \gamma \le y(k) \\ -\varphi^T(k)\theta - r\|\varphi(k)\| - \gamma \le -y(k), k = 1, 2 \cdots N \end{cases} \qquad (2.52)$$

According to the linear programming problem (2.52), there are no any references on how to solve it, so the main contributions of the next two sections are to solve this linear programming problem and choose an appropriate fixed non-negative number.

2.3.2   One choice of a fixed non-negative number

In linear programming problem (2.52), as the optimization variables r and γ denote the radius of sphere Δ and magnitude bound on noise $e(k)$, so these two optimization variables must satisfy that

$$r \ge 0 \text{ and } \gamma \ge 0 \qquad (2.53)$$

Combining linear programming problem (2.52) and inequality constraints (2.53), we rewrite the new linear programming problem as

$$\begin{cases} \min_{\theta,\gamma,r} \quad \gamma + ar \\ \text{subject to} \;\; \varphi^T(k)\theta - r\|\varphi(k)\| - \gamma \le y(k) \\ -\varphi^T(k)\theta - r\|\varphi(k)\| - \gamma \le -y(k), k = 1,2\cdots N \\ r \ge 0 \text{ and } \gamma \ge 0 \end{cases} \tag{2.54}$$

Define the Lagrangian function L corresponding to the above linear programming problem by

$$L\left(\theta,\gamma,r,\lambda_1,\lambda_2,\mu_k^+,\mu_k^-\right) = \gamma + ar - \lambda_1\gamma - \lambda_2 r$$

$$-\sum_{k=1}^{N}\mu_k^+\left(y(k) - \varphi^T(k)\theta + r\|\varphi(k)\| + \gamma\right)$$

$$-\sum_{k=1}^{N}\mu_k^-\left(-y(k) + \varphi^T(k)\theta + r\|\varphi(k)\| + \gamma\right) \tag{2.55}$$

We refer to $\lambda_1,\lambda_2,\left\{\mu_k^+,\mu_k^-\right\}_{k=1}^{N}$ as the Lagrangian multipliers. By applying the optimality KKT sufficient and necessary condition on Lagrangian function, then some equality relations for the optimal solution hold.

$$\frac{\partial L}{\partial \theta} = \sum_{k=1}^{N}\left(\mu_k^+ - \mu_k^-\right)\varphi(k) = 0 \tag{2.56}$$

$$\frac{\partial L}{\partial \gamma} = 1 - \lambda_1 - \sum_{k=1}^{N}\left(\mu_k^+ + \mu_k^-\right) = 0 \tag{2.57}$$

$$\frac{\partial L}{\partial r} = a - \lambda_2 - \sum_{k=1}^{N}\left(\mu_k^+ + \mu_k^-\right)\|\varphi(k)\| = 0 \tag{2.58}$$

$$\lambda_1\gamma = 0, \quad \lambda_2 r = 0 \tag{2.59}$$

$$\mu_k^+\left(y(k) - \varphi^T(k)\theta + r\|\varphi(k)\| + \gamma\right) = 0 \tag{2.60}$$

$$\mu_k^-\left(-y(k) + \varphi^T(k)\theta + r\|\varphi(k)\| + \gamma\right) = 0 \tag{2.61}$$

Also as optimization variables r and γ denote the radius of sphere Δ and magnitude bound on noise $e(k)$ respectively, if $\gamma = 0$, then that $|e(k)| = 0$ means no noise exists in the standard auto-regressive structure (2.46). If $r = 0$, then sphere Δ reduces to its center θ, so here for interval predictor model, we want to satisfy that

$$r > 0 \text{ and } \gamma > 0 \tag{2.62}$$

When r and γ are all equal to zero, then $I(\varphi(k))$ is not an interval, but a fixed output value. The midpoint of the interval predictor model is a central estimate here.

$$I\left(\varphi(k)\right) = \frac{\varphi^T\left(k\right)\theta - \left(r\left\|\varphi(k)\right\| + \gamma\right) + \varphi^T\left(k\right)\theta + \left(r\left\|\varphi(k)\right\| + \gamma\right)}{2} = \varphi^T\left(k\right)\theta$$

Comparing (2.59) and (2.62), we see that equation (2.59) holds unless Lagrangian multipliers $\{\lambda_1, \lambda_2\}$ must satisfy by

$$\lambda_1 = \lambda_2 = 0 \tag{2.63}$$

Further in equation (2.63), assume regression vectors $\varphi(1), \varphi(2)\cdots\varphi(N)$ at different instants are linearly independent. So in order to let equation (2.56) hold, the $\left\{\mu_k^+, \mu_k^-\right\}_{k=1}^N$ need to satisfy by

$$\mu_k^+ = \mu_k^- = \mu_k, k = 1, 2\cdots N \tag{2.64}$$

Substituting (2.63) and (2.64) into (2.57) and (2.58), we obtain the following simplified forms

$$\begin{cases} \displaystyle\sum_{k=1}^N \mu_k = \frac{1}{2} \\ \displaystyle\sum_{k=1}^N \mu_k \left\|\varphi(k)\right\| = \frac{a}{2} \end{cases} \tag{2.65}$$

From the idea of equation (2.65), one choice of this fixed non-negative number a is given here. As $\displaystyle\sum_{k=1}^N \mu_k = \frac{1}{2}$ holds, we set $\mu_k = \frac{1}{2N}$. Then after substituting $\mu_k = \frac{1}{2N}$ into equality $\displaystyle\sum_{k=1}^N \mu_k \left\|\varphi(k)\right\| = \frac{a}{2}$, we have

$$\frac{1}{2N}\left[\left\|\varphi(1)\right\| + \left\|\varphi(2)\right\| + \cdots + \left\|\varphi(N)\right\|\right] = \frac{a}{2} \tag{2.66}$$

It means that

$$a = \frac{1}{N}\left[\left\|\varphi(1)\right\| + \left\|\varphi(2)\right\| + \cdots + \left\|\varphi(N)\right\|\right] \tag{2.67}$$

Equation (2.66) is one of the choices of that fixed non-negative number a. As regressor vectors $\left\{\varphi(k)\right\}_{t=1}^N$ and the number of data are given, so the form (2.67) of a can be computed easily.

2.3.3 Newton method for interval predictor model

Because interval predictor model (2.50) is dependent on the linear programming problem (2.54) with respect to three kinds of optimization variables $\{\theta, \gamma, r\}$, so the important step in constructing the interval predictor model (2.50) is to solve that linear programming problem (2.54). First we rewrite the linear programming problem (2.54) as its standard form. Define a new vector $x \in R^{n+2}$ as $x = \{\theta, \gamma, r\}$. Based on the new optimization vector $x \in R^{n+2}$, the cost function in (2.54) can be rewritten as

$$\gamma + ar = \begin{pmatrix} 0 & 1 & a \end{pmatrix} \begin{pmatrix} \theta \\ \gamma \\ r \end{pmatrix} = C^T x; C^T = \begin{pmatrix} 0 & 1 & a \end{pmatrix} \tag{2.68}$$

Also each inequality can be rewritten as

$$\left\{ \begin{array}{l} \varphi^T(k)\theta - r\|\varphi(k)\| - \gamma \le y(k) \Leftrightarrow \begin{pmatrix} \varphi^T(k) & -1 & -\|\varphi(k)\| \end{pmatrix} \begin{pmatrix} \theta \\ \gamma \\ r \end{pmatrix} \le y(k) \\[3mm] -\varphi^T(k)\theta - r\|\varphi(k)\| - \gamma \le -y(k) \Leftrightarrow \begin{pmatrix} -\varphi^T(k) & -1 & -\|\varphi(k)\| \end{pmatrix} \begin{pmatrix} \theta \\ \gamma \\ r \end{pmatrix} < -y(k) \\[3mm] r \ge 0 \Leftrightarrow \begin{pmatrix} 0 & 0 & -1 \end{pmatrix} \begin{pmatrix} \theta \\ \gamma \\ r \end{pmatrix} \le 0, \quad \gamma \ge 0 \Leftrightarrow \begin{pmatrix} 0 & -1 & 0 \end{pmatrix} \begin{pmatrix} \theta \\ \gamma \\ r \end{pmatrix} \le 0 \end{array} \right. \tag{2.69}$$

Define some matrices to merge all inequities in equation (2.69).

$$A = \begin{bmatrix} 0 & -1 & 0 \\ 0 & 0 & -1 \\ \varphi^T(1) & -1 & -\|\varphi(1)\| \\ \vdots & \vdots & \vdots \\ \varphi^T(N) & -1 & -\|\varphi(N)\| \\ -\varphi^T(1) & -1 & -\|\psi(1)\| \\ \vdots & \vdots & \vdots \\ -\varphi^T(N) & -1 & -\|\varphi(N)\| \end{bmatrix}, \quad B = \begin{bmatrix} 0 \\ 0 \\ y(1) \\ \vdots \\ y(N) \\ y(1) \\ \vdots \\ -y(N) \end{bmatrix} \tag{2.70}$$

where $A \in R^{(2N+2)\times 3}, B \in R^{(2N+2)\times 1}$. Given regression vector $\{\varphi(k)\}_{k=1}^{N}$ and output measured data $\{y(k)\}_{k=1}^{N}$, the above two matrices A, B are known.

Applying two matrices A, B, all inequities in equation (2.69) are obtained in a simplified form.

$$Ax \le B \tag{2.71}$$

Then based on (2.68) and (2.71), the formal linear programming problem (2.54) can be formulated into a standard linear programming form.

$$\begin{cases} \min_{x} \ C^T x \\ \text{subject to } Ax \le B \end{cases} \tag{2.72}$$

Now we propose a Newton method to solve the above standard linear programming problem. Our Newton method introduces the active set approach into the classical Newton approach. The active set approach introduced here is based on a transformation by means of which the optimality KKT conditions are converted into a system of nonlinear equations.

Firstly for a fixed scalar $c > 0$, consider the open set $S_c^* \subset R^{2N+2} \times R^{2N+2}$ defined by

$$S_c^* = \left\{ (x,m) / m_j + cA_j x, j = 1, 2 \cdots 2N + 2 \right\} \tag{2.73}$$

where A_j is the j column of matrix A, and the system of equations on S_c^*.

$$\begin{cases} C + \nabla_x g^+ (x,m,c) m = 0 \\ g^+ (x,m,c) = 0 \end{cases} \tag{2.74}$$

where the function $g^+ (x,m,c)$ is defined by

$$g^+ (x,m,c) = \begin{bmatrix} g_1^+ (x,m_1,c) \\ \vdots \\ g_{2N+2}^+ (x,m_{2N+2},c) \end{bmatrix}; \ g_j^+ (x,m_j,c)$$

$$= \max \left\{ A_j x, -\frac{m_j}{c} \right\}, j = 1, 2 \cdots 2N + 2 \tag{2.75}$$

Note that $g^+(x, m, c)$ is differentiable on S_c^* as $Ax - B$, so equations (2.74), (2.75) are well defined. $g^+(x, m, c)$ appears in the definition of the augmented Lagrangian function, which takes the form as

$$L_c (x,m,c) = C^T x + m' g^+ (x,m,c) + \frac{1}{2} c \left| g^+ (x,m,c) \right|^2$$

Secondly consider the implementation of Newton approach. Define for $(x,m) \in S_c^*$.

$$L^+ (x,m,c) = C^T x + m' g^+ (x,m,c)$$

$$A_c (x,m) = \left\{ j / A_j x - B_j > -\frac{m_j}{c}, j = 1, 2 \cdots 2N + 2 \right\} \tag{2.76}$$

Assume without loss of generality that $A_c(x,m) = \{1 \cdots p\}$ for one integer p. This integer p depends on x and m. We view $A_c(x, m)$ as the active index set, in the sense that indices in $A_c(x, m)$ are predicted by the algorithm to be active at the solution. By differentiation in equation (2.74), we propose that the Newton method consists of the iteration

$$\bar{x} = x + \Delta x, \quad \bar{m} = m + \Delta m \tag{2.77}$$

where $(\Delta x, \Delta m)$ is the solution of the following system:

$$\begin{bmatrix} \nabla^2_{xx}\{m'g^+(x,m,c)\} & N(x,m,c) & 0 \\ N(x,m,c) & 0 & 0 \\ 0 & 0 & -\frac{1}{c}I \end{bmatrix} \begin{bmatrix} \Delta x \\ \Delta m_1 \\ \vdots \\ \Delta m_p \\ \Delta m_{p+1} \\ \vdots \\ \Delta m_{2N+2} \end{bmatrix} = - \begin{bmatrix} C + \nabla_x\{m'g^+(x,m,c)\} \\ g_1^+(x,m,c) \\ \vdots \\ g_p^+(x,m,c) \\ g_{p+1}^+(x,m,c) \\ \vdots \\ g_{2N+2}^+(x,m,c) \end{bmatrix} \tag{2.78}$$

where $N(x,m,c)$ is the $(2N+2) \times p$ matrix having as columns the gradients $A_j, j \in A_c(x,m)$, I is the $(2N+2-p) \times (2N+2-p)$ identity matrix, and the zero matrices have appropriate dimension. Since we see that

$$g_j^+(x,m,c) = -\frac{m_j}{c}, \forall j \notin A_c(x,m)$$

It follows that

$$\bar{m}_j = 0, \forall j \notin A_c(x,m)$$

It follows from equation (2.78) that the remaining variables Δx and $\Delta m_1 \cdots \Delta m_p$ are obtained by solving the reduced system.

$$\begin{bmatrix} \nabla^2_{xx}\{m'g^+(x,m,c)\} & N(x,m,c) \\ N(x,m,c) & 0 \end{bmatrix} \begin{bmatrix} \Delta x \\ \Delta m_1 \\ \vdots \\ \Delta m_p \end{bmatrix} = - \begin{bmatrix} C + \nabla_x\{m'g^+(x,m,c)\} \\ A_1 x - B_1 \\ \vdots \\ A_p x - B_p \end{bmatrix} \tag{2.79}$$

where we make use of the fact that

$$g_j^+(x,m,c) = g_j(x), \forall j \in A_c(x,m)$$
$$\nabla_x L^+(x,m,c) = C + \sum_{j \in A_c(x,m)} (A_j - B_j) m_j \tag{2.80}$$

From above equations, we see that the proposed Newton iteration can be described in a simpler manner.

2.3.4 Application interval predictor model into model predictive control

In this section, we start to apply the above obtained interval predictor model into model predictive control strategy. Set the optimal variables corresponding to the linear programming problem (2.52) as follows: $\hat{\theta}, \hat{\gamma}, \hat{r}$

Then based on these three optimal variables $\left(\hat{\theta}, \hat{\gamma}, \hat{r}\right)$, the interval predictor model $I\left(\varphi(k)\right)$ is defined as

$$I\left(\varphi(k)\right) = \left[\varphi^T(k)\hat{\theta} - \left(\hat{r}\|\varphi(k)\| + \hat{\gamma}\right), \ \varphi^T(k)\hat{\theta} + \left(\hat{r}\|\varphi(k)\| + \hat{\gamma}\right)\right] \quad (2.81)$$

Then from the probability theory, the predictor of output value will be included in this confidence interval with one guaranteed accuracy, i.e.

$$y(k) \in I\left(\varphi(k)\right), \forall\, k = 0,1,2\cdots N \quad (2.82)$$

2.3.4.1 Model predictive control problem

As the goal of model predictive control is to control the system in order to track a desired output reference and reject disturbances from $k = 0$ up to some finite time step N, where this time step N can be very large.

Assuming that the control input $u(l), l = -1, -2, \cdots$ are known, then a control objective can be formalized by one optimization problem.

$$\min_{u(0),u(1)\cdots u(N)} \sum_{k=0}^{N} \left[y(k) - y_{des}(k)\right]^T Q_1 \left[y(k) - y_{des}(k)\right] + u^T(k) S_1 u(k) \quad (2.83)$$

where $y_{des}(k)$ is the desired output reference, Q_1, S_1 are positive semi-definite weighting matrices selected by the designer. Observing (2.82) and (2.83), the predictor of output value $y(k)$, $k = 0, 1, 2 \cdots N$ exists in cost function (2.83), but the only knowledge about the predictor of output value is that $y(k)$ is included in confidence interval $I(\varphi(k))$, so there are two cases to be considered here. One case is to expand the cost function (2.83), then a numerical value about $y(k)$ is needed to substitute into cost function, i.e. the midpoint of that confidence interval can be used as the prediction of output value $y(k)$.

$$I\left(\varphi(k)\right) = \frac{\varphi^T(k)\hat{\theta} - \left(\hat{r}\|\varphi(k)\| + \hat{\gamma}\right) + \varphi^T(k)\hat{\theta} + \left(\hat{r}\|\varphi(k)\| + \hat{\gamma}\right)}{2} = \varphi^T(k)\hat{\theta} \quad (2.84)$$

Replacing $y(k)$ as $\varphi^T(k)\hat{\theta}$ into cost function (2.83), we obtain an explicit form corresponding to optimization problem.

$$\min_{u(0),u(1)\cdots u(N)} \sum_{k=0}^{N} \left[\varphi^T(k)\hat{\theta} - y_{des}(k) \right]^T Q_1 \left[\varphi^T(k)\hat{\theta} - y_{des}(k) \right] + u^T(k)S_1 u(k)$$

$$(2.85)$$

The second case is to use confidence interval $I(\varphi(k))$ into the optimization problem directly, but another max operation is added, i.e. a robust model predictive control problem is obtained.

$$\min_{u(0),u(1)\cdots u(N)} \max_{y(k)\in I(\varphi(k))} \sum_{k=0}^{N} \left[y(k) - y_{des}(k) \right]^T Q_1 \left[y(k) - y_{des}(k) \right] + u^T(k)S_1 u(k)$$

$$(2.86)$$

Here, we only consider the first case (2.84) and the second case (2.85) corresponding to a robust model predictive control problem will be solved later. Further, the controller will satisfy input and output constraints as

$$\begin{cases} u_{min} \le u(i) \le u_{max} \\ y_{min} \le y(i) \le y_{max} \end{cases} \quad i = 0,1\cdots N \qquad (2.87)$$

Combining (2.85) and (2.87), the model prediction control problem is formalized by the following optimization problem with inequality constrained condition.

$$\min_{u(0),u(1)\cdots u(N)} \sum_{k=0}^{N} \left[\varphi^T(k)\hat{\theta} - y_{des}(k) \right]^T Q_1 \left[\varphi^T(k)\hat{\theta} - y_{des}(k) \right] + u^T(k)S_1 u(k)$$

subject to $\begin{cases} u_{min} \le u(i) \le u_{max} \\ y_{min} \le y(i) \le y_{max} \end{cases} \quad i = 0,1\cdots N$

$$(2.88)$$

where u_{min}, u_{max} and y_{min}, y_{max} denote the lower and upper bounds on input and output respectively. Before solving the optimization problem (2.88), we analyze the cost function and inequality constrained condition in (2.88). The second term in cost function can be rewritten as

$$\sum_{k=0}^{N} u^T(k)S_1 u(k) = \begin{bmatrix} u(0) & u(1) & \cdots & u(N) \end{bmatrix} \begin{bmatrix} S_1 & 0 & \cdots & 0 \\ 0 & S_1 & \cdots & 0 \\ \vdots & \vdots & \vdots & \vdots \\ 0 & 0 & \cdots & S_1 \end{bmatrix} \begin{bmatrix} u(0) \\ u(1) \\ \vdots \\ u(N) \end{bmatrix} = u^T S u$$

$$(2.89)$$

where vector u and matrix S are defined as

$$u = \begin{bmatrix} u(0) & u(1) & \cdots & u(N) \end{bmatrix}^T, S = diag\left(S_1 \quad S_1 \quad \cdots \quad S_1 \right)$$

It means that vector u includes all optimal input at all time step i, $i = 0, 1\cdots$ N. The first term in cost function can be also reformulated as

$$\min_{u(0),u(1)\cdots u(N)} \sum_{k=0}^{N} \left[\varphi^T(k)\hat{\theta} - y_{des}(k) \right]^T Q_1 \left[\varphi^T(k)\hat{\theta} - y_{des}(k) \right]$$

$$= \sum_{k=0}^{N} \varphi^T(k)\hat{\theta}Q_1\varphi^T(k)\hat{\theta} - 2\varphi^T(k)\hat{\theta}Q_1 y_{des}(k) + \sum_{k=0}^{N} y_{des}(k)^T Q_1 y_{des}(k)$$

As no input $u(i)$ exists in term $\sum_{k=0}^{N} y_{des}(k)^T Q_1 y_{des}(k)$, then this term can be omitted. For clarity of presentation, set the control inputs $u(l), l = -1, -2, \cdots N$ be zero, if not, we use coordinate transformation to satisfy it. Due to $\hat{\theta} \in R^n$, then regressor vector $\varphi(k)$ and $\hat{\theta}$ can be defined as follows.

$$\begin{cases} \varphi(k) = \left[u(k-1) \quad u(k-2) \quad \cdots \quad u(k-N) \right]^T \\ \hat{\theta} = \left[\hat{\theta}_1 \quad \hat{\theta}_2 \quad \cdots \quad \hat{\theta}_N \right]^T \end{cases}$$

From above descriptions, the following relations hold:

$$\varphi^T(0)\hat{\theta} = \left[u(-1) \quad u(-2) \quad \cdots \quad u(-N) \right] \begin{bmatrix} \hat{\theta}_1 \\ \hat{\theta}_2 \\ \vdots \\ \hat{\theta}_N \end{bmatrix} = 0$$

$$\varphi^T(1)\hat{\theta} = \left[u(0) \quad u(-1) \quad \cdots \quad u(1-N) \right] \begin{bmatrix} \hat{\theta}_1 \\ \hat{\theta}_2 \\ \vdots \\ \hat{\theta}_N \end{bmatrix} = \left[u(0) \quad 0 \quad \cdots \quad 0 \right] \begin{bmatrix} \hat{\theta}_1 \\ \hat{\theta}_2 \\ \vdots \\ \hat{\theta}_N \end{bmatrix}$$

$$= \left[u(0) \quad u(1) \quad \cdots \quad u(N) \right] \begin{bmatrix} 1 & 0 & \cdots & 0 \\ 0 & 0 & \cdots & 0 \\ \vdots & & \vdots & \\ 0 & 0 & \cdots & 0 \end{bmatrix} \begin{bmatrix} \hat{\theta}_1 \\ \hat{\theta}_2 \\ \vdots \\ \hat{\theta}_N \end{bmatrix} = u^T I_1 \hat{\theta}$$

$$\varphi^T(2)\hat{\theta} = \left[u(1) \quad u(0) \quad \cdots \quad 0 \right] \begin{bmatrix} \hat{\theta}_1 \\ \hat{\theta}_2 \\ \vdots \\ \hat{\theta}_N \end{bmatrix}$$

$$= \begin{bmatrix} u(0) & u(1) & \cdots & u(N) \end{bmatrix} \begin{bmatrix} 0 & 1 & \cdots & 0 \\ 1 & 0 & \cdots & 0 \\ \vdots & & \vdots \\ 0 & 0 & \cdots & 0 \end{bmatrix} \begin{bmatrix} \hat{\theta}_1 \\ \hat{\theta}_2 \\ \vdots \\ \hat{\theta}_N \end{bmatrix} = u^T I_2 \hat{\theta}$$

$$\varphi^T(N)\hat{\theta} = \begin{bmatrix} u(N-1) & u(N-2) & \cdots & u(N-n) \end{bmatrix} \begin{bmatrix} \hat{\theta}_1 \\ \hat{\theta}_2 \\ \vdots \\ \hat{\theta}_N \end{bmatrix}$$

$$= \begin{bmatrix} u(0) & u(1) & \cdots & u(N) \end{bmatrix} \begin{bmatrix} 0 & 0 & \cdots & 1 & 0 \\ 0 & 0 & \cdots & 0 & 0 \\ \vdots & & \vdots & & \\ 0 & 1 & \cdots & 0 & 0 \\ 1 & 0 & \cdots & 0 & 0 \\ 0 & 0 & \cdots & 0 & 0 \end{bmatrix} \begin{bmatrix} \hat{\theta}_1 \\ \hat{\theta}_2 \\ \vdots \\ \hat{\theta}_N \end{bmatrix} = u^T I_N \hat{\theta}$$

Substituting above relations into the first term of the cost function, we obtain

$$\sum_{k=0}^{N} \varphi^T(k)\hat{\theta} Q_1 \varphi^T(k)\hat{\theta} - 2\varphi^T(k)\hat{\theta} Q_1 y_{des}(k)$$

$$= \sum_{k=0}^{N} u^T \left[I_k \hat{\theta} Q_1 I_k^T \right] u + u^T Su - 2 \left[I_k Q_1 y_{des}(k) \right] u \tag{2.90}$$

Set

$$\sum_{k=0}^{N} \left[I_k \hat{\theta} Q_1 I_k^T + S \right] = \frac{1}{2}Q, \quad 2 \left[I_k Q_1 y_{des}(k) \right] = b$$

Then the cost function in the optimization problem (2.88) can be simplified as

$$\min_{u} \frac{1}{2} u^T Q u - bu \tag{2.91}$$

Those inequality constraints on input and output can be also reformulated as

$$u(i) \leq u_{max}, i = 0, 1 \cdots N \Rightarrow \begin{bmatrix} 1 & 0 & \cdots & 0 \\ 0 & 1 & \cdots & 0 \\ \vdots & & \vdots & \\ 0 & 0 & \cdots & 1 \end{bmatrix}$$

$$\begin{bmatrix} u(0) & u(-1) & \cdots & u(1-N) \end{bmatrix} \le \begin{bmatrix} u_{\max} \\ u_{\max} \\ \vdots \\ u_{\max} \end{bmatrix} \Rightarrow Iu \le U_{\max}$$

$$-u(i) \le u_{\min}, i = 0,1 \cdots N \Rightarrow - \begin{bmatrix} 1 & 0 & \cdots & 0 \\ 0 & 1 & \cdots & 0 \\ \vdots & & \vdots & \\ 0 & 0 & \cdots & 1 \end{bmatrix}$$

$$\begin{bmatrix} u(0) & u(-1) & \cdots & u(1-N) \end{bmatrix} \le \begin{bmatrix} -u_{\min} \\ -u_{\min} \\ \vdots \\ -u_{\min} \end{bmatrix} \Rightarrow -Iu \le -U_{\min}$$

$$y(i) \le y_{\max}, i = 0,1 \cdots N \Rightarrow \varphi^T(i)\theta \le y_{\max}$$

$$\Rightarrow \theta^T \underbrace{\begin{bmatrix} I_1 & 0 & \cdots & 0 \\ 0 & I_2 & \cdots & 0 \\ \vdots & & \vdots & \\ 0 & 0 & \cdots & I_N \end{bmatrix}}_{I_\Delta} u \le \begin{bmatrix} y_{\max} \\ y_{\max} \\ \vdots \\ y_{\max} \end{bmatrix} \Rightarrow I_\Delta u \le Y_{\max}$$

Similarly, we have that

$$-y(i) \le -y_{\min}, i = 0,1 \cdots N \Rightarrow -I_\Delta u \le -Y_{\min}$$

where I denotes identify matrix, I_i, $i = 1,2 \cdots N$ is from cost function, I_Δ is defined as

$$I_\Delta = diag[I_1 \quad I_2 \quad \dots \quad I_N]\theta$$

For notational clarity, combining above four inequalities to obtain

$$\underbrace{\begin{bmatrix} I \\ -I \\ I_\Delta \\ -I_\Delta \end{bmatrix}}_{E} u \le \underbrace{\begin{bmatrix} U_{\max} \\ -U_{\min} \\ Y_{\max} \\ -Y_{\min} \end{bmatrix}}_{F} \Rightarrow Eu \le F \qquad (2.92)$$

Merging (2.91) and (2.92), the following quadratic programming problem is considered:

$$\min_{u} \frac{1}{2} u^T Q u - bu$$

$$\text{subject to } Eu \le F \tag{2.93}$$

In order to obtain the optimal control input u by solving the above quadratic programming problem (2.93), the following Gauss-Seidel algorithm is used.

2.3.4.2 Gauss-Seidel algorithm

To solve that quadratic programming problem (2.93) with inequality constraint, as the dual of the quadratic programming problem is an unconstrained optimization problem [18], so it is easy to solve its dual problem.

The dual function is defined as

$$q(z) = \inf_{u} \left\{ \frac{1}{2} u^T Q u - bu + z(Eu - F) \right\}$$

The infimum is attained for $u = Q^{-1}(b - Ez)$ and after substitution of this expression in the preceding relation for $q(z)$, a straightforward calculation yields

$$q(z) = \frac{1}{2} Q^{-1} (b - Ez)^T QQ^{-1} (b - Ez)$$

$$= \frac{1}{2} (b - Ez)^T Q^{-1} (b - Ez) - bQ^{-1} (b - Ez) + z \left[EQ^{-1} (b - Ez) - F \right]$$

$$= -\frac{1}{2} z^T EQ^{-1} E^T z - zF + zEQ^{-1} b \tag{2.94}$$

Then the dual of the quadratic programming problem is given by

$$\min_{z} \frac{1}{2} z^T G z + dz$$

$$\text{subject to } z \ge 0 \tag{2.95}$$

where

$$G = EQ^{-1}E^T, \quad d = f - EQ^{-1}b \tag{2.96}$$

If z^* solve the dual problem (2.95), then $u^* = Q^{-1}(b - Ez^*)$ solves the primal problem [18]. Let g_j denote the jth column of E, assume that g_j is nonzero for all j. Since Q is symmetric and positive define, the first partial derivative of the dual cost function with respect to z_j is given by

$$d_j + \sum_{k=1}^{N} g_{jk} z_k \tag{2.97}$$

where g_{jk} and d_j are the corresponding elements of the matrix G and the vector d, respectively. Set the derivative to be zero, the unconstrained minimum of the dual cost function along the *j*th coordinate starting from *z* is attained at $\tilde{z}_j$ given by

$$\tilde{z}_j = -\frac{1}{g_{jj}}\left(d_j + \sum_{k \neq j} g_{jk} z_k\right) = z_j - \frac{1}{g_{jj}}\left(d_j + \sum_{k=1} g_{jk} z_k\right)$$

Taking into account the non-negativity constraint $z_j \geq 0$, when the *j*th coordinate is updated, the Gauss-Seidel iteration has the form

$$\begin{cases} z_j = \max\{0, \tilde{z}_j\} = \max\left\{0, z_j - \frac{1}{g_{jj}}\left(d_j + \sum_{k=1} g_{jk} z_k\right)\right\} \\ z_i = z_i, \quad \forall i \neq j \end{cases} \tag{2.98}$$

Consider a linear projection Jacobi method, it is a special case of the Gauss-Seidel algorithm. Taking into account the first partial derivation of the dual cost function with respect to z_j, another iteration formula is given by

$$z_j(t+1) = \max\left\{0, z_j(t) - \frac{\eta}{g_{jj}}\left(d_j + \sum_{k=1} g_{jk} z_k(t)\right)\right\} \tag{2.99}$$

where $\eta > 0$ is the stepsize parameter. The above iteration is more suitable for parallelization than the Gauss-Seidel iteration. On the other hand, for convergence, the stepsize η should be chosen sufficiently small, and some experimentations may be needed to obtain the appropriate range for η. In the dual problem (2.95), let k_G be the largest eigenvalue of G, and assume that $k_G > 0$, then the Lipschitz condition corresponding to the dual function $q(z) = \frac{1}{2}z^T G z + dz$ is written as

$$\|\nabla q(z_1) - \nabla q(z_2)\| \leq k_G \|z_1 - z_2\|$$

or equivalently

$$\|G(z_1 - z_2)\| \leq k_G \|z_1 - z_2\|$$

Then from the parallel distribution algorithm [19], we obtain the following Theorem 2.6 easily.

Theorem 2.6 (Convergence of the Gauss-Seidel algorithm): Suppose that dual function $q(z)$ satisfies the Lipchitz condition, if $0 < \eta < \frac{1}{k_G} g_{jj}$ and if z^* is a limit point of the sequence $\{z(t)\}$ generated by the Gauss-Seidel algorithm, then $(z - z^*)\nabla q(z^*) \geq 0$ for all $z \geq 0$. Moreover if $q(z)$ is convex on the set $\{z/z \geq 0\}$, then z^* minimizes $q(z)$ over the set $\{z/z \geq 0\}$.

2.3.5   Simulation examples

Now, we propose a simulation example to illustrate the nature of the above results. In this simulation example, the unknown regressive structure is assumed as follows

$$y(k) = 0.1u(k-1) + 0.2u(k-2) + e(k) = \left(u(k-1) \quad u(k-2) \right) \begin{pmatrix} 0.1 \\ 0.2 \end{pmatrix} + e(k)$$

Setting regression vector as

$$\varphi^T(k) = \left(u(k-1) \quad u(k-2) \right)$$

Seeking an explanatory interval predictor model of the form

$$y(k) = \varphi^T(k)\theta(k) + e(k), |e(k)| \le \gamma$$

In order to fit this above interval predictor model to the measured data, we choose $u(k) = \cos(k)$ and collect $N = 500$ observations as the data sequence $D_N = \{\varphi(k), y(k)\}_{k=1}^N$. Choosing that size measurement as $\mu = \gamma + 0.8r$, and solving the linear programming problem by the Newton method on the basis of our measured data $D_N = \{\varphi(k), y(k)\}_{k=1}^N$, we obtain one optimal center $\theta = \begin{pmatrix} 0.1 \\ 0.18 \end{pmatrix}$ with bounded radius $r = 0.2$, and level of magnitude bound $\gamma = 0.1$. The resulting iterative estimations of the unknown parameter is shown in Fig. 2.2, with the iterative estimations shown clustering around at the point $\theta = \begin{pmatrix} 0.1 \\ 0.18 \end{pmatrix}$.

In Fig. 2.2, the center of the circle is the optimal or true value, denoted by the red triangle. The black triangles are denoted as iterative estimations, obtained by the Newton method. From Fig.2.1, we see that the iterative estimations will converge to their optimal or true value with increasing iteration steps. After substituting the optimal center into equation (2.50), our interval predictor model for output value is obtained. This interval predictor model contains the output of the parametric model with some guaranteed probability. The whole output frequency response curves are shown in Fig. 2.3, based on estimated model parameters. The red curve is the actual true amplitude curve from the Bode plot tool. When the estimated parameters are contained in the uncertainty bound with probability level 0.99, the amplitude curves lie above or below the red curve. From Fig. 2.3, we can see that these three curves are very close and the red amplitude curve lies between two confidence amplitude curves with probability level 0.99.

When using a Matlab simulation tool to simulate the output response of Bode plot in a closed loop, the phase plot is acquired with the amplitude plot, simultaneously. The confidence interval phase plot is given in Fig. 2.4 and the red phase curve also lies between two confidence phase curves with the probability level 0.99. This is similar to the derivation of Fig. 2.4. The output error between

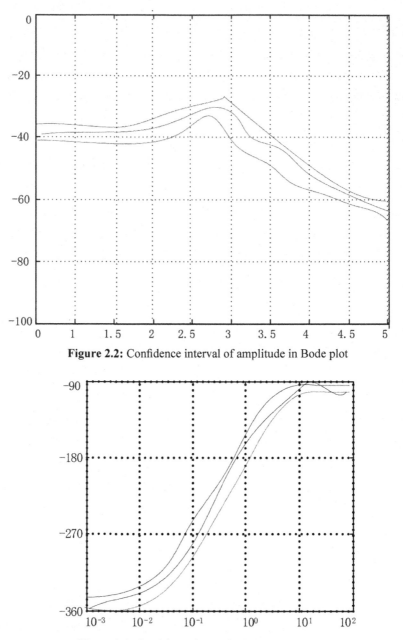

Figure 2.2: Confidence interval of amplitude in Bode plot

Figure 2.3: Confidence interval of phase in Bode plot

true output value and actual output value will converge to zero with time increases in Fig. 2.4, then this model error can be neglected in our next process in designing controllers.

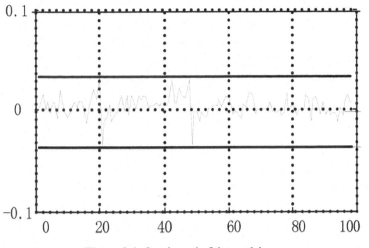

Figure 2.4: One bound of the model error

2.4   Stability analysis in cooperative distributed model predictive control

Due to the success of the MPC strategy in the process industry from the 1980s, large scale complex plants were increasingly controlled by several MPC controllers in a decentralized way [16]. Further unconstrained decentralized control had been a field of active research for several years and numerous analysis and design methods were readily available, except systemic methods for constrained controllers in MPC. In the decentralized control scheme, all control variables are computed by a single regulator. Large scale systems are always thought of as the result of many subsystems interacting through the coupling of physical variables or the transmission of information over one communication network. So all subsystems transmit their outputs to the decentralized controllers to design the control inputs, which will be sent back to actuators collected with subsystems. But the decentralized control scheme for the large scale systems will suffer from many obvious problems, such as

(1) Computational complexity, a decentralized regulator needs a considerable amount of computing power and memory to compute control inputs within a sampling interval,

(2) Communication networks, a decentralized control requires a star like topology of the communication network that impacts the cost of the control system, and

(3) Reliability, a failure in a single subsystem or a link could comprise the proper function of the overall controlled large scale system.

With decentralized MPC being readily used in large scale processes, some authors pointed out that while a completely decentralized MPC was not realistic due to stability and fault tolerance reason, some levels of coordination between the decentralized MPC could be beneficial. To avoid the above mentioned drawbacks, each subsystem is equipped with a local controller to receive the outputs from the corresponding subsystem and compute the control input. This idea is called the distributed control scheme, and it means that each subsystem is regulated by one local controller. This local controller transmits or receives quantities from other subsystems. In a distributed control scheme, the design process for a local controller uses only information from the parent subsystem. The communication flow process at the design stages has the same topology as the coupling graph. After each parent subsystem sends the required knowledge to its children, the design for a local controller can be achieved in parallel form. The complexity of synthesizing a local controller for a subsystem scales only with the number of its parents rather than the total number of subsystems. Furthermore, if a subsystem joins an existing network, then at the most, subsystems that are influenced by it can retune their controllers. Similarly, if a subsystem leaves the network, then at the most, its children can retune their controllers too.

In particular, we hope that by such a distributed MPC approach, the efficiency and operating range of the plant could be increased, while its operational cost could be decreased. Now many control engineers think that a distributed MPC approach could be beneficial for a variety of applications. All of these applications are concentrated on large scale networks of systems, which are subject to communication constraints. Generally, the distributed MPC scheme is an extension to decentralize MPC. Same as in a decentralized approach, a globally defined system is controlled by several local controllers. However, some controllers in the network exchange information. Within this general definition of distributed MPC, existing methods available in references vary considerably in the ways local controllers are operated and information is shared and utilized.

Using the above descriptions and advantages of distributed MPC scheme, in this short note, we continue to study the stability analysis for the cooperative distributed MPC scheme. As the previous work was focused on optimization algorithms for designing the MPC controller, there are a few references that can be used for stability analysis of cooperative distributed MPC scheme. After reviewing some concepts such as the discrete time constrained linear time invariant system (LTI), MPC and distributed idea, set invariance and Lyapunov stability condition are introduced, which can be extended to the cooperative distributed scheme. Based on these mentioned Lyapunov stability conditions, all cost function terms are considered as quadratic function forms, then the Lyapunov stability condition is equivalent to the set of linear matrix inequalities (LMIs). This equivalent property is derived from the Schur complement and S-procedure strategy. Furthermore, considering the local state and input constraint sets, more complex LMIs are obtained by combining all state-input constraint sets and our previous LMIs. These two sets of LMIs are regarded as sufficient conditions to guarantee stability for the cooperative distributed MPC scheme. The classical Lyapunov stability condition is sufficient too.

2.4.1   Preliminaries

Consider the following discrete time constrained LTI system:

$$\begin{cases} x(k+1) = Ax(k) + Bu(k) + w(k) \\ y(k) = Cx(k), \forall k \geq 0 \end{cases} \tag{2.100}$$

where $x(k) \in R^n$ is the state of the system at time instant k, $u(k) \in R^m$ is the input, $y(k) \in R^p$ its output and $w(k) \in R^n$ an additive disturbance. System matrix $A \in R^{n \times n}$, and input map $B \in R^{n \times m}$, output matrix $C \in R^{p \times n}$. Both the input and state $x(k), u(k)$ are subject to constraints

$$x(k) \in X, \quad u(k) \in U, \forall k \geq 0 \tag{2.101}$$

Both $X \subseteq R^n$, and $U \subseteq R^m$ are convex sets which contain the origin in their interior. If no disturbance acts on the system, the corresponding system is denoted as nominal, i.e.

$$x(k+1) = Ax(k) + Bu(k), \forall k \geq 0 \tag{2.102}$$

If the input of the above nominal system is given as a function $K((u)k) : R^n \rightarrow R^m$ of the current state, the resulting closed loop system is that

$$x(k+1) = Ax(k) + BK(u(k)), \forall k \geq 0 \tag{2.103}$$

All eigenvalues of the matrix $A + BK$ lie strictly within the unit circle of the complex plane.

2.4.2   Distributed discrete time constrained LTI system

For a network of constrained linear systems, a set of coupled dynamic systems is denoted with local constraints. The whole network system is divided into M non-overlapping subsystems, each of which has a local state $x_i \in R^{n_i}$, a local input $u_i \in R^{m_i}$ and a local output $y_i \in R^{p_i}$ and local disturbance $w_i \in R^{n_i}$. The dimensions of these local vectors are such that

$$\sum_{i=1}^{M} n_i = n, \sum_{i=1}^{M} m_i = m, \sum_{i=1}^{M} p_i = p \tag{2.104}$$

The global state vector and global input are defined as

$$\begin{cases} x = col_{i \in \{1 \cdots M\}}(x_i), \\ u = col_{i \in \{1 \cdots M\}}(u_i) \end{cases} \tag{2.105}$$

Similarly, the global output and global disturbance are

$$\begin{cases} y = col_{i \in \{1 \cdots M\}} (y_i), \\ w = col_{i \in \{1 \cdots M\}} (w_i) \end{cases} \tag{2.106}$$

The global system (2.100) can be equivalently described by the ith subsystem, given by

$$\begin{cases} x_i \left(k+1 \right) = \sum_{j=1}^{M} A_{ij} x_j \left(k \right) + B_i u_i \left(k \right) + w_i \left(k \right) \\ y_i \left(k \right) = \sum_{j=1}^{M} C_{ij} x_j \left(k \right), \forall k \geq 0 \end{cases} \tag{2.107}$$

The local constraint set and disturbance have the following forms:

$$x_i \in X_i, \quad u_i \in U_i, \quad w_i \in W_i, \forall i = \{1 \cdots M\}$$

Similarly, all X_i and U_i are convex sets which contain the origin in their interior, and all W_i are convex sets which contain the origin too.

Using the definition of neighboring systems from [20], the local system (2.107) is equivalently written as one compact form.

$$\begin{cases} x_i \left(k+1 \right) = A_{N_i} x_{N_i} \left(k \right) + B_i u_i \left(k \right) \\ y_i \left(k \right) = C_{N_i} x_{N_i} \left(k \right), \forall k \geq 0, i = \{1 \cdots M\} \end{cases} \tag{2.108}$$

where N_i is the set of all neighbors of i, the states and outputs of all systems $j \in N_i$ are denoted as

$$\begin{cases} x_{N_i} = col_{i \in N_i} \left(x_i \right) \in R^{n_{N_i}}, \\ y_{N_i} = col_{i \in N_i} \left(y_i \right) \in R^{p_{N_i}} \end{cases}$$

For all networks of nominal LTI systems, we make the following assumption.

Assumption 2.2: The pair (A, B) is stabilizable by a structured linear state feedback control law of the form

$$K \left(x \right) = Kx = col_{\{i \in 1 \cdots M\}} \left(K_{N_i} x_{N_i} \right)$$

where

$$K \in R^{m \times n} \text{ and } K_{N_i} \in R^{m_i \times n_{N_i}}, \forall i = \{1 \cdots M\}$$

As this section is related to the stability of constrained networks LTI systems, so set invariance is very important in stability analysis. The definition for positively invariance set is introduced.

Definition 2.1 (Positively invariance set): If set $X_{PI} \subseteq R^n$ is positively invariance for one closed loop system $x(k+1) = Ax(k) + BK(x(k))$, $\forall k \geq 0$, if

$$x(k) \in X_{PI} \Rightarrow x(k+1) \in X_{PI}$$

This definition for positively invariance set will be applied in a later stability analysis for our cooperative distributed MPC scheme.

2.4.3 MPC scheme

MPC is a well established control methododology for constrained systems. The control law is defined through a finite horizon optimal control problem, i.e. at every point in time, the solution of the finite horizon problem for the current initial state if found, but only the first element of the optimal input trajectory is applied to the plant. The MPC control law is defined as follows through the finite horizon optimal control problem.

$$V(x) = \min_u V_f(x(N)) + \sum_{k=0}^{N-1} l(x(k), u(k))$$

$$\text{subject to} \begin{cases} x(0) = x_0 \\ x(k+1) = Ax(k) + Bu(k) \\ (x(k), u(k)) \in X \times U, \ \forall k \in \{0 \cdots N-1\} \\ x(N) \in X_f \end{cases}$$

(2.109)

where $l(x(k), u(k))$ is the state cost, the terminal cost $V_f(x(N))$ are positive definite convex functions, and the terminal set $X_f \subseteq R^n$ is convex. $u = \{u(0) \cdots u(N-1)\}$ denotes an input trajectory over N steps. The input trajectory $u^*(x) = \{u^*(0) \cdots u^*(N-1)\}$ denotes the minimizer of problem (2.109) and its first element is used to define the state feedback control law $K_{MPC}(x) = u^*(0)$, that leads to one nonlinear closed loop system.

$$x(k+1) = Ax(k) + BK_{MPC}(x(k))$$

(2.110)

In a closed loop operation, the systems optimize their inputs in a sequential fashion, keeping the variance of the neighboring systems constant. At each timestep, either one system optimizes its inputs or several systems do so in a sequential fashion.

We are only concerned with a cooperative distributed MPC scheme whose global MPC problem is solved online by distributed optimization at every time step. In order to achieve the cooperative distributed property, a decomposition of the terminal cost is needed into exactly one unit per system in the network, i.e.

$$V_f(x) = \sum_{i=1}^{M} V_{f,i}(x_i) \tag{2.111}$$

where for each $i = \{1 \cdots M\}$, $V_{f,i}(\ ): R^{n_i} \Rightarrow R$ is the local terminal cost contribution. Further, as proposed in [20], it is desirable to allow a local terminal cost contribution to increase as long as at the same time the global terminal cost decreases.

Theorem 2.7: [20] Let $X_f \subseteq R^n$ be a positively invariance set for system (2.107) under the control law $K_f(x) = col_{i \in \{1 \cdots M\}} \left(K_{N_i}(x_{N_i}) \right)$. If there exist functions $V_{f,i}(x_i)$, $\gamma_i(x_{N_i})$, $K_{N_i}(x_{N_i})$ and $l_i \left(x_{N_i}, K_{N_i}(x_{N_i}) \right)$, as well as K_∞ class functions $\beta_{1,i}(\ ), \beta_{2,i}(\ )$ and $\beta_{3,i}(\ )$ such that $\forall x = col_{i \in \{1 \cdots M\}} \left(x_i \in X_f \right)$, it holds that

$$\begin{cases} \beta_{1,i} \left(\|x_i\| \right) \leq V_{f,i}(x_i) \leq \beta_{2,i} \left(\|x_i\| \right), \forall i = \{1 \cdots M\} \\ \beta_{3,i} \left(\|x_{N_i}\| \right) \leq l_i \left(x_{N_i}, K_{N_i}(x_{N_i}) \right) \\ V_{f,i} \left(x_i(k+1) \right) - V_{f,i} \left(x_i(k) \right) \leq -l_i \left(x_{N_i}, K_{N_i}(x_{N_i}) \right) + \gamma_i(x_{N_i}) \\ \sum_{i=1}^{M} \gamma_i(x_{N_i}) \leq 0 \end{cases} \tag{2.112}$$

Then the function $V_f(x) = \sum_{i=1}^{M} V_{f,i}(x_i)$ is a Lyapunov function for system (2.107) and this system is asymptotically stable on X_f. Furthermore, if the conditions (2.112) hold, then the functions $V_{f,i}(x_i), \forall i = \{1 \cdots M\}$ can be used as local terminal cost in the cooperative distributed MPC scheme.

From Theorem 2.7, the main essence of this short note is to derive a set of LMIs, so that the inequalities in Theorem 2.7 can be replaced by our derived LMIs. Moreover, Theorem 2.7 is a sufficient condition for stability in the cooperative distributed MPC scheme, so our derived set of LMIs is also a sufficient condition.

2.4.4   Stability analysis based on LMIs

Consider local state and input constraint sets of the following forms:

$$\begin{cases} X_i = \left\{ x_i \in R^{n_i} / G_i x_i \leq g_i \right\} \\ U_i = \left\{ u_i \in R^{m_i} / H_i u_i \leq h_i \right\}, \forall i = \{1 \cdots M\} \end{cases} \tag{2.113}$$

where $G_i \in R^{l_{x_i} \times n_i}, g_i \in R^{l_{x_i}}, H_i \in R^{l_{u_i} \times m_i}$ and $h_i \in R^{l_{u_i}}$. And all cost function terms are considered as quadratic function forms, hence

$$\begin{cases} l_i\left(x_{N_i}, u_i\right) = x_{N_i}^T Q_{N_i} x_{N_i} + u_i^T R_i u_i \\ V_{f,i}\left(x_i\right) = x_i^T P_{f,i} x_i \\ \gamma_i\left(x_{N_i}\right) = x_{N_i}^T \Gamma_{N_i} x_{N_i} \end{cases} \tag{2.114}$$

where Q_{N_i}, R_i and $P_{f,i}$ are positive definite matrices, further the terminal control law is structured and linear, i.e.

$$K_f\left(x\right) = col_{i \in \{1 \cdots M\}}\left(K_{N_i} x_{N_i}\right) \tag{2.115}$$

Substituting (2.114) and (2.115) into the third inequality of (2.112), then expanding it as

$$\left(A_i x_i + B_i K_{N_i} x_{N_i}\right)^T P_{f,i}\left(A_i x_i + B_i K_{N_i} x_{N_i}\right) - x_{N_i}^T P_{f,i} x_{N_i}$$
$$\leq -x_{N_i}^T Q_{N_i} x_{N_i} - u_i^T R_i u_i + x_{N_i}^T \Gamma_{N_i} x_{N_i} \tag{2.116}$$

It means that

$$x_{N_i}^T\left(A_i + B_i K_{N_i}\right)^T P_{f,i}\left(A_i + B_i K_{N_i}\right) x_{N_i} - x_{N_i}^T P_{f,i} x_{N_i}$$
$$\leq -x_{N_i}^T Q_{N_i} x_{N_i} - u_i^T R_i u_i + x_{N_i}^T \Gamma_{N_i} x_{N_i} \tag{2.117}$$

Equation (2.117) holds if and only if

$$\underbrace{\left(A_i + B_i K_{N_i}\right)^T P_{f,i}\left(A_i + B_i K_{N_i}\right) - P_{f,i} \leq -Q_{N_i} - K_{N_i}^T R_i K_{N_i} + \Gamma_{N_i}}$$
$$\Updownarrow$$

$$P_{f,i} + \Gamma_{N_i} - \left(Q_{N_i} + K_{N_i}^T R_i K_{N_i}\right) \geq \left(A_i + B_i K_{N_i}\right)^T P_{f,i}\left(A_i + B_i K_{N_i}\right) \tag{2.118}$$

Using Schur complement, we obtain one following LMI.

$$\begin{bmatrix} P_{f,i} + \Gamma_{N_i} - \left(Q_{N_i} + K_{N_i}^T R_i K_{N_i}\right) & \left(A_i + B_i K_{N_i}\right)^T \\ \left(A_i + B_i K_{N_i}\right) & P_{f,i}^{-1} \end{bmatrix} \geq 0 \tag{2.119}$$

Set $M_{f,i} = P_{f,i}^{-1}$, and multiplying $M_{f,i} = P_{f,i}^{-1}$ on both sides of equation (2.118) to get

$$P_{f,i}^{-1}\left(A_i + B_i K_{N_i}\right)^T P_{f,i}\left(A_i + B_i K_{N_i}\right) P_{f,i}^{-1} - P_{f,i}^{-1} \leq -P_{f,i}^{-1}$$
$$\left(Q_{N_i} + K_{N_i}^T R_i K_{N_i}\right) P_{f,i}^{-1} + P_{f,i}^{-1} \Gamma_{N_i} P_{f,i}^{-1} \tag{2.120}$$

Substituting (2.119) into (2.118), then it holds that

$$\left(A_i + B_i K_{N_i} \right)^T \left(A_i + B_i K_{N_i} \right) P_{f,i}^{-1} - P_{f,i}^{-1} \le -P_{f,i}^{-1}$$

$$\left(Q_{N_i} + K_{N_i}^T R_i K_{N_i} \right) P_{f,i}^{-1} + P_{f,i}^{-1} \Gamma_{N_i} P_{f,i}^{-1} \tag{2.121}$$

Using $M_{f,i} = P_{f,i}^{-1}$ to simplify above inequality to get

$$\left(A_i + B_i K_{N_i} \right)^T \left(A_i + B_i K_{N_i} \right) M_{f,i} - M_{f,i} \le -M_{f,i}$$

$$\left(Q_{N_i} + K_{N_i}^T R_i K_{N_i} \right) M_{f,i} + M_{f,i} \Gamma_{N_i} M_{f,i} \tag{2.122}$$

Expanding equation (2.122), we have that

$$A_i^T M_{f,i} A_i + 2 A_i^T M_{f,i} B_i K_{N_i} + K_{N_i}^T B_i B_i M_{f,i} K_{N_i} - M_{f,i} \le -M_{f,i}$$

$$\left(Q_{N_i} + K_{N_i}^T R_i K_{N_i} \right) M_{f,i} + M_{f,i} \Gamma_{N_i} M_{f,i} \tag{2.123}$$

Formulating above equation (2.123), we see that

$$A_i^T M_{f,i} A_i + K_{N_i}^T \left(B_i^T B_i M_{f,i} + M_{f,i} R_i M_{f,i} \right) K_{N_i} + K_{N_i}$$

$$\left(2 A_i^T M_{f,i} B_i \right) + M_{f,i} Q_{N_i} M_{f,i} + M_{f,i} \Gamma_{N_i} M_{f,i} - M_{f,i} \le 0 \tag{2.124}$$

Separating the gain matrix K_{N_i}, one LMI is obtained.

$$\begin{bmatrix} B_i^T B_i M_{f,i} + M_{f,i} R_i M_{f,i} & A_i^T M_{f,i} B_i \\ B_i^T M_{f,i} A_i & A_i^T M_{f,i} A_i + M_{f,i} Q_{N_i} M_{f,i} + M_{f,i} \Gamma_{N_i} M_{f,i} - M_{f,i} \end{bmatrix} \le 0$$

$$\tag{2.125}$$

Observing (2,2) constant term, the following relation holds:

$$\begin{bmatrix} B_i^T B_i M_{f,i} + M_{f,i} R_i M_{f,i} & A_i^T M_{f,i} B_i \\ B_i^T M_{f,i} A_i & -M_{f,i} \end{bmatrix} - \begin{bmatrix} 0 & 0 \\ A_i^T M_{f,i}^{\frac{1}{2}} & M_{f,i} \left(Q_{N_i} + \Gamma_{N_i} \right)^{\frac{1}{2}} \end{bmatrix}$$

$$
\begin{bmatrix} 0 & M_{f,i}^{\frac{1}{2}} A_i \\ 0 & \left(Q_{N_i} + \Gamma_{N_i}\right)^{\frac{1}{2}} M_{f,i} \end{bmatrix} = \begin{bmatrix} B_i^T B_i M_{f,i} + M_{f,i} R_i M_{f,i} & A_i^T M_{f,i} B_i \\ B_i^T M_{f,i} A_i & -M_{f,i} \end{bmatrix} -
$$

$$
\begin{bmatrix} 0 & 0 \\ 0 & -A_i^T M_{f,i} A_i - M_{f,i} Q_{N_i} M_{f,i} - M_{f,i} \Gamma_{N_i} M_{f,i} \end{bmatrix} =
$$

$$
\begin{bmatrix} B_i^T B_i M_{f,i} + M_{f,i} R_i M_{f,i} & A_i^T M_{f,i} B_i \\ B_i^T M_{f,i} A_i & A_i^T M_{f,i} A_i + M_{f,i} Q_{N_i} M_{f,i} + M_{f,i} \Gamma_{N_i} M_{f,i} - M_{f,i} \end{bmatrix}
$$

$$(2.126)$$

Combining the above relation and Schur complement, all of our derivation can be formulated as Theorem 2.8.

Theorem 2.8 (Stability condition 1): Consider for each $\forall i = \{1 \cdots M\}$, the function $V_f(x) = \sum_{i=1}^{M} V_{f,i}(x_i)$ is a Lyapunov function and system is asymptotically stable, then the condition (2.112) is equivalent to the set of LMIs.

$$
\begin{bmatrix} B_i^T B_i M_{f,i} + M_{f,i} R_i M_{f,i} & A_i^T M_{f,i} B_i & 0 & 0 \\ B_i^T M_{f,i} A_i & -M_{f,i} & M_{f,i}^{\frac{1}{2}} A_i & M_{f,i} \left(Q_{N_i} + \Gamma_{N_i}\right)^{\frac{1}{2}} \\ 0 & A_i^T M_{f,i}^{\frac{1}{2}} & I & 0 \\ 0 & \left(Q_{N_i} + \Gamma_{N_i}\right)^{\frac{1}{2}} M_{f,i} & 0 & I \end{bmatrix}
$$
$$
\le 0; \forall i = \{1 \cdots M\} \qquad (2.127)
$$

Theorem 2.8 gives one sufficient condition for asymptotically stable of a cooperative distributed MPC scheme when local state and input constraint sets (2.113) are not considered. Now we want to derive sufficient conditions for asymptotically stable, while satisfying local state and input constraint sets. In this more complex case, the sufficient condition can be also formulated as Theorem 2.9.

Theorem 2.9 (Stability condition 2): Similarly, consider for each $\forall i = \{1 \cdots M\}$ and local state and input constraint sets

$$
\begin{cases} X_i = \{x_i \in R^{n_i} \mid G_i x_i \le g_i\} \\ U_i = \{u_i \in R^{m_i} \mid H_i u_i \le h_i\}, \forall i = \{1 \cdots M\} \end{cases}
$$

The function $V_f(x) = \sum_{i=1}^{M} V_{f,i}(x_i)$ is a Lyapunov function and the constrained system is asymptotically stable, then one sufficient condition is given as the set of LMIs, i.e. there exist two scalars $\tau_1, \tau_2 \geq 0$ such that

$$
\begin{bmatrix}
B_i^T B_i M_{f,i} + M_{f,i} R_i M_{f,i} & A_i^T M_{f,i} B_i & \frac{\tau_2}{2} H_i^T h_i^{\frac{1}{2}} \\
B_i^T M_{f,i} A_i & -M_{f,i} - \frac{\tau_1}{4} G_i^T g_i G_i & 0 \\
\frac{\tau_2}{2} H_i^T h_i^{\frac{1}{2}} & 0 & I \\
0 & M_{f,i}^{\frac{1}{2}} A_i & 0 \\
0 & \left(Q_{N_i} + \Gamma_{N_i} \right) M_{f,i} & 0 \\
& 0 & 0 \\
& A_i^T M_{f,i}^{\frac{1}{2}} & M_{f,i} \left(Q_{N_i} + \Gamma_{N_i} \right)^{\frac{1}{2}} \\
& 0 & 0 \\
& I & 0 \\
& 0 & I
\end{bmatrix} \leq 0
$$

$\forall i = \{1 \cdots M\}$ \hfill (2.128)

Proof: Rewrite these local state and input constraint sets X_i, U_i as

$$
\begin{cases}
G_i x_i \leq g_i \Rightarrow \begin{pmatrix} x_i \\ 1 \end{pmatrix}^T \begin{pmatrix} 0 & \frac{1}{2} G_i \\ \frac{1}{2} G_i & -g_i \end{pmatrix} \begin{pmatrix} x_i \\ 1 \end{pmatrix} \leq 0 \\
H_i u_i \leq h_i \Rightarrow H_i K_i x_i \leq h_i \Rightarrow \begin{pmatrix} x_i \\ 1 \end{pmatrix}^T \begin{pmatrix} 0 & \frac{1}{2} H_i K_i \\ \frac{1}{2} H_i K_i & -h_i \end{pmatrix} \begin{pmatrix} x_i \\ 1 \end{pmatrix} \leq 0
\end{cases} \quad (2.129)
$$

Due to gain matrix, K_{N_i} is separated in deriving equation (2.125), so we do the same separations in (2.129) to separate gain matrix K_{N_i}

$$
\left\{
\begin{aligned}
& 0 + \frac{1}{4} G_i^T g_i G_i \le 0 \Rightarrow
\begin{bmatrix}
0 & 0 \\
0 & \dfrac{1}{4} G_i^T g_i G_i
\end{bmatrix}
\le 0; 0 + \frac{1}{4} H_i^T K_i h_i K_i H_i \le 0 \\
& \qquad\qquad \Rightarrow
\begin{bmatrix}
\dfrac{1}{4} H_i^T h_i H_i & 0 \\
0 & 0
\end{bmatrix}
\le 0
\end{aligned}
\right.
\tag{2.130}
$$

Combining LMIs (2.125) and (2.130) and applying S-procedure strategy, there exist two scalars $\tau_1, \tau_2 \ge 0$ such that

$$
\begin{bmatrix}
B_i^T B_i M_{f,i} + M_{f,i} R_i M_{f,i} & A_i^T M_{f,i} B_i \\
B_i^T M_{f,i} A_i & A_i^T M_{f,i} A_i + M_{f,i} Q_{N_i} M_{f,i} + M_{f,i} \Gamma_{N_i} M_{f,i} - M_{f,i}
\end{bmatrix}
$$
$$
- \tau_1
\begin{bmatrix}
0 & 0 \\
0 & \dfrac{1}{4} G_i^T g_i G_i
\end{bmatrix}
- \tau_2
\begin{bmatrix}
\dfrac{1}{4} H_i^T h_i H_i & 0 \\
0 & 0
\end{bmatrix}
\le 0
\tag{2.131}
$$

Expanding equation (2.131) to get

$$
\left[
\begin{array}{l}
B_i^T B_i M_{f,i} + M_{f,i} R_i M_{f,i} - \dfrac{\tau_2}{4} H_i^T h_i H_i \\[2mm]
\qquad\qquad B_i^T M_{f,i} A_i \\
\end{array}
\right.
$$
$$
\left.
\begin{array}{r}
A_i^T M_{f,i} B_i \\[2mm]
A_i^T M_{f,i} A_i + M_{f,i} \left(Q_{N_i} + \Gamma_{N_i} - I \right) M_{f,i} - \dfrac{\tau_1}{4} G_i^T g_i G_i
\end{array}
\right]
\le 0
\tag{2.132}
$$

Applying Schur complement to the element (2, 2), we have

$$
\begin{bmatrix}
B_i^T B_i M_{f,i} + M_{f,i} R_i M_{f,i} - \dfrac{\tau_2}{4} H_i^T h_i H_i & A_i^T M_{f,i} B_i & 0 & 0 \\[2mm]
B_i^T M_{f,i} A_i & -M_{f,i} - \dfrac{\tau_1}{4} G_i^T g_i G_i & A_i^T M_{f,i}^{\frac{1}{2}} & M_{f,i} \left(Q_{N_i} + \Gamma_{N_i} \right)^{\frac{1}{2}} \\[2mm]
0 & M_{f,i}^{\frac{1}{2}} A_i & I & 0 \\[2mm]
0 & \left(Q_{N_i} + \Gamma_{N_i} \right) M_{f,i} & 0 & I
\end{bmatrix}
\le 0
\tag{2.133}
$$

Then applying Schur complement to the element (1, 1) again, the result (2.128) can be obtained, thus concluding the proof. These two sets of LMIs show the indications of asymptotically stable for the cooperative distributed MPC scheme. On the contrary, they are two sufficient conditions, not the necessary and sufficient conditions. For the problem of how to solving these sets of LMIs, the classical interior point algorithm can be applied.

2.4.5 Conclusion

In this short section, we study the problem of stability analysis in cooperative distributed MPC. Based on the classical Lyapunov stability condition, all cost function terms are chosen as quadratic function forms. After substituting these quadratic function forms into the given Lyapunov stability condition, a set of LMIs can be obtained to replace the former inequalities. Furthermore, by combining local state and input constraint sets while guaranteeing the asymptotically stable property simultaneously, a set of more complex LMIs is derived to achieve our goal. All of our derivations are based on the Schur complement and S-procedure strategy, which lead to the interior point optimization algorithm.

2.5   Summary

Data driven strategy is from the system identification field, and model predictive control is one mature control strategy in recent research. This chapter aims to introduce the idea of data drive into model predictive control strategy, then the goal of identification for control is achieved while combining the system identification and model predictive control.

References

[1] Campi, M.C. and Calafiore, G. 2009. Interval predictor models: Identification and reliability, Automatica, 45(2): 382-393.
[2] Campi, M.C. and Vidyasagar, M. 2001. Learning with prior information, IEEE Transaction on Automatic Control, 46(11): 1682-1694.
[3] Vidyasagar, M. and Karandikar, R.L. 2008. A learning theory approach to system identification and stochastic adaptive control, Journal of Process Control, 18(3): 421-430.
[4] Campi, M.C. and Kumar, P.R. 1998. Learning dynamical systems in a stationary environment, Systems & Control Letters, 34(3): 125-132.
[5] Calafiore, G. and Campi, M.C. 2005. Uncertain convex programs: Randomized solutions and confidence levels, Mathematical Programming, 102(11): 25-46.
[6] Milanese, M. and Novara, C. 2004. Set membership identification of nonlinear systems, Automatica, 40(6): 957-975.
[7] Alamo, T., Bravo, J.M. and Camacho, E.F. 2005. Guaranteed state estimation by zonotopes, Automatica, 41(6): 1035-1043.

[8] Bravo, J.M., Alamo, T. and Camacho, E.F. 2006. Bounded error identification of systems with time varying parameters, IEEE Transaction on Automatic Control, 51(7): 1144-1150.

[9] Bravo, J.M., Suarez, A. and Vasallo, M. 2016. Slide window bounded error tome varying systems identification, IEEE Transaction on Automatic Control, 61(8): 2282-2287.

[10] Bravo, J.M., Alamo, T. and Vasallo, M. 2017. A general framework for predictions based on bounding techniques and local approximation, IEEE Transaction on Automatic Control, 62(7): 3430-3435.

[11] Tanaskovic, M., Fagiano, L. and Smith, R. 2014. Adaptive receding horizon control for constrained MIMO systems, Automatica, 50(12): 3019-3029.

[12] Tanaskovic, M., Fagiano, L. and Novara, C. 2017. Data driven control of nonlinear systems: An on line direct approach, Automatica, 75(1): 1-10.

[13] Novara, C., Formentin, S. and Savaresi, S.M. 2016. Data driven design of two degrees of freedom nonlinear controllers, Automatica, 72(10): 19-27.

[14] Casini, M., Garulli, A. and Vicino, A. 2014. Feasible parameter set approximation for linear models with bounded uncertain regressions, IEEE Transaction on Automatic Control, 59(11): 2910-2920.

[15] Cerone, V., Lasserre, J.B. and Piga, D. 2014. A unified framework for solving a general class of conditional and robust set membership estimation problems, IEEE Transaction on Automatic Control, 59(11): 2897-2909.

[16] Casini, M., Garulli, A. and Vicino, A. 2017. A linear programming approach to online set membership parameter estimation for linear regression models, International Journal of Adaptive Control and Signal Processing, 31(3): 360-378.

[17] Zhang, X., Kamgarpour, M. and Georghiou, A. 2017. Robust optimal control with adjustable uncertainty sets, Automatica, 75(1): 249-259.

[18] Zeilinger, M., Jones, C.N. and Morari, M. 2011. Real time suboptimal model predictive control using a combination of explicit MPC and online optimization, IEEE Transaction on Automatic Control, 56(7): 1524-1534.

[19] Zeilinger, M., Jones, C.N. and Morari, M. 2014. Soft constrained model predictive control with robust stability guarantees, IEEE Transaction on Automatic Control, 59(5): 1190-1202.

[20] Richter, S., Jones, C.N. and Morari, M. 2012. Computational complexity certification for real time MPC with input constraints based on fast gradient method, IEEE Transaction on Automatic Control, 57(6): 1391-1403.

Data Driven Identification for Closed Loop System

3.1   Introduction

Nowadays many systems operate under feedback control situations, due to the required safety of operation or to unstable behavior of the plant, as occurs in many industrial processes such as paper production, glass production, separation process like crystallization, etc. As feedback in an open loop structure does not exist , the plant output affects the input less. In a closed loop system, a feedback controller is added to return the collected output to the excited input. Then one error signal between the excited input and feedback output can be imposed on the plant to generate one correction action, which makes the output converge to a given value. The essences of a closed loop system are to decrease the error using the negative feedback controller, and to correct the deviation from the given value automatically. As the closed loop system can suppress the errors coming from the internal or external disturbances and achieve the specified control goal, the closed loop system is most needed in all engineering fields. In this chapter 3, we continue to study some interesting topics as follows.

(1)   From the idea of a direct approach, we introduce stealth identification for a closed loop system to tackle the problem of an unknown controller. Stealth identification makes use of a prior experiment from which an initial estimate is used to replace the true parameter vector θ_0 in the prediction error and inverse covariance matrix expression. Stealth identification involves modifying the feedback mechanism in a classical closed loop identification scheme in such a way that the unknown controller does not sense the excitation signal. In this modified feedback mechanism, the sensitivity function becomes a constant 1, then the new prediction error and inverse covariance matrix will be very simple. These simple forms can simplify the later optimization problems, which are used to devise the parameter estimation or optimal input. If this modified feedback mechanism is used, to improve the convergence of classical prediction error identification, a candidate domain of attraction for

the new objective function is introduced and one convergence condition is derived to assume that a given set is a candidate domain of attraction.

(2) Here using the framework of a tailor made parameterization for a closed loop system, we study the performance analysis problem where a closed loop transfer function is parameterized using the parameters of an open loop plant model and utilizing knowledge of a feedback controller. When the plant model and feedback controller are all polynomial forms, a recursive least squares method with forgetting schemes is proposed to verify that this recursive method can be regarded as a regularization least squares problem. Based on the parameter vector, one uncertainty bound about the parameter vector is constructed to reflect the identification accuracy by using the statistical probability theory. Using a tailor made parameterization form, some results from the robust control theory and related stability property are used to give a preliminary performance analysis corresponding to the closed loop transfer function. Generally, this preliminary performance analysis is extended to a transfer function matrix form which is constituted by three transfer functions. The worst case performance at frequencies is analyzed by solving one standard convex optimization problem involving some linear matrix inequality constraints.

(3) To design one feedback controller in a closed loop system, the idea of minimum variance control is used to realize this goal. Two explicit forms corresponding to the closed loop system are considered, i.e. its general form and rational transfer function form respectively. Firstly, one closed form solution of the minimum variance controller is derived in the general form of the closed loop system, and an optimization algorithm is proposed to obtain the controller in practice. Secondly, in the rational transfer function form of the closed loop system, the minimum variance controller is determined, while guaranteeing the modified variances of output and input are as small as possible.

3.2   Stealth identification strategy for closed loop linear time invariant system

There are many subjects in the closed loop system, for example, closed loop system identification, closed loop controller design, closed loop performance monitoring, diagnosis, etc. There are three identification methods for a closed loop system, i.e. direct approach, indirect approach, and joint input-output approach, where the feedback is neglected in the direct approach, then the plant model is identified directly using only input-output data. For the indirect approach, the feedback effect is considered and the input-output data from the whole closed loop condition are used to identify the plant model. The joint input-output approach is very similar to the indirect approach. The joint input-output approach requires two separate steps, (1) identification of the closed loop system and (2) recalculation of the open loop model. For the problem of how to design a feedback controller in a closed loop system, generally, two strategies are used, i.e. model based design and direct data driven design. The primary step of model based design is to construct the plant model in a closed loop system using the system identification theory and apply this mathematical model in designing a feedback controller. Conversely, for the direct data driven design, the modeling process is not needed and the controller is directly designed using only the observed input-output data under closed loop conditions. By comparing these two strategies, the direct data driven method is worth studying intensely in the future. But now as the first model based design strategy is applied more widely, we need to do more research on closed loop system identification. Closed loop performance monitoring and diagnosis comprise a crucial step in the maintenance of a model based control system. In the event of performance degradation, diagnostic tools allow us to verify if the unsatisfactory closed loop operation results from the idea of plant-model mismatch. As above, all subjects need the mathematical model corresponding to the closed loop system, so it is necessary to propose a new identification strategy for the closed loop system under different situations. Some references on closed loop system identification are given as follows.

In [1], the above three methods are presented to identify a closed loop system. In [2], research on the detailed system identification theory is introduced in the time domain. Similarly, the frequency domain system identification is given in [3], where lots of other methods are proposed to solve closed loop system identification, such as maximum likelihood estimation, prediction error method, bias correction method, subspace method, etc. A new virtual closed loop method for closed loop identification is proposed in [4]. In [5], one projection algorithm is proposed based on the recursive prediction error method. In [6], when many inputs exist in a closed loop system, can a closed loop system be identified with parts of the inputs controlled? The closed relationship between closed loop system identification and closed loop control is obtained in [7]. In [8], the linear matrix inequality is used to describe the problem of optimal input design in a closed loop system. Further, the least cost identification experiment problem is analyzed in [9]. The power spectral of the input signal is considered to be an objective function and the accuracy of the parameter estimations is the constraint [10]. In [11], the H infinity norm from robust control is introduced to the objective function in the optimal input design problem. Based on the H infinity norm, the uncertainty between the identified model and the nominal model is measured and the optimal input is chosen by minimizing this uncertainty [12]. The selection of the optimal input can also be determined from the point of asymptotic behavior corresponding to the parameter estimation [13]. The persistent excitation input in a closed loop system is analyzed and we obtain some conditions about how to achieve persistent excitation [14]. In [15], the problem of how to apply closed loop system identification into adaptive control is solved, so that bias and covariance terms are isolated separately. All the above results hold when the number of the observed signals will converge to infinity. In our paper [16], we consider the problem of model structure validation for closed loop system identification, and two probabilistic model uncertainties are derived from some statistical properties of the parameter estimation. The probabilistic bounds and optimum input filter are based on an asymptotic normal distribution of the parameter estimator and its covariance matrix, which was estimated from infinite sampled data. Using some derived results from our paper [16], a new technique for estimating bias and variance contributions to the model error is suggested in [17], then one bound described as an inequality corresponds to a condition on the model error.

The above mentioned references used to identify the closed loop system often assume that an open loop system is closed by a feedback mechanism containing a known, linear time invariant controller $C(z)$. As a result, the prediction error and the inverse covariance matrix are the functions of the true parameter vector θ_0, but this true parameter vector is unknown. This difficulty can be resolved by replacing θ_0 with an initial estimate θ_{init} that is obtained from a previous experiment. Further, the inverse covariance matrix is a function of the sensitivity function of the closed loop system. Thus the sensitivity function must also be known, it means that when the system is in a closed loop with a known linear time invariant controller, the sensitivity function is trivially calculated. However, there are many systems that are closed loop systems with an unknown controller $C(z)$. For instance, industrial process controller by model predictive control, which is not linear time invariant, and a dynamic network, in which the feedback from the output to the input consists of many unknown linear time invariant transfer functions that are connected in parallel or in series. In such complex systems, the two expressions for the prediction error and inverse covariance matrix cannot be calculated, as the sensitivity function is unknown. Consequently, this unknown property limits the applicability of the classical prediction error identification for a closed loop system with an unknown controller. It is well known that an advantage of the direct approach for closed loop system is that the identification procedure does not require the knowledge of the controller during the identification experiment.

3.2.1 Closed loop system description

Consider the following linear time invariant controller $C(z)$ regulating a single input single output linear time invariant system, consisting of output $y(t)$ and input $u(t)$ (see Fig. 3.1).

In Fig. 3.1, $G_0(z)$ is a true plant model, $H_0(z)$ is a noise filter, they are all stable, discrete time transfer functions, and $H_0(z)$ is monic and minimum phase. $C(z)$ is a stable linear time invariant controller. The excitation signal $r(t)$ and external disturbance or noise $e(t)$ are uncorrelated, $e(t)$ is a white noise with zero mean value and variance λ_0. $v(t)$ is a colored noise obtained by passing white noise $e(t)$ through the noise filter $H_0(z)$. z is the delay operator, it means that $zu(t) = u(t + 1)$.

$$v(t) = H_0(z)\, e(t) \tag{3.1}$$

In above closed loop system structure, after some computations, we derive one transfer function from.

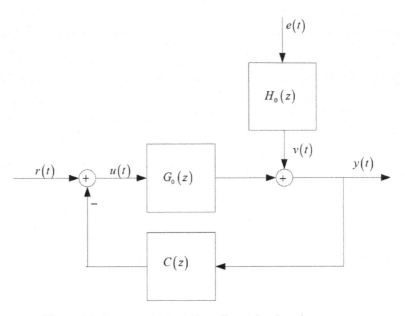

Figure 3.1: Structure of closed loop linear time invariant system

$$\begin{cases} y(t) = G_0(z)u(t) + H_0(z)e(t) \\ u(t) = r(t) - C(z)y(t) \end{cases} \tag{3.2}$$

Continuing to do some computations and we get

$$\begin{cases} y(t) = \dfrac{G_0(z)}{1 + G_0(z)C(z)} r(t) + \dfrac{H_0(z)}{1 + G_0(z)C(z)} e(t) \\ u(t) = \dfrac{1}{1 + G_0(z)C(z)} r(t) - \dfrac{C(z)H_0(z)}{1 + G_0(z)C(z)} e(t) \end{cases} \tag{3.3}$$

To simplify the latter analysis process, define the sensitivity function as

$$S_0(z) = \frac{1}{1 + G_0(z)C(z)} \tag{3.4}$$

Then input and output relations corresponding to the closed loop system may be written as

$$\begin{cases} y(t) = G_0(z)S_0(z)r(t) + H_0(z)S_0(z)e(t) \\ u(t) = S_0(z)r(t) - C(z)H_0(z)S_0(z)e(t) \end{cases} \tag{3.5}$$

As now our goal is how to identify this closed loop system without any prior knowledge about controller $C(z)$, so in next section the identification problem is formulated and the classical prediction error identification is used to solve our identification process, as the classical prediction error identification is similar to the direct approach.

3.2.2 Classical prediction error identification

Introduce the unknown parameter vector θ in closed loop system, the parameterized form are given by

$$
\begin{cases}
y(t,\theta) = \dfrac{G(z,\theta)}{1+G(z,\theta)C(z)}r(t) + \dfrac{H(z,\theta)}{1+G(z,\theta)C(z)}e(t) \\
u(t,\theta) = \dfrac{1}{1+G(z,\theta)C(z)}r(t) - \dfrac{C(z)H(z,\theta)}{1+G(z,\theta)C(z)}e(t)
\end{cases}
\tag{3.6}
$$

where θ denotes the unknown parameter vector, it exists in the parameterized plant model $G(z, \theta)$ and noise filter $H(z, \theta)$ respectively. The goal of closed loop identification is to identify the unknown parameter vector $\hat{\theta}_N$ from one collected input-output data set $Z^N = \{y(t), u(t)\}_{t=1}^N$, where N denotes the number of total observed data.

According to (3.6), the prediction of output $y(t, \theta)$ can be calculated as the one step ahead prediction.

$$
\begin{aligned}
\hat{y}(t,\theta) &= \frac{1+G(z,\theta)C(z)}{H(z,\theta)} \times \frac{G(z,\theta)}{1+G(z,\theta)C(z)}r(t) + \left[1 - \frac{1+G(z,\theta)C(z)}{H(z,\theta)}\right]y(t) \\
&= \frac{G(z,\theta)}{H(z,\theta)}r(t) + \frac{H(z,\theta)-1-G(z,\theta)C(z)}{H(z,\theta)}y(t)
\end{aligned}
\tag{3.7}
$$

Computing the one step ahead prediction error or residual, it becomes

$$
\varepsilon(t,\theta) = y(t) - \hat{y}(t,\theta) = \frac{1+G(z,\theta)C(z)}{H(z,\theta)}\left[y(t) - \frac{G(z,\theta)}{1+G(z,\theta)C(z)}r(t)\right]
\tag{3.8}
$$

In the standard prediction error identification algorithm, when using the input-output data set $Z^N = \{y(t), u(t)\}_{t=1}^N$ with the data number N, the unknown parameter vector θ is identified by

$$
\hat{\theta}_N = \arg\min_{\theta} V_N\left(\theta, Z^N\right) = \arg\min_{\theta} \frac{1}{N}\sum_{t=1}^N \varepsilon^2(t,\theta)
$$

$$
V_N\left(\theta, Z^N\right) = \frac{1}{N}\sum_{t=1}^N \varepsilon^2(t,\theta)
\tag{3.9}
$$

In the common identification process, assume that there always exists one true parameter vector θ_0 such that

$$G(z,\theta_0) = G_0(z),\ H(z,\theta_0) = H_0(z)$$

This assumption shows that the identified model is contained in the considered model set. To show what make us propose stealth identification for a closed loop system, we concentrate on that prediction error $\varepsilon(t,\theta)$ in (3.8). Due to equation (3.8) it is implicit and it cannot be expanded easily, we use another expression, which is similar to direct approach.

$$\varepsilon(t,\theta) = H^{-1}(z,\theta)\left[y(t) - G(z,\theta)u(t)\right] \tag{3.10}$$

Substituting (3.3) into (3.10), we obtain that

$$y(t) - G(z,\theta)u(t) = \frac{G_0(z)}{1 + G_0(z)C(z)}r(t) + \frac{H_0(z)}{1 + G_0(z)C(z)}e(t)$$

$$-\frac{G(z,\theta)}{1 + G_0(z)C(z)}r(t) - \frac{G(z,\theta)C(z)H_0(z)}{1 + G_0(z)C(z)}e(t)$$

$$= \frac{G_0(z) - G(z,\theta)}{1 + G_0(z)C(z)}r(t) + \frac{H_0(z)\big(1 - G(z,\theta)C(z)\big)}{1 + G_0(z)C(z)}e(t) \tag{3.11}$$

After substituting (3.11) into prediction error (3.10), it holds that

$$\varepsilon(t,\theta) = \frac{G_0(z) - G(z,\theta)}{H(z,\theta)} \times \frac{1}{1 + G_0(z)C(z)}r(t) + \frac{H_0(z)}{H(z,\theta)}\frac{1 - G(z,\theta)C(z)}{1 + G_0(z)C(z)}e(t)$$

$$= \frac{G_0(z) - G(z,\theta)}{H(z,\theta)} \times S_0(z)r(t) + \frac{H_0(z)}{H(z,\theta)}\frac{1 - G(z,\theta)C(z)}{1 + G_0(z)C(z)}e(t) \tag{3.12}$$

Based on above prediction error expression, we get the asymptotic covariance matrix $P_{N,\theta}$ corresponding to the parameter estimate $\hat{\theta}_N$ as that

$$P_{N,\theta} = \lambda_0 \left\langle \varphi, \varphi \right\rangle^{-1} \tag{3.13}$$

where $\left\langle \varphi, \varphi \right\rangle$ denotes some inter product operator, φ is the negative gradient of the prediction error, i.e. it can be computed from

$$\varphi(t,\theta) = -\frac{\partial \varepsilon(t,\theta)}{\partial \theta} \tag{3.14}$$

The inverse covariance matrix $P_{N,\theta}^{-1}$ can be given in the frequency domain.

$$P_{N,\theta}^{-1} = \frac{N}{2\pi\lambda_0}\int_{-\pi}^{\pi} F_r\left(e^{iw},\theta_0\right)F_r^{H}\left(e^{iw},\theta_0\right)\phi_r(w)\,dw + \frac{N}{2\pi}$$

$$\int_{-\pi}^{\pi} F_v\left(e^{iw},\theta_0\right)F_v^{H}\left(e^{iw},\theta_0\right)dw \tag{3.15}$$

where $\phi_r(w)$ is the power spectrum of the excitation signal $r(t)$, the subscript H is the Hermitian conjugate, and the vector functions $F_r(z)$ and $F_v(z)$ are defined by

$$\begin{cases} F_r(z) = H_0^{-1}(z)S_0(z)\left[\nabla_\theta G(z,\theta)\right]_{\theta=\theta_0} \\ F_v(z) = H_0^{-1}(z)\left[\nabla_\theta H(z,\theta)\right]_{\theta=\theta_0} - C(z)S_0(z)\left[\nabla_\theta G(z,\theta)\right]_{\theta=\theta_0} \end{cases} \quad (3.16)$$

where ∇_θ is the gradient operation.

Observe that in order to identify the unknown parameter vector from optimization problem (3.8), controller $C(z)$ needs to be known explicitly. Further, when constructing an asymptotic confidence region for an unknown parameter vector, the inverse covariance matrix must be computed. Indeed, the two terms $F_r\left(e^{iw},\theta_0\right)$ and $F_v\left(e^{iw},\theta_0\right)$ in the inverse covariance matrix are dependent on a closed form expression of the controller $C(z)$. Consequently, the prediction error identification is only possible for a closed loop system when the feedback controller $C(z)$ is known. However, in the engineering field, there are many closed loop systems with an unknown controller, so in this circumstance stealth identification is proposed to solve this difficulty.

3.2.3 Stealth identification

As stealth identification adapts the classical closed loop system such that the controller no longer senses the excitation signal, it can be applied to linear systems with an unknown controller. Stealth identification involves modifying the feedback mechanism in the classical closed loop identification scheme in such a way that the unknown controller does not sense the excitation signal. It means that the controller $C(z)$ will not exist in term $F_r(z)$, i.e. the dependence of inverse covariance matrix $P_{N,\theta}^{-1}$ on $\phi_r(w)$ becomes independent of the expression of the controller.

We now address stealth identification for a closed loop system with unknown controller, see Fig. 3.2. The goal of this modified structure is to ensure that the excitation signal $r(t)$ is not affected by the controller.

The feedback term to the controller is adjusted from $y(t)$ to $y(t) - G_{id}r(t) - H_{id}v(t)$, where G_{id} and H_{id} are initial estimate of $G_0(z)$ and $H_0(z)$ respectively.

$$\begin{cases} G_{id} = G(z,\theta_{init}) \\ H_{id} = H(z,\theta_{init}) \end{cases} \quad (3.17)$$

where θ_{init} is an initial estimate corresponding to parameter vector θ. To obtain an initial estimate θ_{init}, a coarse identification experiment is performed prior to the actual optimal identification experiment. For example, we apply an excitation signal $\{r_{init}(t)\}_t^N$ and collect the input-output data set $\{u_{init}(t),y_{init}(t)\}_t^N$, then use optimization problem (3.9) to obtain an initial estimate θ_{init}.

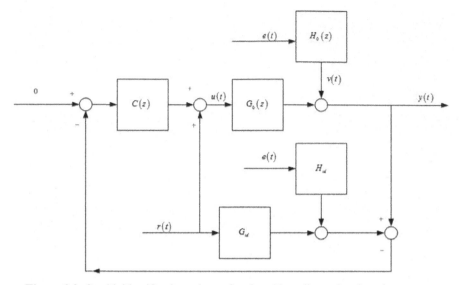

Figure 3.2: Stealth identification scheme for closed loop linear time invariant system

Observe Fig. 3.2 carefully, after simple but tedious calculation, the input-output relations are defined by

$$\begin{cases} u(t) = C(z)\big[-y(t) + G_{id}r(t) + H_{id}e(t)\big] + r(t) \\ y(t) = G_0(z)u(t) + H_0(z)e(t) \end{cases} \quad (3.18)$$

Continuing to do some computations and we obtain

$$y(t) = \frac{G_0(z)\big[1 + C(z)G_{id}\big]}{1 + G_0(z)C(z)}r(t) + \frac{G_0(z)C(z)H_{id} + H_0(z)}{1 + G_0(z)C(z)}e(t)$$

$$y(t) - H_0(z)e(t) = \frac{G_0(z)\big[1 + C(z)G_{id}\big]}{1 + G_0(z)C(z)}r(t)$$

$$+ \left[\frac{G_0(z)C(z)H_{id} + H_0(z)}{1 + G_0(z)C(z)} - H_0(z)\right]e(t)$$

$$= \frac{G_0(z)\big[1 + C(z)G_{id}\big]}{1 + G_0(z)C(z)}r(t) + \frac{G_0(z)C(z)\big[H_{id} - H_0(z)\big]}{1 + G_0(z)C(z)}e(t)$$

$$u(t) = \frac{1 + C(z)G_{id}}{1 + G_0(z)C(z)}r(t) + \frac{C(z)\big[H_{id} - H_0(z)\big]}{1 + G_0(z)C(z)}e(t) \quad (3.19)$$

substituting the above two input-output relations into equation (3.9) to obtain a new expression of prediction error $\varepsilon_1(t,\theta)$ in our modified closed loop structure.

$$\varepsilon_1(t,\theta) = H^{-1}(z,\theta)\big[y(t) - G(z,\theta)u(t)\big]$$

$$y(t) - G(z,\theta)u(t) = \frac{G_0(z)\big[1 + C(z)G_{id}\big]}{1 + G_0(z)C(z)}r(t)$$

$$+\frac{G_0(z)C(z)H_{id} + H_0(z)}{1 + G_0(z)C(z)}e(t) - \frac{G(z,\theta)\big[1 + C(z)G_{id}\big]}{1 + G_0(z)C(z)}r(t)$$

$$+\frac{G(z,\theta)C(z)\big[H_{id} - H_0(z)\big]}{1 + G_0(z)C(z)}e(t)$$

$$=\big[G_0(z) - G(z,\theta)\big]\frac{1 + C(z)G_{id}}{1 + G_0(z)C(z)}r(t)$$

$$+\frac{G_0CH_{id} + H_0 - G(\theta)CH_{id} + G(\theta)CH_0}{1 + G_0(z)C(z)}e(t)$$

$$=\big[G_0(z) - G(z,\theta)\big]\frac{1 + C(z)G_{id}}{1 + G_0(z)C(z)}r(t)$$

$$+\frac{\big[G_0(z) - G(z,\theta)\big]C(z)G_{id} + \big[1 + G(z,\theta)C(z)\big]H_0(z)}{1 + G_0(z)C(z)}e(t) \qquad (3.20)$$

Then the new prediction error $\varepsilon_1(t,\theta)$ is given as

$$\varepsilon_1(t,\theta) = \frac{1}{H(z,\theta)}\left[\begin{array}{c} \big[G_0(z) - G(z,\theta)\big]\dfrac{1 + C(z)G_{id}}{1 + G_0(z)C(z)}r(t) \\[2ex] + \left(\begin{array}{c}\dfrac{\big[G_0(z) - G(z,\theta)\big]C(z)H_{id}}{1 + G_0(z)C(z)} \\[2ex] +\dfrac{\big[1 + G(z,\theta)C(z)\big]H_0(z)}{1 + G_0(z)C(z)}\end{array}\right)e(t) \end{array}\right]$$

$$(3.21)$$

If the following assumptions hold,

$$\begin{cases} G_{id} = G(z,\theta_{init}) = G_0(z) \\ H_{id} = H(z,\theta_{init}) = H_0(z) \end{cases}$$

Then above input-output relations become

$$\begin{cases} u(t) = r(t) \\ y(t) = G_0(z)u(t) + H_0(z)e(t) \end{cases} \qquad (3.22)$$

Consequently, the new prediction error $\varepsilon_1(t,\theta)$ becomes

$$\varepsilon_1\left(t,\theta\right)=\frac{1}{H\left(z,\theta\right)}\left[\begin{array}{l}\left[G_0\left(z\right)-G\left(z,\theta\right)\right]r\left(t\right)\\+\left[\dfrac{H_0\left(z\right)+G_0\left(z\right)C\left(z\right)H_{id}}{1+G_0\left(z\right)C\left(z\right)}\right]\\+\dfrac{\left[H_0\left(z\right)-H_{id}\right]G\left(z,\theta\right)C\left(z\right)}{1+G_0\left(z\right)C\left(z\right)}\end{array}\right]e\left(t\right)$$

$$=\frac{1}{H\left(z,\theta\right)}\left[G_0\left(z\right)r\left(t\right)+H_0\left(z\right)e\left(t\right)-G\left(z,\theta\right)r\left(t\right)\right]$$

$$=\frac{1}{H\left(z,\theta\right)}\left[\left[G_0\left(z\right)-G\left(z,\theta\right)\right]r\left(t\right)+H_0\left(z\right)e\left(t\right)\right] \qquad (3.23)$$

Comparing the new prediction error $\varepsilon_1\left(t,\theta\right)$ (3.23) with that original prediction error $\varepsilon\left(t,\theta\right)$ (3.11), we found that the sensitivity function $S_0\left(z\right)$ does not exist in $\varepsilon_1\left(t,\theta\right)$, i.e. the new prediction error $\varepsilon_1\left(t,\theta\right)$ is independent of the unknown controller $C(z)$. It is the idea of the name stealth identification. To further study the advantage of stealth identification in inverse covariance matrix $P_{N,\theta}^{-1}$, using equation (3.13), the new inverse covariance matrix $P_{N,\theta,1}^{-1}$ is obtained by

$$P_{N,\theta,1}^{-1}=\frac{1}{\lambda_0}E\left[\varphi\left(t,\theta\right)\varphi^T\left(t,\theta\right)\right]_{\theta=\theta_0} \qquad (3.24)$$

where E denotes the expectation operator, and the negative gradient of the new prediction error $\varepsilon_1\left(t,\theta\right)$ is computed from

$$\varphi\left(t,\theta\right)=-\frac{\partial\varepsilon_1\left(t,\theta\right)}{\partial\theta} \qquad (3.25)$$

After simple but tedious calculation, the new inverse covariance matrix $P_{N,\theta,1}^{-1}$ is obtained by

$$P_{N,\theta,1}^{-1}=\frac{N}{2\pi\lambda_0}\int_{-\pi}^{\pi}F_{r1}\left(e^{iw},\theta_0\right)F_{r1}^H\left(e^{iw},\theta_0\right)\phi_r\left(w\right)dw$$

$$+\frac{N}{2\pi}\int_{-\pi}^{\pi}F_{v1}\left(e^{iw},\theta_0\right)F_{v1}^H\left(e^{iw},\theta_0\right)dw$$

$$F_{r1}\left(z\right)=H_0^{-1}\left(z\right)\left[\nabla_\theta G\left(z,\theta\right)\right]_{\theta=\theta_0}$$

$$F_{v1}\left(z\right)=-H_0^{-1}\left(z\right)\left(H_0\left(z\right)+G_0\left(z\right)C\left(z\right)H_{id}\right)\times H_0^{-1}\left(z\right)\left[\nabla_\theta H\left(z,\theta\right)\right]_{\theta=\theta_0}$$

$$+H_0^{-1}\left(z\right)\left(H_0\left(z\right)-H_{id}\right)C\left(z\right)\left[\nabla_\theta G\left(z,\theta\right)\right]_{\theta=\theta_0}$$

$$-H_0^{-1}\left(z\right)\left(H_0\left(z\right)-H_{id}\right)G_0\left(z\right)C\left(z\right)\left[\nabla_\theta H\left(z,\theta\right)\right]_{\theta=\theta_0} \qquad (3.26)$$

The above vector functions $F_{r1}(z)$ and $F_{v1}(z)$ are general forms, if the condition of equation (3.22) holds then they will show two very simple forms such as

$$F_{r1}(z) = H_0^{-1}(z)\left[\nabla_\theta G(z,\theta)\right]_{\theta=\theta_0}$$
$$F_{v1}(z) = H_0^{-1}(z)\left[\nabla_\theta H(z,\theta)\right]_{\theta=\theta_0} \tag{3.27}$$

Comparing equation (3.27) and (3.16), the new inverse covariance matrix $P_{N,\theta,1}^{-1}$ is easier than $P_{N,\theta}^{-1}$. This simpler form can simplify the other fields, such as model structure validation and optimal signal design, etc.

3.2.4　Convergence condition in stealth identification

Under the Stealth identification scheme for the closed loop system, we now start to introduce a candidate domain of attraction for the objective function $V_N(\theta, Z^N)$ and derive one convergence condition, which ensures that a given set is a candidate domain of attraction.

Define the asymptotic limit parameter estimate θ^* as

$$\theta^* = \arg\min_\theta \lim_{N\to\infty} E\left\{V_N\left(\theta, Z^N\right)\right\} = \arg\min_\theta \lim_{N\to\infty} E\left\{\varepsilon_1^2\left(t,\theta\right)\right\}$$

$$= \arg\min_\theta \lim_{N\to\infty} \frac{1}{N}\sum_{t=1}^{N}\varepsilon_1^2\left(t,\theta\right) \tag{3.28}$$

For notation clarity, we define

$$V(\theta) = EV_N\left(\theta, Z^N\right)$$

$$V_N(\theta) = V_N\left(\theta, Z^N\right) = \frac{1}{N}\sum_{t=1}^{N}\varepsilon_1^2\left(t,\theta\right) \tag{3.29}$$

To solve the optimization problem (3.28), the least squares method applied to the optimization problem (3.28) can be reformulated as an iterative algorithm. For example, the classical steepest descent method is used. The iterative formula is given along the Newton search direction.

$$\hat{\theta}_N^{(j+1)} = \hat{\theta}_N^{(j)} - \mu^{(j)}\left[\hat{R}_N\left(\hat{\theta}_N^{(j)}\right)\right]^{-1} V_N'\left(\hat{\theta}_N^{(j)}\right) \tag{3.30}$$

where $\mu^{(j)}$ is the step size, its appropriate choice must guarantee the new quadratic objective function $V_N(\theta)$ is decreasing at every iteration.

Denote the estimations of gradient and Hessian matrix about the quadratic objective function $V_N(\theta)$ as $V_N'(\theta)$ and $\hat{R}_N(\theta)$ respectively. This iterative algorithm has the property that under mild conditions the parameter estimate $\hat{\theta}_N$ converges, for $N \to +\infty$, to a set

$$\Theta^* = \left\{ \theta^* = \arg\min_{\theta} V(\theta) \right\} \tag{3.31}$$

where $V(\theta)$ is defined in equation (3.29).

The goal of estimation theory is to obtain the global minimum, but not the local minimum. So the initial condition that is used to start the iterative algorithm must be close enough to the global minimum. A set of initial conditions for which the iterative algorithm will converge to the global minimum is called a domain of attraction of the algorithm.

Now two definitions from system identification theory are given as follows:

Definition 3.1: Assume θ^* to be the unique global minimum of the function $V(\theta): R^n \rightarrow R$. A set $\Omega \subset R^n$ is a domain of attraction of an iterative algorithm, if

$$\lim_{N \rightarrow \infty} \hat{\theta}_N = \theta^*, \quad \forall \hat{\theta}_1 \in \Omega$$

A good initialization for the iterative algorithm can guarantee that the parameter estimate will approach to the global minimum.

Definition 3.2: Let θ^* be the global minimum of the objective function $V_N(\theta)$. A set Λ is a candidate domain of attraction for the objective function $V_N(\theta)$, if

$$\left(\theta - \theta^*\right) \nabla V_N(\theta) > 0, \forall \theta \in \Lambda \text{ and } \theta \neq \theta^* \tag{3.32}$$

where $\nabla V_N(\theta)$ denotes the gradient of the objective function.

Now the properties of the candidate domain of attraction are explored to obtain a convergence condition which ensures that a given set is a candidate domain of attraction for stealth identification structure.

Theorem 3.1: Consider the input-output relations (3.18) and the objective function $V_N(\theta)$ in the stealth identification structure. A set $\theta_0 \in \Lambda$ is a candidate domain of attraction if and only if

$$\left(\theta - \theta_0\right) \frac{1}{\pi} \int_{-\pi}^{\pi} \phi_r \left[H^{-1}(\theta)\left(G_0 - G(\theta)\right) \right]^T \times$$

$$\left[\nabla H^{-1}(\theta)\left(G_0 - G(\theta)\right) - H^{-1}(\theta)\nabla G(\theta) \right] + \lambda_0 |H_0|^2 \left| \nabla H^{-1}(\theta) \right|^2 dw \geq 0 \tag{3.33}$$

For any $\theta \in \Lambda$, such that $\theta \neq \theta_0$, where θ_0 is defined in section 3.2.3.

Proof: The corresponding objective function in stealth identification for closed loop system is given by

$$V_N(\theta) = V_N\left(\theta, Z^N\right) = \frac{1}{N} \sum_{t=1}^{N} \varepsilon_1^2(t, \theta) \tag{3.34}$$

where the new prediction error $\varepsilon_1(t, \theta)$ is defined in (3.23) and (3.24).

By the application of Parseval's theorem, $V_N(\theta)$ can be rewritten in frequency domain under our modified closed loop structure.

$$V_N(\theta) = \frac{1}{2\pi}\int_{-\pi}^{\pi}\left|H^{-1}\left(e^{iw},\theta\right)\left(G_0\left(e^{iw}\right)-G\left(e^{iw},\theta\right)\right)\right|^2\phi_r(w)$$

$$+\left|H^{-1}\left(e^{iw},\theta\right)\left(a_1+\frac{\left(H_0-H_{id}\right)C\left(e^{iw}\right)}{1+G_0\left(e^{iw}\right)C\left(e^{iw}\right)}G\left(e^{iw},\theta\right)\right)\right|^2\lambda_0 \, dw$$

$$= \frac{1}{2\pi}\int_{-\pi}^{\pi}\left|H^{-1}\left(e^{iw},\theta\right)\left(G_0\left(e^{iw}\right)-G\left(e^{iw},\theta\right)\right)\right|^2\phi_r(w)$$

$$+\left|H^{-1}\left(e^{iw},\theta\right)\left(a_1+a_2 G\left(e^{iw},\theta\right)\right)\right|^2 dw \tag{3.35}$$

where a_1 and a_2 are defined as

$$a_1 = \frac{H_0\left(e^{iw}\right)G_0\left(e^{iw}\right)C\left(e^{iw}\right)H_{id}C\left(e^{iw}\right)}{1+G_0\left(e^{iw}\right)C\left(e^{iw}\right)}, a_2 = \frac{\left(H_0-H_{id}\right)C\left(e^{iw}\right)}{1+G_0\left(e^{iw}\right)C\left(e^{iw}\right)} \tag{3.36}$$

Now compute the gradient of the objective function $V_N(\theta)$ with respect to θ.

$$\nabla V_N(\theta) = \frac{1}{\pi}\int_{-\pi}^{\pi}\phi_r\left[H^{-1}(\theta)\left(G_0-G(\theta)\right)\right]^T$$

$$\times\left[\nabla H^{-1}(\theta)\left(G_0-G(\theta)\right)-H^{-1}(\theta)\nabla G(\theta)\right]$$

$$+\lambda_0\left[H^{-1}\left(e^{iw},\theta\right)\left(a_1+a_2 G\left(e^{iw},\theta\right)\right)\right]$$

$$\times H^{-1}(\theta)\nabla G(\theta)+\nabla H^{-1}(\theta)\left(a_1+a_2 G\left(e^{iw},\theta\right)\right)dw \tag{3.37}$$

By using definition 3.2, we compute that

$$(\theta-\theta_0)\nabla V_N(\theta) = (\theta-\theta_0)\frac{1}{\pi}\int_{-\pi}^{\pi}\phi_r\left[H^{-1}(\theta)\left(G_0-G(\theta)\right)\right]^T$$

$$\times\left[\nabla H^{-1}(\theta)\left(G_0-G(\theta)\right)-H^{-1}(\theta)\nabla G(\theta)\right]$$

$$+\lambda_0\left[H^{-1}\left(e^{iw},\theta\right)\left(a_1+a_2 G\left(e^{iw},\theta\right)\right)\right]$$

$$\times H^{-1}(\theta)\nabla G(\theta)+\nabla H^{-1}(\theta)\left(a_1+a_2 G\left(e^{iw},\theta\right)\right)dw \tag{3.38}$$

when the following assumption holds

$$\begin{cases}G_{id} = G\left(z,\theta_{init}\right) = G_0(z)\\ H_{id} = H\left(z,\theta_{init}\right) = H_0(z)\end{cases}$$

Then some simplified results are obtained by

$$\varepsilon_1(t,\theta) = \frac{1}{H(z,\theta)}\big[\big[G_0(z) - G(z,\theta)\big]r(t) + H_0(z)e(t)\big] \tag{3.39}$$

Then

$$V_N(\theta) = \frac{1}{2\pi}\int_{-\pi}^{\pi}\big|H^{-1}(e^{iw},\theta)\big(G_0(e^{iw}) - G(e^{iw},\theta)\big)\big|^2\phi_r(w)$$
$$+ \big|H^{-1}(e^{iw},\theta)\big\|H_0(e^{iw})\big|^2\lambda_0\,dw \tag{3.40}$$

Its gradient form is

$$\nabla V_N(\theta) = \frac{1}{\pi}\int_{-\pi}^{\pi}\phi_r\big[H^{-1}(\theta)\big(G_0 - G(\theta)\big)\big]^T$$
$$\times\big[\nabla H^{-1}(\theta)\big(G_0 - G(\theta)\big) - H^{-1}(\theta)\nabla G(\theta)\big]$$
$$+ \lambda_0|H_0|^2\big|\nabla H^{-1}(\theta)\big|^2\,dw \tag{3.41}$$

By using the condition that

$$(\theta - \theta_0)\nabla V_N(\theta) > 0, \quad \forall\theta\in\Lambda \text{ and } \theta\neq\theta_0 \tag{3.42}$$

which concludes that the set Λ is a candidate domain of attraction.

3.2.5   Simulation example

In this section, we apply the stealth identification on a single input and single output system controlled by a model predictive controller. A true data generating system considered here is given as

$$\begin{cases} G_0(z) = \dfrac{0.25z^{-1} + 0.12z^{-2}}{1 - 1.6z^{-1} + 0.8z^{-2} - 0.64z^{-3} + 0.65z^{-4}} \\ H_0(z) = \dfrac{1 + 0.2z^{-1}}{1 + 0.5z^{-1}} \end{cases} \tag{3.43}$$

Its parameterized form is given as

$$\begin{cases} G(z,\theta) = \dfrac{a_5z^{-1} + a_6z^{-2}}{1 + a_1z^{-1} + a_2z^{-2} + a_3z^{-3} + a_4z^{-4}} \\ H(z,\theta) = \dfrac{1 + b_2z^{-1}}{1 + b_1z^{-1}} \end{cases} \tag{3.44}$$

A Gaussian white noise signal $e(t)$ with variance $\lambda_0 = 0.5$ is added through the noise filter $H_0(z)$. The sample times $T_s = 1$ second, the true parameter vector θ_0 is that

$$\theta_0 = (-1.6 \quad 0.8 \quad -0.64 \quad 0.65 \quad 0.25 \quad 0.12 \quad 0.5 \quad 0.2)^T$$

The data generating system is operated in the closed loop system with a model predictive control based on our commissioning model (G_{init}, H_{init}), where θ_{init} is chosen as

$$\theta_{init} = (-1.7 \quad 0.7 \quad -0.4 \quad 0.8 \quad 0.15 \quad 0.1 \quad 0.4 \quad 0.1)^T$$

The model predictive control is tuned so that we get sufficient performance for the commissioning model (G_{init}, H_{init}). The model predictive control is set to have a prediction horizon of 100 and a control horizon of 100. The output variable $y(t)$ has a weight of 10, and the input variable $u(t)$ has a weight of 1. An excited signal $r(t)$ has a bound, i.e. $-1 \le r(t) \le 1$. The applied input signal is given in Fig. 3.3, and we measure the output signal $y(t)$ by some measuring devices, where the observed output signal is plotted in Fig. 3.4.

When using the prediction error identification method to identify unknown parameters, if estimation error $\delta = \left\| \hat{\theta}(t) - \theta \right\| / \| \theta \|$ is less than one very small value $\varepsilon = 0.005$, i.e. $\delta = \left\| \hat{\theta}(t) - \theta \right\| / \| \theta \| \le 0.05$, then it terminates the recursive methods. The parameter estimations can be seen in Table 3.1 with iterative step increases. From Table 3.1, we see that when external noise $e(t)$ is a white noise, the parameter estimations will converge to their corresponding true values with iterative step approaches to 3000.

To verify the efficiency of the identified model $G\left(\hat{\theta}_N\right)$ by our stealth identification strategy and make sure that this identified model can be used to replace the true model, we compare the Bode responses among the true plant model $G_0(q)$, our identified model $G\left(\hat{\theta}_N\right)$, and the classical plant model respectively in Fig. 3.5. From Fig.3.5, we see that these three Bode response curves coincide with each other, and our identified model $G\left(\hat{\theta}_N\right)$ can converge to its true plant model quicker than the classical identified plant model. This means that very quickly, our model error $\tilde{G}(q)$ will converge to zero with time increases.

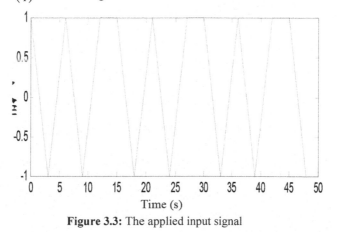

Figure 3.3: The applied input signal

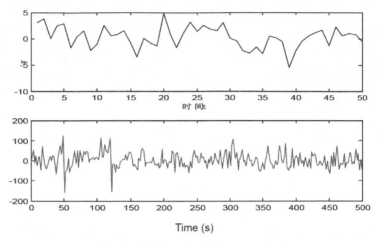

Figure 3.4: The observed output signal

Table 3.1: Parameter estimations identified by stealth identification framework

t	a_1	a_2	a_3	a_4	a_5	a_6
100	-1.66010	0.71224	-0.36960	0.76926	0.15739	0.10412
500	-1.61858	0.71936	-0.57727	0.75419	0.22146	0.12554
1000	-1.63117	0.71189	-0.56187	0.68829	0.24662	0.10467
1500	-1.63560	0.70161	-0.58364	0.68297	0.24484	0.10423
2000	-1.63341	0.70037	-0.60951	0.65304	0.24476	0.10680
2500	-1.64755	0.70421	-0.62808	0.65519	0.24388	0.10213
3000	-1.64664	0.73523	-0.64690	0.64788	0.24975	0.10434
True	−1.6	0.8	−0.64	0.65	0.25	0.12

Furthermore, the Bode responses among the true noise filter, our identified noise filter, and the classical noise filter are also given in Fig. 3.6, where two Bode response curves between the true noise filter and our identified noise filter coincide with each other. But another Bode response curve corresponding to the classical noise filter deviates from the true Bode response curve. By comparing the Bode response curves in Fig. 3.5 and Fig. 3.6, we see that our identified plant model and noise filter, obtained by our stealth identification strategy are more accurate, as excited signal and unknown controller do not affect the identification accuracy.

3.2.6   Conclusion

In this section, a new stealth identification strategy is proposed to identify the closed loop linear time invariant system. Stealth identification modified the

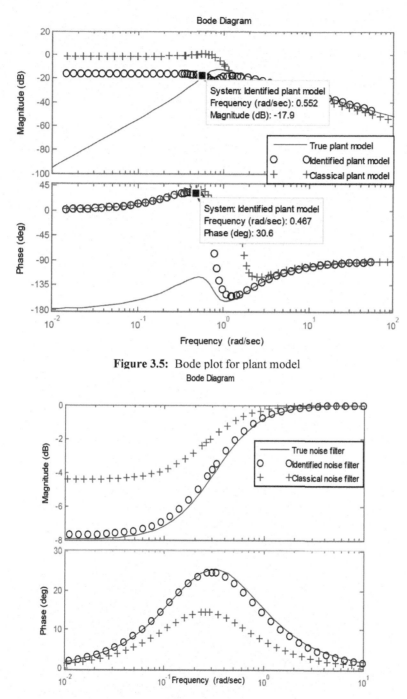

Figure 3.5: Bode plot for plant model

Figure 3.6: Bode plot for noise filter

closed loop feedback mechanism so that the unknown controller does not affect the excitation signal. The stealth identification strategy makes it possible for the classical prediction error method to identify the closed loop system with an unknown controller. One convergence condition to ensure the global minimum in the process of stealth identification is given. In the future, we want to apply this stealth identification strategy to model structure validation for the closed loop linear time invariant system.

3.3   Performance analysis of closed loop system with a tailor made parameterization

In this section, we concentrate on the performance analysis of the closed loop system in depth. All theories are based on a tailor made parameterization used to parameterize the closed loop system. A tailor made parameterization combines those two separate steps from the indirect closed loop identification approach. This means that knowledge of the closed loop system and knowledge of the controller are employed in the parameterization of the closed loop system. This method applies knowledge of the controller and minimizes an error between the true closed loop transfer function and identified closed loop using a parameterization model of the open loop model only. Here we study the tailor made parameterization method in a linear framework, where the plant model and controller are all parameterized as polynomials. To identify the closed loop parameter vector, a recursive least squares method with forgetting schemes is proposed. This recursive least squares method with forgetting schemes achieves the reformulation of the classical recursive least squares with forgetting schemes as a regularized least squares problem. To reflect the identification accuracy, we apply the statistical probability framework to derive the variance matrix of the unknown parameters. This variance matrix is decomposed into one inter product form which is used to construct one uncertainty bound about the unknown parameter estimation. This uncertainty bound is called the confidence interval and it constitutes the guaranteed confidence region test with respect to the model parameter estimation under closed loop conditions. Using only one plant model, we define a Vinnicombe distance between its true and identified plant model. Then we use the results of some robust control theories such as the Vinnicombe gap between the plant and its related stability property to give a preliminary performance analysis. Generally, this preliminary performance analysis is extended to a transfer function matrix form which is constituted by three transfer functions. The worst case performance at frequencies is analyzed by solving one standard convex optimization problem involving some linear matrix inequality constraints.

3.3.1   Closed loop system description

Consider the following closed loop system configuration in Fig. 3.1. again. The closed loop system configuration in the figure appears in many practical engineering problems, for example, flight simulation. Flight simulation is a speed servo system with a high precision position. The driven element of flight simulation is an electric motor, and the essence of the control structure in flight simulation is a closed loop system corresponding to the position or speed of that electric motor.

According to the analysis of the servo control system, one negative feedback part is added to reduce the sensitivity in the closed loop system, while the cascade regulator is introduced in each feedback control structure to reduce the dependence on the electric motor's parameter. Here, we give an example of the pitch position tracking loop from flight simulation to verify the feasibility of our iterative correlation tuning control approach in the precision servo control system. In the closed loop system of flight simulation, the photoelectric encoder is mounted on the outer pitch frame, and the angular position signal collected at the outer pitch frame is regarded as the position feedback part. After the difference between two angular positions goes through the position correlation part and power amplifier part, then this difference will make the electric motor start to rotate. The pitch position tracking loop from flight simulation is simplified in Fig. 3.7.

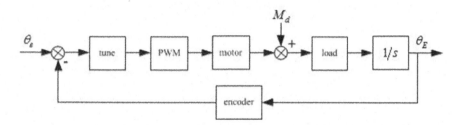

Figure 3.7: The simplified pitch position tracking loop

In Fig. 3.7 the input signal is the relative angular signal of the inner pitch loop, and this input signal is collected by one photoelectric encoder located in the inner pitch frame. It means that one photoelectric encoder collects the angular position signal to send one position feedback part. The transfer function model of that simplified pitch position tracking loop can be seen in Fig. 3.8.

In Fig. 3.8 we regard the encoder as a constant and merge it in the power amplifier, then the close loop system is an unit feedback. θ_{me} is the input signal with respect to the electric motor, the controller in this position tracking loop is the classical PID controller.

Observing the closed loop system configuration, we obtain the following transfer function form:

$$y(t) = G_0(q)r(t) - G_0(q)C(q)y(t) + H_0(q)e(t) \qquad (3.45)$$

Continuing to do some simple computations, we get

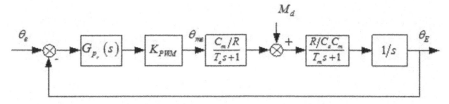

Figure 3.8: The transfer function model of that simplified pitch position tracking loop

$$\begin{cases} y(t) = \dfrac{G_0(q)}{1+G_0(q)C(q)} r(t) + \dfrac{H_0(q)}{1+G_0(q)C(q)} e(t) \\[3mm] u(t) = \dfrac{1}{1+G_0(q)C(q)} r(t) - \dfrac{C(q)H_0(q)}{1+G_0(q)C(q)} e(t) \end{cases} \qquad (3.46)$$

To simplify the analysis process, one sensitivity function is defined as

$$S_0(q) = \frac{1}{1+G_0(q)C(q)}$$

Applying the above defined sensitivity function, the output of closed loop system can be rewritten as

$$y(t,\theta) = G_0(q)S_0(q)r(t) + H_0(q)S_0(q)e(t)$$

Introduce one unknown parameter vector θ into the closed loop system, the parameterized form corresponding to equation (3.46) is given by

$$y(t) = \frac{G(q,\theta)}{1+G(q,\theta)C(q)} r(t) + \frac{H(q,\theta)}{1+G(q,\theta)C(q)} e(t) \qquad (3.47)$$

where θ denotes the unknown parameter vector, it exists in the parameterized plant model $G(q,\theta)$ and noise filter $H(q,\theta)$ respectively. The goal of closed loop identification is to identify the unknown parameter vector from one given data set $Z^N = \{r(t), y(t)\}_{t=1}^N$ and priori known controller $C(q)$, where N denotes the total number of observed data.

According to equation (3.47), the prediction of output $y(t, \theta)$ can be calculated as the one step ahead prediction.

$$\begin{aligned} \hat{y}(t,\theta) &= \frac{1+G(q,\theta)C(q)}{H(q,\theta)} \times \frac{G(q,\theta)}{1+G(q,\theta)C(q)} r(t) + \left[1 - \frac{1+G(q,\theta)C(q)}{H(q,\theta)}\right] y(t) \\[2mm] &= \frac{G(q,\theta)}{H(q,\theta)} r(t) + \frac{H(q,\theta)-1-G(q,\theta)C(q)}{H(q,\theta)} y(t) \end{aligned} \qquad (3.48)$$

Constructing one step ahead prediction error or residual as

$$\varepsilon(t,\theta) = y(t) - \hat{y}(t,\theta) = \frac{1 + G(q,\theta)C(q)}{H(q,\theta)}\left[y(t) - \frac{G(q,\theta)}{1 + G(q,\theta)C(q)}r(t)\right] \quad (3.49)$$

In the standard prediction error algorithm, using input-output data set $Z^N = \{r(t), y(t)\}_{t=1}^{N}$ with the number N, the unknown parameter vector is identified by solving one optimization problem.

$$\hat{\theta}_N = \arg\min_{\theta} V_N\left(\theta, Z^N\right) = \arg\min_{\theta}\frac{1}{N}\sum_{t=1}^{N}\varepsilon^2(t,\theta) \quad (3.50)$$

The above equation (3.50) is similar to the classical prediction error algorithm and direct approach. In the next section, it will be made clear that a tailor made parameterization is used. The parameterized plant model $G(q, \theta)$ and feedback controller $C(q)$ are all assumed to be polynomials. Then we propose a recursive least squares method with forgetting schemes to identify the unknown parameter vector θ. Based on this identified parameter vector, one confidence interval of an unknown parameter vector is constructed under closed loop condition.

3.3.2 Confidence interval analysis with a tailor made parameterization

Let the plant model $G(q, \theta)$ be parameterized as one polynomial.

$$G(q,\theta) = \frac{B(q,\theta)}{A(q,\theta)} = \frac{b_1 q^{-1} + \cdots b_{n_b} q^{-n_b}}{1 + a_1 q^{-1} + \cdots a_{n_a} q^{-n_a}} \quad (3.51)$$

where $\theta = [a_1 \cdots a_n, b_1 \cdots b_n]^T$. Similarly, the feedback controller is as parameterized as

$$C(q) = \frac{N_c(q)}{D_c(q)} = \frac{n_0 + n_1 q^{-1} + \cdots n_{n_N} q^{-n_N}}{1 + d_1 q^{-1} + \cdots d_{n_D} q^{-n_D}} \quad (3.52)$$

where $N_c(q)$ and $D_c(q)$ are coprime polynomials. Based on these two polynomial forms (3.51), (3.52), the parameterization of the output predictor is given by

$$\hat{y}(t/t-1,\theta) = \frac{D_c(q)B(q,\theta)}{D_c(q)A(q,\theta) + N_c(q)B(q,\theta)}r(t) \quad (3.53)$$

The denominator of the closed loop transfer function can be written as a function of the open loop unknown parameter vector θ.

$$D_c(q)A(q,\theta) + N_c(q)B(q,\theta) = 1 + \begin{bmatrix} q^{-1} & q^{-2} & \cdots & q^{-n} \end{bmatrix}\theta_{cl} \quad (3.54)$$

The order of the closed loop polynomial is given by

$$n = \max\left(n_a + n_D, n_b + n_N\right)$$

The closed loop parameter vector θ_{cl} is given as

$$\theta_{cl} = S\theta + \rho \tag{3.55}$$

Matrix S and vector ρ are parameterized as

$$\rho = \begin{bmatrix} d_1 & \cdots & d_{n_D} & 0 & \cdots & 0 \end{bmatrix}^T \in R^n, \; S = \begin{bmatrix} P_D & P_N \\ 0 & 0 \end{bmatrix}$$

$$P_D = \begin{bmatrix} 1 & 0 & \cdots & 0 \\ d_1 & 1 & \cdots & \vdots \\ d_2 & d_1 & \ddots & \vdots \\ \vdots & & & 1 \\ d_{n_D} & & \ddots & d_1 \\ 0 & & \ddots & d_2 \\ \vdots & & & \vdots \\ 0 & \cdots & 0 & d_{n_D} \end{bmatrix}, \; P_N = \begin{bmatrix} n_0 & 0 & \cdots & 0 \\ n_1 & n_0 & \cdots & \vdots \\ n_2 & n_1 & \cdots & \vdots \\ \vdots & \vdots & \cdots & n_0 \\ n_{n_N} & & \cdots & n_1 \\ \vdots & & \cdots & \vdots \\ 0 & \cdots & & n_{n_N} \end{bmatrix} \tag{3.56}$$

when the feedback controller $C(q)$ is priori known, then matrix S and vector ρ can be constructed by using parameters from coprime polynomials.

Rearranging equation (3.53), we obtain

$$\left[D_c(q)A(q,\theta) + N_c(q)B(q,\theta)\right]y(t) = \left[D_c(q)B(q,\theta)\right]r(t) \tag{3.57}$$

Substituting (3.54), (3.55) and (3.56) into (3.57), it yields

$$\left(1 + \begin{bmatrix} q^{-1} & q^{-2} & \cdots & q^{-n} \end{bmatrix}\left[\begin{pmatrix} P_D & P_N \end{pmatrix}\theta + \rho\right]\right)y(t)$$
$$= \begin{bmatrix} q^{-1} & q^{-2} & \cdots & q^{-n} \end{bmatrix}\begin{bmatrix} 0 & P_D \end{bmatrix}\theta r(t) \tag{3.58}$$

Expanding above equation, we see that

$$y(t) + \begin{bmatrix} y(t-1) & y(t-2) & \cdots & y(t-n) \end{bmatrix}\left[\begin{pmatrix} P_D & P_N \end{pmatrix}\theta + \rho\right]$$
$$= \begin{bmatrix} r(t-1) & r(t-2) & \cdots & r(t-n) \end{bmatrix}\begin{bmatrix} 0 & P_D \end{bmatrix}\theta$$

It means that

$$y(t) = \left(\begin{bmatrix} r(t-1) & r(t-2) & \cdots & r(t-n) \end{bmatrix}\begin{bmatrix} 0 & P_D \end{bmatrix} - \begin{bmatrix} y(t-1) & y(t-2) & \cdots & y(t-n) \end{bmatrix}\begin{pmatrix} P_D & P_N \end{pmatrix}\right)\theta$$

Define one vector $\varphi(t)$ as

$$\varphi^T(t) = \left(\begin{bmatrix} r(t-1) & r(t-2) & \cdots & r(t-n) \end{bmatrix} \begin{bmatrix} 0 & P_D \end{bmatrix} - \\ \begin{bmatrix} y(t-1) & y(t-2) & \cdots & y(t-n) \end{bmatrix} \begin{pmatrix} P_D & P_N \end{pmatrix} \right)$$

Then output of the closed loop system can be written as

$$y(t) = \varphi^T(t)\theta \tag{3.59}$$

Vector $\varphi(t)$ is similar to the classical regression vector. A common way to identify the unknown parameter vector θ in (3.59) relies on the recursive least squares with forgetting schemes, where the parameter vector estimation $\hat{\theta}_t$ is given as

$$\hat{\theta}_t = \arg\min_{\theta} V_1(\theta) \tag{3.60}$$

where the loss function is defined as

$$V_1(\theta) = \sum_{s=1}^{t} \lambda^{t-s} \left(y(s) - \varphi^T(s)\theta \right) \tag{3.61}$$

The forgetting factor $\lambda \in [0, 1]$ operates as an exponential weight which decreases with more remote data. Optimization problem (3.60) admits the recursive solution.

$$\begin{cases} R_t = \lambda R_{t-1} + \varphi(t)\varphi^T(t) \\ \hat{\theta}_t = \hat{\theta}_{t-1} + R_t^{-1}\varphi(t)\left(y(t) - \varphi^T(t)\hat{\theta}_{t-1} \right) \end{cases} \tag{3.62}$$

Define $P_t = R_t^{-1}$, then one equivalent recursion is obtained.

$$\hat{\theta}_t = \hat{\theta}_{t-1} + K_t \left(y(t) - \varphi^T(t)\hat{\theta}_{t-1} \right)$$

$$K_t = \frac{P_{t-1}\varphi(t)}{\lambda + \varphi^T(t)P_{t-1}\varphi(t)} ; P_t = \frac{1}{\lambda}\left(I - K_t\varphi^T(t) \right) P_{t-1} \tag{3.63}$$

Observing optimization problem (3.60) again, let $Q_t = diag(1 \ \cdots \ \lambda^{t-1})$ and consider

$$\hat{\theta}_t = \arg\min_{\theta} \sum_{i=1}^{t} \left(y(i) - \varphi^T(i)\theta \right)^2 \lambda^{t-i} = \arg\min_{\theta} \left(y(t) - \varphi^T(t)\theta \right)^2$$

$$+ \lambda \sum_{i=1}^{t-1} \left(y(i) - \varphi^T(i)\theta \right)^2 \lambda^{t-i-1} = \arg\min_{\theta} \left(y(t) - \varphi^T(t)\theta \right)^2$$

$$+ \lambda \sum_{i=1}^{t-1} \left[\left(\varphi^T(i)\left(\theta - \hat{\theta}_{t-1} \right) \right)^2 - 2\left(y(i) - \varphi^T(i)\hat{\theta}_{t-1} \right)\varphi^T(t)\left(\theta - \hat{\theta}_{t-1} \right) \right] \lambda^{t-i-1}$$

$$\tag{3.64}$$

where we use the following relation:

$$\hat{\theta}_{t-1} = \arg\min_{\theta} \sum_{i=1}^{t-1} \left(y(i) - \varphi^T(i)\theta \right)^2 \lambda^{t-i-1}$$

By using optimality condition, it holds that

$$\hat{\theta}_t = \arg\min_{\theta} \left(y(t) - \varphi^T(t)\theta \right)^2 + \lambda \left(\theta - \hat{\theta}_{t-1} \right)^T R_{t-1} \left(\theta - \hat{\theta}_{t-1} \right) \tag{3.65}$$

where the updating law can be seen in equation (3.63). Equation (3.65) shows that the recursive least squares with forgetting scheme can be regarded as regularization least squares problem.

Based on optimization problem (3.50), define the asymptotic limit parameter estimate θ^* as

$$\theta^* = \arg\min_{\theta} \lim_{N \to \infty} E\left\{ V_N \left(\theta, Z^N \right) \right\}$$

where E denotes the expectation operator. In the common identification process, assume that there always exists one true parameter vector θ_0 such that

$$G(q, \theta_0) = G_0(q), \, H(q, \theta_0) = H_0(q)$$

This above assumption shows that the identified model is contained in the considered model set, then the asymptotic covariance matrix of the parameter vector is obtained.

$$P_\theta = \text{cov}\,\hat{\theta}_N = \sigma_0^2 \left\langle \varphi(t), \varphi(t) \right\rangle^{-1} \tag{3.66}$$

where $\left\langle \varphi(t), \varphi(t) \right\rangle$ denotes some inter product operator, φ is the negative gradient of the predictor error, i.e. it can be computed from equation (3.59).

$$\varphi(t, \theta) = -\frac{\partial \varepsilon(t, \theta)}{\partial \theta} = \frac{\partial \hat{y}(t, \theta)}{\partial \theta}$$

On basis of (3.66), the following asymptotic result can be got:

$$\hat{\theta}_N \xrightarrow{N \to \infty} \theta_0$$

It shows that the parameter estimator $\hat{\theta}_N$ will converge to its limit θ_0, and further $\hat{\theta}_N$ will asymptotically converge ($N \to \infty$) to normally distributed random variable with mean θ_0 and variance P_θ.

$$\sqrt{N} \left(\hat{\theta}_N - \theta_0 \right) \to \mathbb{N}(0, P_\theta), \text{ as } N \to \infty$$

This asymptotic result can be rewritten in a quadratic form, and then we get one λ^2 distribution.

$$N\left(\hat{\theta}_N - \theta_0\right)^T P_\theta^{-1}\left(\hat{\theta}_N - \theta_0\right) \xrightarrow{N\to\infty} \lambda_n^2 \qquad (3.67)$$

where n is the number of degrees of freedom in the λ^2 distribution, being equal to the dimension of the parameter vector. Equation (3.67) implies that the random variable $\hat{\theta}_N$ satisfies one uncertainty bound.

$$\hat{\theta}_N \in D\left(\alpha,\theta_0\right); D\left(\alpha,\theta_0\right) = \left\{\theta \, / \, N\left(\theta-\theta_0\right)^T P_\theta^{-1}\left(\theta-\theta_0\right) \leq \lambda_{n,\alpha}^2\right\} \qquad (3.68)$$

with $\lambda_{n,\alpha}^2$ corresponding to a probability level α in λ_n^2 distribution, But now in order to quantity the uncertainty on θ_0 rather than on $\hat{\theta}_N$. For every realization of $\hat{\theta}_N$, it holds that

$$\hat{\theta}_N \in D\left(\alpha,\theta_0\right) \Leftrightarrow \theta_0 \in D\left(\alpha,\hat{\theta}_N\right)$$

It signifies that

$$\theta_0 \in D\left(\alpha,\hat{\theta}_N\right) \text{ with probability } \alpha$$

$$D\left(\alpha,\hat{\theta}_N\right) = \left\{\theta \, / \, N\left(\hat{\theta}_N - \theta\right)^T P_\theta^{-1}\left(\hat{\theta}_N - \theta\right) \leq \lambda_{n,\alpha}^2\right\} \qquad (3.69)$$

Equations (3.68) and (3.69) give the confidence intervals of the unknown parameter vector under the closed loop condition. The probability level of the event $\hat{\theta}_N \in D\left(\alpha,\theta_0\right)$ holds is at least α.

From the above statistical derivation, we obtain one kind of performance analysis corresponding to the unknown parameter vector in the closed loop system. Also when considered in a tailor made parameterization, the negative gradient of the prediction (3.66) is exactly the regression vector (3.59).

3.3.3 Performance analysis only on one transfer function

Here, we give some robust control theories to give a preliminary performance analysis. Combining equation (3.59) and (3.54), the closed loop transfer function can be reformulated as that

$$\frac{D_c\left(q\right)B\left(q,\theta\right)}{D_c\left(q\right)A\left(q,\theta\right)+N_c\left(q\right)B\left(q,\theta\right)} = \frac{\left[q^{-1} \quad q^{-2} \quad \cdots \quad q^{-n}\right]\left[0 \quad P_D\right]\theta}{1+\left[q^{-1} \quad q^{-2} \quad \cdots \quad q^{-n}\right]\left(S\theta+\rho\right)}$$

$$= \frac{Z_2\theta}{1+\left[q^{-1} \quad q^{-2} \quad \cdots \quad q^{-n}\right]\rho+\left[q^{-1} \quad q^{-2} \quad \cdots \quad q^{-n}\right]\theta} = \frac{Z_2\theta}{a+Z_1\theta} \qquad (3.70)$$

where we apply the parameterized plant model $G(q, \theta)$, and column vectors Z_1, Z_2 are defined as following respectively:

$$\begin{cases} Z_1 = \begin{bmatrix} q^{-1} & q^{-2} & \cdots & q^{-n} \end{bmatrix} \\ Z_2 = \begin{bmatrix} q^{-1} & q^{-2} & \cdots & q^{-n} \end{bmatrix} \begin{bmatrix} 0 & P_D \end{bmatrix} \\ a = 1 + \begin{bmatrix} q^{-1} & q^{-2} & \cdots & q^{-n} \end{bmatrix} \rho \end{cases} \qquad (3.71)$$

To simplify the following mathematical derivation, we use $T(\theta)$ to denote the above closed loop transfer function. As the closed loop system is considered here, so we use the closed loop transfer function $T(\theta)$ in our performance analysis, not the former open loop transfer function $G(q, \theta)$. After substituting the true parameter vector θ_0 into the above closed loop transfer function $T(\theta)$, we obtain the true closed loop transfer function $T(\theta)$ as

$$T_0 = T(\theta_0) = \frac{Z_2 \theta_0}{a + Z_1 \theta_0}$$

Remark: In reality, the true closed loop transfer function T_0 does not exist, here it is used in our performance analysis. Ideally when the number N tends to ∞, then we have the following asymptotic result.

$$T(\hat{\theta}_N) \xrightarrow{N \to \infty} T_0 = T(\theta_0)$$

As the above asymptotic result is an ideal case, we define one measure to quality the quantity between the parameterized closed loop transfer function and its true value. An alternative measure from robust control theory is the Vinnicombe distance as that

$$\delta_v(T(\theta), T_0) = \begin{cases} \max_w k(T(\theta), T_0) = \max_w \dfrac{|T(\theta) - T_0|}{\sqrt{1 + |T_0|^2} \sqrt{1 + |T(\theta)|^2}} & \text{if (3.69) is satisfied} \\ 1 & \text{otherwise} \end{cases}$$

$$(3.72)$$

The condition to be satisfied in order to have $\delta_v(T(\theta), T_0) < 1$ is

$$\begin{cases} 1 + T_0^* T(\theta)(jw) \neq 0 & \text{for all } w \\ wno\left(1 + T_0^* T(\theta)(jw)\right) + \eta(T(\theta)) - \tilde{\eta}(T_0^*) = 0 \end{cases} \qquad (3.73)$$

where $T^*(q) = T(-q)$, $\tilde{\eta}(T)$ denotes the number of the closed right half plane of T, while $\eta(T)$ denotes the number of open right half plane poles of T, $wno(T)$ denotes the winding number of the origin of $T(q)$ as q follows the standard Nyquist D-contour. Here $\delta_v(T(\theta), T_0)$ is the Vinnicombe distance between $T(\theta)$ and T_0.

From the robust control theory, the worst case Vinnicombe distance is at least one optimal value $\sqrt{\gamma}$. This requirement is equivalent to the following inequality.

$$\delta_\nu\left(T(\theta),T_0\right) = \max_w \frac{\left|T(\theta)-T_0\right|}{\sqrt{1+\left|T_0\right|^2}\sqrt{1+\left|T(\theta)\right|^2}} \le \sqrt{\gamma} \tag{3.74}$$

Taking square operation on both sides, one inequality is easily obtained.

$$\left(\frac{\left|T(\theta)-T_0\right|}{\sqrt{1+\left|T_0\right|^2}\sqrt{1+\left|T(\theta)\right|^2}}\right)^2 \le \gamma$$

Expanding the above inequality, we obtain

$$\left(\left|T(\theta)-T_0\right|\right)^2 \le \gamma\left(1+\left|T_0\right|^2\right)\left(1+\left|T(\theta)\right|^2\right)$$

$$\Updownarrow$$

$$T_0^* T_0 + T^*(\theta)T(\theta) - 2T_0^* T(\theta) \le \gamma\left(1+T_0^* T_0\right)\left(1+T^*(\theta)T(\theta)\right)$$

$$\Updownarrow$$

$$T_0^* T_0 + T^*(\theta)T(\theta) - T_0^* T(\theta) - T_0 T^*(\theta) - \gamma\left(1+T_0^* T_0\right)$$

$$-\gamma\left(1+T_0^* T_0\right)T^*(\theta)T(\theta) \le 0 \tag{3.75}$$

We regard $T^*(\theta)\,T(\theta)$ as a free variable and formulate a quadratic function corresponding to $T^*(\theta)\,T(\theta)$, then one linear matrix inequality can be get

$$\begin{pmatrix}T^*(\theta) \\ 1\end{pmatrix}\begin{pmatrix}1-\gamma\left(1+T_0^* T_0\right) & -T_0 \\ -T_0^* & T_0^* T_0 - \gamma\left(1+T_0^* T_0\right)\end{pmatrix}\begin{pmatrix}T(\theta) \\ 1\end{pmatrix} \le 0 \tag{3.76}$$

Substituting $T(\theta) = \dfrac{Z_2\theta}{a+Z_1\theta}$ into above linear matrix inequality, we obtain

$$\begin{pmatrix}\dfrac{Z_2\theta}{a+Z_1\theta} \\ 1\end{pmatrix}^*\begin{pmatrix}1-\gamma\left(1+T_0^* T_0\right) & -T_0 \\ -T_0^* & T_0^* T_0 - \gamma\left(1+T_0^* T_0\right)\end{pmatrix}\begin{pmatrix}\dfrac{Z_2\theta}{a+Z_1\theta} \\ 1\end{pmatrix} \le 0 \tag{3.77}$$

By pre-multiplying (3.77) by $(a+Z_1\theta)^*$ and post-multiplying it by $(a+Z_1\theta)$, we have

$$\begin{pmatrix}Z_2\theta \\ a+Z_1\theta\end{pmatrix}^*\begin{pmatrix}1-\gamma\left(1+T_0^* T_0\right) & -T_0 \\ -T_0^* & T_0^* T_0 - \gamma\left(1+T_0^* T_0\right)\end{pmatrix}\begin{pmatrix}Z_2\theta \\ a+Z_1\theta\end{pmatrix} \le 0 \tag{3.78}$$

which is equivalent to the following constraint after complex mathematical derivation and with $Q = (1 + T_0{}^* T_0)$:

$$\begin{pmatrix} \theta \\ 1 \end{pmatrix}^* \begin{pmatrix} \left(1 - \gamma Q\right) Z_2^2 - T_0^* Z_1 Z_2 - T_0 Z_1 Z_2 + \left(T_0^* T_0 - \gamma Q\right) Z_1^2 & \\ -T_0^* a Z_2 + \left(T_0^* T_0 - \gamma Q\right) a Z_1 & \\ & -T_0 a Z_2 + \left(T_0^* T_0 - \gamma Q\right) a Z_1 \\ & \left(T_0^* T_0 - \gamma Q\right) a^2 \end{pmatrix} \begin{pmatrix} \theta \\ 1 \end{pmatrix} \leq 0 \qquad (3.79)$$

To simplify the above expression, we introduce three variables as

$$\begin{cases} a_{11} = \left(1 - \gamma Q\right) Z_2^2 - T_0^* Z_1 Z_2 - T_0 Z_1 Z_2 + \left(T_0^* T_0 - \gamma Q\right) Z_1^2 \\ a_{12} = -T_0 a Z_2 + \left(T_0^* T_0 - \gamma Q\right) a Z_1 \\ a_{22} = \left(T_0^* T_0 - \gamma Q\right) a^2 \end{cases}$$

Then we obtain Theorem 3.2.

Theorem 3.2: In the performance analysis process of the closed loop system, to quality the distance between the parameterized closed loop transfer function and its true value, the requirement that the worst case Vinnicombe distance is equal to one optimal value $\sqrt{\gamma}$ can be reformulated as an optimization problem with linear matrix inequality constraints.

$$\min_{\gamma} \gamma$$

$$\text{subject to} \quad \begin{pmatrix} a_{11} & a_{12} \\ a_{12}^* & a_{22} \end{pmatrix} \leq 0 \qquad (3.80)$$

This above optimization problem can be solved directly by the Matlab Toolbox. After one optimal value γ is solved, then the worst case Vinnicombe distance is equal to this optimal value.

3.3.4   Performance analysis on one transfer function matrix

The above section studies the performance analysis of closed loop system only on one closed loop transfer function, but from equation (3.46), there are four closed loop transfer functions which are used to constitute one closed loop transfer function matrix form. Its parameterized form is that

$$H(G,C) = \begin{pmatrix} \dfrac{G(q,\theta)}{1+G(q,\theta)C(q)} & \dfrac{H(q,\theta)}{1+G(q,\theta)C(q)} \\ \dfrac{1}{1+G(q,\theta)C(q)} & -\dfrac{C(q)H(q,\theta)}{1+G(q,\theta)C(q)} \end{pmatrix}$$

where we consider the case that $H(q,\ \theta) \equiv 1$. If $H(q,\ \theta) \neq 1$, we also assume $H(q,\ \theta)$ can be parameterized as a polynomial. Then the following mathematical derivation is similar. Let $H(q,\ \theta) \equiv 1$, the above closed loop transfer function matrix $H(G,\ C)$ is reduced to

$$H(G,C) = \begin{pmatrix} \dfrac{G(q,\theta)}{1+G(q,\theta)C(q)} & \dfrac{1}{1+G(q,\theta)C(q)} \\ \dfrac{1}{1+G(q,\theta)C(q)} & -\dfrac{C(q)}{1+G(q,\theta)C(q)} \end{pmatrix} \qquad (3.81)$$

From equation (3.51), we rewrite open loop plant model $G(q,\ \theta)$ as

$$G(q,\theta) = \frac{B(q,\theta)}{A(q,\theta)} = \frac{b_1 q^{-1} + \cdots b_{n_b} q^{-n_b}}{1 + a_1 q^{-1} + \cdots a_{n_a} q^{-n_a}} = \frac{Z_4 \theta}{1 + Z_3 \theta}$$

$$Z_3 = \begin{pmatrix} q^{-1} & \cdots & q^{-n_a} & 0 & \cdots & 0 \end{pmatrix}$$

$$Z_4 = \begin{pmatrix} 0 & \cdots & 0 & q^{-1} & \cdots & q^{-n_b} \end{pmatrix} \qquad (3.82)$$

Substituting (3.52) and (3.72) into each element of matrix $H(G,\ C)$, we obtain respectively

$$\begin{cases} \dfrac{G(q,\theta)}{1+G(q,\theta)C(q)} = \dfrac{\dfrac{Z_4\theta}{1+Z_3\theta}}{1+\dfrac{Z_4\theta}{1+Z_3\theta}\times\dfrac{N_c}{D_c}} = \dfrac{Z_4\theta D_c}{(1+Z_3\theta)D_c + Z_4\theta N_c} \\[4mm] \dfrac{1}{1+G(q,\theta)C(q)} = \dfrac{1}{1+\dfrac{Z_4\theta}{1+Z_3\theta}\times\dfrac{N_c}{D_c}} = \dfrac{(1+Z_3\theta)D_c}{(1+Z_3\theta)D_c + Z_4\theta N_c} \\[4mm] -\dfrac{C(q)}{1+G(q,\theta)C(q)} = -\dfrac{N_c}{D_c}\times\dfrac{(1+Z_3\theta)D_c}{(1+Z_3\theta)D_c + Z_4\theta N_c} = -\dfrac{(1+Z_3\theta)N_c}{(1+Z_3\theta)D_c + Z_4\theta N_c} \end{cases}$$

$$(3.83)$$

In a tailor made parameterization case, after substituting (3.83) into matrix $H(G,\ C)$, then its parameterized form is

$$H(\theta) = \begin{pmatrix} \dfrac{Z_4\theta D_c}{(1+Z_3\theta)D_c + Z_4\theta N_c} & \dfrac{(1+Z_3\theta)D_c}{(1+Z_3\theta)D_c + Z_4\theta N_c} \\ \dfrac{(1+Z_3\theta)D_c}{(1+Z_3\theta)D_c + Z_4\theta N_c} & -\dfrac{(1+Z_3\theta)N_c}{(1+Z_3\theta)D_c + Z_4\theta N_c} \end{pmatrix} \tag{3.84}$$

Based on some of the above definitions, given a plant model $G(q, \theta)$ and a stabilizing controller $C(q)$, the performance of a closed loop system $[C, G]$ is defined as the following frequency function.

$$J(G, W_l, W_r, \Omega) = \sigma_1(W_l H(\theta) W_r) \tag{3.85}$$

where Ω denotes the frequency point, W_l, W_r are diagonal weights.

$$W_l = \begin{pmatrix} W_{l1} & 0 \\ 0 & W_{l2} \end{pmatrix}, W_r = \begin{pmatrix} W_{r1} & 0 \\ 0 & W_{r2} \end{pmatrix}$$

$\sigma_1(A)$ denotes the largest singular value of matrix A. This matrix A can be computed as

$$A = W_l H(\theta) W_r = \begin{pmatrix} W_{l1} & 0 \\ 0 & W_{l2} \end{pmatrix} \begin{pmatrix} \dfrac{Z_4\theta D_c}{(1+Z_3\theta)D_c + Z_4\theta N_c} & \dfrac{(1+Z_3\theta)D_c}{(1+Z_3\theta)D_c + Z_4\theta N_c} \\ \dfrac{(1+Z_3\theta)D_c}{(1+Z_3\theta)D_c + Z_4\theta N_c} & -\dfrac{(1+Z_3\theta)N_c}{(1+Z_3\theta)D_c + Z_4\theta N_c} \end{pmatrix}$$

$$\begin{pmatrix} W_{r1} & 0 \\ 0 & W_{r2} \end{pmatrix}$$

$$= \begin{pmatrix} \dfrac{W_{l1}Z_4\theta D_c}{(1+Z_3\theta)D_c + Z_4\theta N_c} & \dfrac{W_{l1}(1+Z_3\theta)D_c}{(1+Z_3\theta)D_c + Z_4\theta N_c} \\ \dfrac{W_{l2}(1+Z_3\theta)D_c}{(1+Z_3\theta)D_c + Z_4\theta N_c} & -\dfrac{W_{l2}(1+Z_3\theta)N_c}{(1+Z_3\theta)D_c + Z_4\theta N_c} \end{pmatrix} \begin{pmatrix} W_{r1} & 0 \\ 0 & W_{r2} \end{pmatrix}$$

$$= \begin{pmatrix} \dfrac{W_{l1}Z_4\theta D_c W_{r1}}{(1+Z_3\theta)D_c + Z_4\theta N_c} & \dfrac{W_{l1}(1+Z_3\theta)D_c W_{r2}}{(1+Z_3\theta)D_c + Z_4\theta N_c} \\ \dfrac{W_{l2}(1+Z_3\theta)D_c W_{r1}}{(1+Z_3\theta)D_c + Z_4\theta N_c} & -\dfrac{W_{l2}(1+Z_3\theta)N_c W_{r2}}{(1+Z_3\theta)D_c + Z_4\theta N_c} \end{pmatrix} \tag{3.86}$$

From knowledge of matrix theory, we see that the worst case performance at frequency point Ω is equal to one optimal value $\sqrt{\gamma}$ is equivalent to the following inequality:

$$\max_{\Omega} \sigma_1(A_r) = \sqrt{\gamma} \tag{3.87}$$

Equation (3.87) is equivalent to

$$\lambda_1\left(A^*A\right) = \gamma \tag{3.88}$$

where $\lambda_1(A^*A)$ denotes the largest eigenvalue of A^*A. It means that we need to solve one largest eigenvalue problem by using the linear matrix inequality condition.

$$\lambda_1\left(A^*A\right) \le \gamma \tag{3.89}$$

where a new matrix A^*A can be computed through complex matrix product operation. As matrix A is a rank one matrix, then the problem $\lambda_1(A^*A) \le \gamma$ is equivalent to

$$
\begin{pmatrix}
\dfrac{\left[W_{l_1}Z_4\theta D_c W_{r_1}\right]^2}{\left[(1+Z_3\theta)D_c + Z_4\theta N_c\right]^2} & \dfrac{W_{l_1}Z_4\theta D_c W_{r_1}W_{l_1}(1+Z_3\theta)D_c W_{r_2}}{\left[(1+Z_3\theta)D_c + Z_4\theta N_c\right]^2} \\[3ex]
+\dfrac{\left[W_{l_2}(1+Z_3\theta)D_c W_{r_1}\right]^2}{\left[(1+Z_3\theta)D_c + Z_4\theta N_c\right]^2} & -\dfrac{W_{l_2}(1+Z_3\theta)D_c W_{r_1}W_{l_2}(1+Z_3\theta)N_c W_{r_2}}{\left[(1+Z_3\theta)D_c + Z_4\theta N_c\right]^2} \\[3ex]
\dfrac{W_{l_1}(1+Z_3\theta)D_c W_{r_2}W_{l_1}Z_4\theta D_c W_{r_1}}{\left[(1+Z_3\theta)D_c + Z_4\theta N_c\right]^2} & \dfrac{\left[W_{l_1}(1+Z_3\theta)D_c W_{r_2}\right]^2}{\left[(1+Z_3\theta)D_c + Z_4\theta N_c\right]^2} \\[3ex]
-\dfrac{W_{l_2}(1+Z_3\theta)N_c W_{r_2}W_{l_2}(1+Z_3\theta)D_c W_{r_1}}{\left[(1+Z_3\theta)D_c + Z_4\theta N_c\right]^2} & +\dfrac{\left[W_{l_2}(1+Z_3\theta)N_c W_{r_2}\right]^2}{\left[(1+Z_3\theta)D_c + Z_4\theta N_c\right]^2}
\end{pmatrix} \le \gamma I
$$

$$\gamma I = \begin{pmatrix} \gamma & 0 \\ 0 & \gamma \end{pmatrix} \tag{3.90}$$

By pre-multiplying (3.90) by $[(1+Z_3\theta)D_c + Z_4\theta N_c]^2$ and post-multiplying it by $[(1+Z_3\theta)D_c + Z_4\theta N_c]^2$, we have

$$
\begin{pmatrix}
\left[W_{l_1}Z_4\theta D_c W_{r_1}\right]^2 + \left[W_{l_2}(1+Z_3\theta)D_c W_{r_1}\right]^2 & \begin{matrix} W_{l_1}Z_4\theta D_c W_{r_1}W_{l_1}(1+Z_3\theta)D_c W_{r_2} \\ -W_{l_2}(1+Z_3\theta)D_c W_{r_1}W_{l_2}(1+Z_3\theta)N_c W_{r_2} \end{matrix} \\[2ex]
\begin{matrix} W_{l_1}(1+Z_3\theta)D_c W_{r_2}W_{l_1}Z_4\theta D_c W_{r_1} \\ -W_{l_2}(1+Z_3\theta)N_c W_{r_2}W_{l_2}(1+Z_3\theta)D_c W_{r_1} \end{matrix} & \left[W_{l_1}(1+Z_3\theta)D_c W_{r_2}\right]^2 + \left[W_{l_2}(1+Z_3\theta)N_c W_{r_2}\right]^2
\end{pmatrix}
$$

$$\le \gamma\left[(1+Z_3\theta)D_c + Z_4\theta N_c\right]^2 I_2$$

We regard $\theta^*\theta$ as a free variable and formulate a quadratic function corresponding to $\theta^*\theta$. As here the closed loop transfer function matrix is considered, four elements exist in this matrix. Then the four linear matrix inequalities can be easily got.

$$
\begin{pmatrix}\theta\\1\end{pmatrix}^{*}
\begin{pmatrix}
W_{l1}^2 Z_3^2 D_c^2 W_{r2}^2 + W_{l2}^2 Z_3^2 N_c^2 W_{r2}^2 & W_{l1}^2 Z_3 D_c^2 W_{r2}^2 + W_{l1}^2 Z_3 N_c^2 W_{r2}^2 \\
\quad -\gamma\left(Z_3^2 D_c^2 + Z_4^2 N_c^2 + 2Z_3 D_c Z_4 N_c\right) & -\gamma\left(Z_3 D_c^2 + Z_4 D_c N_c\right) \\
* & W_{l1}^2 D_c^2 W_{r2}^2 + W_{l2}^2 N_c^2 W_{r2}^2 - \gamma
\end{pmatrix}
\begin{pmatrix}\theta\\1\end{pmatrix}\le 0
$$

$$\Updownarrow$$

$$\begin{pmatrix}\theta\\1\end{pmatrix}^{*} M_1 \begin{pmatrix}\theta\\1\end{pmatrix}\le 0$$

$$
\begin{pmatrix}\theta\\1\end{pmatrix}^{*}
\begin{pmatrix}
\left[W_{l1} Z_4 D_c W_{r1}\right]^2 + \left[W_{l2} Z_3 D_c W_{r1}\right]^2 & W_{l2}^2 Z_3 D_c^2 W_{r1}^2 - \gamma\left(Z_3 D_c^2 + Z_4 D_c N_c\right) \\
\quad -\gamma\left(Z_3^2 D_c^2 + Z_4^2 N_c^2 + 2Z_3 D_c Z_4 N_c\right) & \\
* & W_{l2}^2 D_c^2 W_{r1}^2 - \gamma
\end{pmatrix}
\begin{pmatrix}\theta\\1\end{pmatrix}\le 0
$$

$$\Updownarrow$$

$$\begin{pmatrix}\theta\\1\end{pmatrix}^{*} M_2 \begin{pmatrix}\theta\\1\end{pmatrix}\le 0$$

$$
\begin{pmatrix}\theta\\1\end{pmatrix}^{*}
\begin{pmatrix}
W_{l1} Z_3 D_c W_{r2} W_{l2} Z_4 D_c W_{r1} - W_{r2} Z_3^2 N_c W_{r2} W_{l2} D_c W_{r1} & \frac{1}{2} W_{l1} D_c W_{r2} W_{l1} Z_4 D_c W_{r1} - W_{l2} Z_3 N_c W_{r2} D_c W_{r1} \\
* & -W_{r2} N_c W_{r2} W_{l1} D_c W_{r1}
\end{pmatrix}
\begin{pmatrix}\theta\\1\end{pmatrix}\le 0
$$

$$\Updownarrow$$

$$\begin{pmatrix}\theta\\1\end{pmatrix}^{*} M_3 \begin{pmatrix}\theta\\1\end{pmatrix}\le 0$$

$$
\begin{pmatrix}\theta\\1\end{pmatrix}^{*}
\begin{pmatrix}
W_{l1} Z_4 D_c W_{r1} W_{l1} Z_3 D_c W_{r2} - W_{l2} Z_3 W_{r1} W_{l2} Z_3 N_c D_c W_{r2} & \frac{1}{2} W_{l1} Z_4 D_c W_{r1} W_{l1} D_c W_{r2} - W_{l2} Z_3 W_{r1} W_{l2} N_c D_c W_{r2} \\
* & -W_{r2} N_c W_{r2} W_{l1} D_c W_{r1}
\end{pmatrix}
\begin{pmatrix}\theta\\1\end{pmatrix}\le 0
$$

$$\Updownarrow$$

$$\begin{pmatrix}\theta\\1\end{pmatrix}^{*} M_4 \begin{pmatrix}\theta\\1\end{pmatrix}\le 0$$

$$(3.91)$$

To simplify the above four linear matrix inequalities, we introduce four matrices M_1, M_2, M_3, M_4 to denote them. The above mathematical derivation is very difficult, if the reader or reviewer want to check them, please ask the first author. Then we obtain another Theorem 3.3.

Theorem 3.3: Consider a closed loop system plotted in Figure 3.1 and the plant model $G(q, \theta)$, a stabilizing controller $C(q)$ are all parameterized as their tailor made parameterization form. The worst case performance at frequency point Ω is equal to one optimal value $\sqrt{\gamma}$. This requirement can be formulated as the following standard convex optimization problem involving linear matrix inequality constraints evaluated at the frequency point.

$$
\begin{aligned}
&\min_{\gamma,\tau_1,\tau_2,\tau_3} \quad \gamma \\
&\text{subject to} \quad \tau_1 \ge 0, \tau_2 \ge 0, \tau_3 \ge 0 \\
&\qquad\qquad M_1 - \tau_1 M_2 - \tau_2 M_3 - \tau_3 M_4 \le 0
\end{aligned}
\qquad (3.92)
$$

Comparing equation (3.92) with the result in [18], we conclude that our results are a generation of [18]. These two optimization problems (3.80) and (3.92) can be solved by many convex optimization algorithms such as fast gradient projection algorithm, active set algorithm, ellipsoidal algorithm, trust region algorithm, etc.

3.3.5 Conclusions

In this section, we consider the problem of performance analysis in a closed loop system where the plant model and controller are all parameterized as their tailor made parameterization forms. Under this framework of a tailor made parameterization, we study the confidence internal analysis corresponding to the parameter vector and the performance analysis only on one transfer function. Then we extend the result to performance analysis on one transfer function matrix.

3.4 Minimum variance control strategy for the closed loop system

Using the above given concise introduction to minimum variance control, we concentrate on designing a minimum variance controller through our derivations. After observing the closed loop system in its general form, the transfer function forms of input and output are given. Based on the idea of minimum variance control, one expectation cost function is constructed to consider input and output simultaneously. If we take the partial derivation with respect to the minimum variance control, a closed form solution for the minimum variance controller is derived. Furthermore, as our considered expectation cost function is a nonlinear function form, the alternating direction method of multipliers is applied to solve that minimum variance controller iteratively. From our derived closed form solution for the minimum variance controller, we see that the minimum variance controller is independent of the plant model, noise filter, input spectrum, and noise spectrum. To be more specific, the rational transfer functions in the closed loop system are studied. The problem is to determine minimum variance control in such a way that the variances of the output and input are as small as possible, our derivations are dependent on the prediction theory and Diophantine equation.

3.4.1 Minimum variance control for general form

The essence of classical minimum variance control is to make the variance of the output as small as possible, but here we modify it to guarantee that the variances of the output and input are small simultaneously. This modification can overcome the sensitivity in the parameters of the transfer function $G(q)$ and $H(q)$. One way to overcome this sensitivity is to change the cost to

$$E\left\{y^2\left(t\right)+u^2\left(t\right)\right\} \tag{3.93}$$

where E is the expectation operator.

To expand equation (3.93), we substitute equation (3.46) into (3.93), and obtain that

$$\begin{cases} y(t)^2 = \left[\dfrac{G(q)}{1+G(q)K(q)}\right]^2 r^2(t) + \left[\dfrac{H(q)}{1+G(q)K(q)}\right]^2 \\[2mm] e^2(t) + 2\dfrac{G(q)H(q)}{\left[1+G(q)K(q)\right]^2} r(t)e(t) \\[2mm] u(t)^2 = \left[\dfrac{1}{1+G(q)K(q)}\right]^2 r^2(t) + \left[\dfrac{K(q)H(q)}{1+G(q)K(q)}\right]^2 \\[2mm] e^2(t) + 2\dfrac{K(q)H(q)}{\left[1+G(q)K(q)\right]^2} r(t)e(t) \end{cases} \tag{3.94}$$

Summing both equations to get

$$y(t)^2 + u(t)^2 = \frac{G^2(q)+1}{\left[1+G(q)K(q)\right]^2} r^2(t) + \frac{H^2(q)+H^2(q)K^2(q)}{\left[1+G(q)K(q)\right]^2} e^2(t)$$

$$+2\frac{G(q)H(q)+K(q)H(q)}{\left[1+G(q)K(q)\right]^2} r(t)e(t) \tag{3.95}$$

Taking expectation operator on above equation (3.95).

$$E\left[y(t)^2 + u(t)^2\right] = \frac{G^2(q)+1}{\left[1+G(q)K(q)\right]^2}\phi_r(w) + \frac{H^2(q)+H^2(q)K^2(q)}{\left[1+G(q)K(q)\right]^2}\lambda \tag{3.96}$$

where we use the uncorrelated property between excited signal $r(t)$ and external noise $e(t)$, $\phi_r(w)$ is input spectrum corresponding to excited signal $r(t)$.

The goal of the minimum variance controller is to make the expectation cost (3.96) as small as possible, i.e. the minimum variance controller $K(q)$ is obtained through minimizing the following optimization problem:

$$\underset{K(q)}{\arg\min}\ \frac{G^2(q)+1}{\left[1+G(q)K(q)\right]^2}\phi_r(w) + \frac{H^2(q)+H^2(q)K^2(q)}{\left[1+G(q)K(q)\right]^2}\lambda \tag{3.97}$$

By differentiation with respect to $K(q)$ and by setting the derivative equal to zero, we have that

$$\left[\left(G^2(q)+1\right)(-2)\left(1+G(q)K(q)\right)^{-3}G(q)\right]\phi_r(w)$$

$$2H^2(q)K(q)\left[1+G(q)K(q)\right]^2 - \left(H^2(q)+H^2(q)K^2(q)\right)2$$

$$+\lambda\frac{\left[1+G(q)K(q)\right]G(q)}{\left[1+G(q)K(q)\right]^4} = 0 \tag{3.98}$$

It means that

$$-G(q)\left(G^2(q)+1\right)\phi_r(w)+H^2(q)K(q)+H^2(q)G(q)K^2$$
$$(q)\lambda-H^2(q)G(q)\lambda-H^2(q)G(q)K^2(q)\lambda=0 \tag{3.99}$$

Then minimum variance controller $K(q)$ is given as the following closed form solution:

$$K(q)=\frac{G(q)\left(G^2(q)+1\right)\phi_r(w)+H^2(q)G(q)\lambda}{H^2(q)} \tag{3.100}$$

From the above closed form solution of minimum variance controller $K(q)$, $K(q)$ is dependent on the plant model $G(q)$, noise filter $H(q)$, input spectrum $\phi_r(w)$, and noise variance λ. But equation (3.100) is a theoretical expression, we can use the iteration method to solve the nonlinear optimization problem (3.97), for example, the more widely used alternating direction method of multipliers. For the sake of simplicity, we rewrite the nonlinear cost function as

$$f\left(K(q)\right)=\frac{G^2(q)+1}{\left[1+G(q)K(q)\right]^2}\phi_r(w)+\frac{H^2(q)+H^2(q)K^2(q)}{\left[1+G(q)K(q)\right]^2}\lambda \tag{3.101}$$

We introduce an additional variable $M(q)$ and reformulate the above problem as

$$\begin{aligned}
\min & \frac{G^2(q)+1}{\left[1+G(q)K(q)\right]^2}\phi_r(w)+H^2(q)\frac{1+K^2(q)}{\left[1+G(q)K(q)\right]^2}\lambda \\[2mm]
= & \frac{G^2(q)+1}{\left[1+G(q)K(q)\right]^2}\phi_r(w)+H^2(q)\frac{1+\dfrac{M^2(q)}{G^2(q)}}{\left[1+M(q)\right]^2}\lambda \\[2mm]
= & \frac{G^2(q)+1}{\left[1+G(q)K(q)\right]^2}\phi_r(w)+\frac{H^2(q)}{G^2(q)}\frac{G^2(q)+M^2(q)}{\left[1+M(q)\right]^2}\lambda \\[2mm]
= & f_1\left(K(q)\right)+f_2\left(M(q)\right) \\[2mm]
& \text{subject to}\quad M(q)=G(q)K(q)
\end{aligned} \tag{3.102}$$

We assign a Lagrange multiplier p to the equality constraint $M(q) = G(q)K(q)$, and consider the following augmented Lagrangian function:

$$L\left(K(q),M(q),p\right)=f_1\left(K(q)\right)+f_2\left(M(q)\right)+p\left(M(q)-G(q)K(q)\right)$$
$$+\frac{c}{2}\left\|G(q)K(q)-M(q)\right\|_2^2 \tag{3.103}$$

Then the alternation direction method of multipliers is formulated by

$$
\begin{cases}
K_{t+1}(q) = \underset{K(q)}{\arg\min} \left\{ f_1\left(K(q)\right) + p(t)G(q)K_t(q) + \frac{c}{2}\left\| G(q)K_t(q) - M(t) \right\|_2^2 \right\} \\[2mm]
M(t+1) = \underset{M(q)}{\arg\min} \left\{ f_2\left(M(q)\right) - p(t)M(q) + \frac{c}{2}\left\| G(q)K_{t+1}(q) - M(q) \right\|_2^2 \right\} \\[2mm]
p(t+1) = p(t) + c\left(G(q)K_{t+1}(q) - M(t+1) \right)
\end{cases}
\tag{3.104}
$$

where $K_{t+1}(q)$ denotes the minimum variance controller $K(q)$ at iteration instant $t + 1$, the parameter c is any positive number, and the initial values $p(0)$ and $M(0)$ are arbitrary. A sequence $\{K_t(q), M(t), p(t)\}$ generated by iteration equation (3.104) is bounded, and every limit point of $\{K_t(q)\}$ is an optimal solution of the original problem $\underset{K(q)}{\arg\min}\, f\left(K(q)\right)$. Furthermore, $\{p(t)\}$ converges to an optimal solution p^* of the dual problem.

$$
\max H_1(p) + H_2(p)
\tag{3.105}
$$

where for all p

$$
\begin{cases}
H_1(p) = \underset{K(q)}{\inf} \left\{ f_1\left(K(q)\right) + pG(q)K(q) \right\} \\[2mm]
H_2(p) = \underset{M(q)}{\inf} \left\{ f_2\left(M(q)\right) - pM(q) \right\}
\end{cases}
\tag{3.106}
$$

In practice or engineering, our above alternative direction method of multipliers (3.104) is more widely used to solve the minimum variance controller $K(q)$ iteratively.

3.4.2 Minimum variance control for rational transfer function form

In equation (3.45), $G(q)$, $H(q)$ and $K(q)$ denote the plant model, noise filter and feedback controller respectively. Here, we only consider their rational transfer function forms as follows.

$$
G(q) = \frac{B(q)}{A(q)}, H(q) = \frac{C(q)}{A(q)}
\tag{3.107}
$$

It means that the above rational transfer function forms $\left\{ \dfrac{B(q)}{A(q)}, \dfrac{C(q)}{A(q)} \right\}$ are called the ARX model, $A(q)$, $B(q)$ and $C(q)$ are rational polynomials. The coefficients of $A(q)$, $B(q)$ and $C(q)$ are denoted as follows.

$$\begin{cases} A(q) = 1 + a_1 q^{-1} + a_2 q^{-2} + \cdots + a_n q^{-n} \\ B(q) = b_1 q^{-1} + b_2 q^{-2} + \cdots + b_n q^{-n} \\ C(q) = c_1 q^{-1} + c_2 q^{-2} + \cdots + c_n q^{-n} \end{cases} \tag{3.108}$$

where n is model order, and it is priori known.

Using the rational transfer functions in equation (3.107), then the considered output is that

$$y(t) = \frac{B(q)}{A(q)} u(t) + \frac{C(q)}{A(q)} e(t) \tag{3.109}$$

3.4.3   Main results

Now our mission is to derive the minimum variance controller $K(q)$ by using the idea of minimum variance control, and the obtained controller $K(q)$ is also a rational transfer function form.

Rewriting equation (3.109) to a slightly different form, namely

$$A(q) y(t) = B(q) u(t) + C(q) e(t) \tag{3.110}$$

The property that the value of $u(t)$ at time t is a function of observed outputs up to and including time t, that is $y(t)$, $y(t-1)$ and of all past control signals $u(t-1)$, $u(t-2) \ldots$. From equation (3.109), we have that

$$y(t+k) = \frac{B(q)}{A(q)} u(t+k) + \frac{C(q)}{A(q)} e(t+k) = \frac{q^{-k} B(q)}{A(q)} u(t) + \frac{C(q)}{A(q)} e(t+k)$$

$$= \frac{B_1(q)}{A(q)} u(t) + \frac{C(q)}{A(q)} e(t+k) \, B_1(q) = q^{-k} B(q) \tag{3.111}$$

The right part consists of terms which can be computed exactly from the observations and terms which are independent of the observations. To separate these groups of terms we write the last term of equation (3.111) as

$$\frac{C(q)}{A(q)} e(t+k) = \frac{1}{A(q)} e(t+k) \left[A(q) F(q) + q^{-k} E(q) \right]$$

$$= F(q) e(t+k) + \frac{q^{-k} E(q)}{A(q)} e(t+k) = F(q) e(t+k) + \frac{E(q)}{A(q)} e(t)$$

$$C(q) = A(q) F(q) + q^{-k} E(q) \tag{3.112}$$

Substituting equation (3.112) into (3.111), we obtain

$$y(t+k) = \frac{B_1(q)}{A(q)} u(t) + F(q) e(t+k) + \frac{E(q)}{A(q)} e(t) \tag{3.113}$$

Observing equation (3.109) again, we get

$$\frac{C(q)}{A(q)}e(t)=y(t)-\frac{B(q)}{A(q)}u(t)$$

$$\updownarrow$$

$$e(t)=\frac{A(q)}{C(q)}\left[y(t)-\frac{B(q)}{A(q)}u(t)\right]=\frac{A(q)}{C(q)}y(t)-\frac{B(q)}{C(q)}u(t) \qquad (3.114)$$

Substituting equation (3.114) into (3.113), then

$$y(t+k)=\frac{B_1(q)}{A(q)}u(t)+F(q)e(t+k)+\frac{E(q)}{A(q)}\left[\frac{A(q)}{C(q)}y(t)-\frac{B(q)}{C(q)}u(t)\right]$$

$$=\frac{B_1(q)}{A(q)}u(t)+F(q)e(t+k)+\frac{E(q)}{C(q)}y(t)-\frac{E(q)B(q)}{A(q)C(q)}u(t)$$

$$=F(q)e(t+k)+\left[\frac{B_1(q)}{A(q)}-\frac{E(q)B(q)}{A(q)C(q)}\right]u(t)+\frac{E(q)}{C(q)}y(t) \qquad (3.115)$$

The second term of the right member can be reduced to

$$\left[\frac{B_1(q)}{A(q)}-\frac{E(q)B(q)}{A(q)C(q)}\right]=\frac{q^{-k}B(q)}{A(q)}-\frac{E(q)B(q)}{A(q)C(q)}=\frac{B(q)}{A(q)}\left[q^{-k}-\frac{E(q)}{C(q)}\right] \qquad (3.116)$$

From the separation of $C(q)$ in equation (3.112), we get

$$C(q)=A(q)F(q)+q^{-k}E(q);$$

$$\updownarrow$$

$$1=\frac{A(q)F(q)}{C(q)}+q^{-k}\frac{E(q)}{C(q)};$$

$$\updownarrow$$

$$1-\frac{A(q)F(q)}{C(q)}=q^{-k}\frac{E(q)}{C(q)};$$

$$\updownarrow$$

$$\frac{E(q)}{C(q)}=q^{k}-q^{k}\frac{A(q)F(q)}{C(q)} \qquad (3.117)$$

Substituting equation (3.117) into (3.116), we get

$$\frac{B_1(q)}{A(q)}-\frac{E(q)B(q)}{A(q)C(q)}=q^{k}\frac{B(q)F(q)}{C(q)} \qquad (3.118)$$

So equation (3.115) can be rewritten as

$$y(t+k) = F(q)e(t+k) + q^k \frac{B(q)F(q)}{C(q)}u(t) + \frac{E(q)}{C(q)}y(t) \qquad (3.119)$$

Taking square and expectation operation, hence

$$E\left[y(t+k)\right]^2 = E\left[F(q)e(t+k)\right]^2 + E\left[q^k \frac{B(q)F(q)}{C(q)}u(t) + \frac{E(q)}{C(q)}y(t)\right]^2$$

$$(3.120)$$

The mixed terms will vanish because $e(t + 1)$, $e(t + 2)$ $\cdots$ $e(t + k)$ are independent of $y(t)$, $y(t{-}1)$ $\cdots$ and $u(t{-}1)$, $u(t{-}2)$ $\cdots$, then

$$E\left[y(t+k)\right] \geq E\left[F(q)e(t+k)\right]^2 = F^2(q)E\left[e(t+k)\right]^2 = F^2(q) \qquad (3.121)$$

where equality holds for

$$q^k \frac{B(q)F(q)}{C(q)}u(t) + \frac{E(q)}{C(q)}y(t) = 0 \qquad (3.122)$$

which gives the following designed control law

$$u(t) = -q^{-k}\frac{C(q)}{B(q)F(q)} \times \frac{E(q)}{C(q)}y(t) = -q^{-k}\frac{E(q)}{B(q)F(q)}y(t) \qquad (3.123)$$

Then we summarize the above analysis process as Theorem 3.4.

Theorem 3.4: Consider one closed output process described by

$$y(t) = \frac{B(q)}{A(q)}u(t) + \frac{C(q)}{A(q)}e(t)$$

where $\{e(t)\}$ is a sequence of independent $(0, \lambda)$ random variables. Let the polynomial $C(q)$ has all its zero inside the unit circle. The minimum variance control law is given by

$$u(t) = -q^{-k}\frac{E(q)}{B(q)F(q)}y(t)$$

where the polynomials $F(q)$ and $E(q)$ respectively are defined by the identity

$$C(q) = A(q)F(q) + q^{-k}E(q)$$

Then the optimal closed loop system is shown in Fig. 3.9.

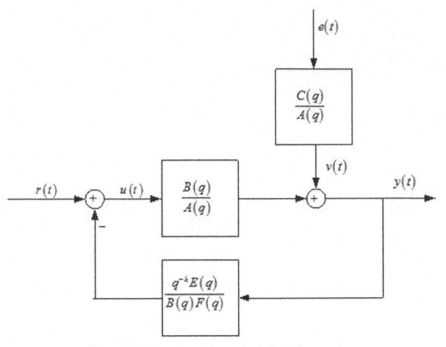

Figure 3.9: Structure of the optimal closed loop system

3.4.4   Extending results

In Theorem 3.4, the variance of output is guaranteed to be as small as possible. The above analysis holds only if the minimum variance control law can be realized through a stable filter $C(q)$. Even if $C(q)$ has its roots outside the unit circle, as some of these roots are near the unit circle, the performance of the minimum variance policy can be very sensitive to variations in the parameters of the polynomials $A(q)$ and $B(q)$. One way to overcome this sensitivity is to modify the cost function to

$$\min_{u(t)} E\left\{ y^2 \left(t+k\right) + u^2 \left(t\right) \right\} \qquad (3.124)$$

From equation (3.109), we obtain

$$u(t) = \frac{A(q)}{B(q)} \left[y(t) - \frac{C(q)}{A(q)} e(t) \right] = \frac{A(q)}{B(q)} y(t) - \frac{C(q)}{B(q)} e(t) \qquad (3.125)$$

Then

$$E\left\{y^2\left(t+k\right)+u^2\left(t\right)\right\}=E\left[F\left(q\right)e\left(t+k\right)\right]^2$$

$$+E\left[q^k\frac{B\left(q\right)F\left(q\right)}{C\left(q\right)}u\left(t\right)+\frac{E\left(q\right)}{C\left(q\right)}y\left(t\right)\right]^2+Eu^2\left(t\right)$$

$$=E\left[F\left(q\right)e\left(t+k\right)\right]^2+E\left[\left(q^k\frac{B\left(q\right)F\left(q\right)}{C\left(q\right)}\right)^2+1\right]u^2\left(t\right)+E\left[\frac{E\left(q\right)}{C\left(q\right)}y\left(t\right)\right]^2$$

$$+2E\left[\frac{E\left(q\right)}{C\left(q\right)}y\left(t\right)u\left(t\right)\right]$$
(3.126)

Applying necessary condition on above modified cost function, it holds that

$$\left[\left(q^k\frac{B\left(q\right)F\left(q\right)}{C\left(q\right)}\right)^2+1\right]u\left(t\right)+\frac{E\left(q\right)}{C\left(q\right)}y\left(t\right)=0$$
(3.127)

It holds that

$$u\left(t\right)=-\frac{\dfrac{E\left(q\right)}{C\left(q\right)}}{\left[\left(q^k\dfrac{B\left(q\right)F\left(q\right)}{C\left(q\right)}\right)^2+1\right]}y\left(t\right)=-\frac{q^k-q^k\dfrac{A\left(q\right)F\left(q\right)}{C\left(q\right)}}{1+\left(q^k\dfrac{B\left(q\right)F\left(q\right)}{C\left(q\right)}\right)^2}y\left(t\right)$$

$$=-\frac{\dfrac{E\left(q\right)}{C\left(q\right)}}{\dfrac{C^2\left(q\right)+q^{2k}B^2\left(q\right)F^2\left(q\right)}{C^2\left(q\right)}}y\left(t\right)=-\frac{E\left(q\right)C\left(q\right)}{C^2\left(q\right)+q^{2k}B^2\left(q\right)F^2\left(q\right)}y\left(t\right)$$

(3.128)

In this modified cost function, the optimal closed loop system is shown in Fig. 3.10.

3.4.5 Conclusion

Here, we study the problem of minimum variance control for the closed loop system through our derivations. Whether the considered closed loop system is in the general form or rational transfer function form, the closed form solution corresponding to the minimum variance controller is given. Furthermore, one alternation direction method of multipliers is proposed to solve the minimum

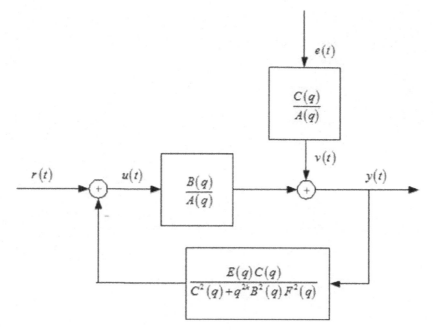

Figure 3.10: Structure of another optimal closed loop system

variance controller in practice. The classical cost function considered in minimum variance control is modified to another new cost function so that the performance of the minimum variance policy will be robust to variations in the parameters of the polynomials. Next, we will study the robust minimum variance control in case of uncertainty.

3.5   Synthesis identification analysis for closed loop system

3.5.1   Closed loop system structure

This section considers the synthesis identification analysis for the closed loop system with a system model and noise model. All of the above derivations on closed loop identification are effective, on the condition that the feedback controller must be known or linear. But this condition does not hold in practice, so we study the closed loop identification with a nonlinear controller. We use one linear feedback controller to replace the original nonlinear controller and establish one equivalent property in the statistical sense. The equivalent linear controller can be constructed by some statistical correlation functions.

Based on our previous results on closed loop system identification, we continue to add some contributions to closed loop system identification, i.e. synthesis identification analysis for closed loop system. We regard this problem from two different points. The first point is about modifying the original cost function to its other simplified form, which is convenient for the later identification process. The second point is to replace the nonlinear feedback controller with one linear feedback controller. Then the existing research on linear controller design or linear system identification can be applied directly.

One closed loop system is given in Fig. 3.11, where $y(t)$ is the closed loop output, and input signal is $u(t)$. $C(z)$ is a simple linear controller.

More specifically, in Fig. 3.11, $G_0(z)$ denotes the unknown plant, $H_0(z)$ is one noise shape model, which is used to deal with the external white noise. External excitation input is $r(t)$, and $e(t)$ is a while noise with zero mean and variance λ_0. Colored noise $v(t)$ is passed through that noise filter z is the backward shift operator.

From Fig. 3.11, we see,

$$v(t) = H_0(z) e(t) \tag{3.129}$$

Using the basic modern control theory and some mathematical equation computations, we have

$$\begin{cases} y(t) = G_0(z) u(t) + H_0(z) e(t) \\ u(t) = r(t) - C(z) y(t) \end{cases} \tag{3.130}$$

Expanding each term in equation (3.130), we get

$$\begin{cases} y(t) = \dfrac{G_0(z)}{1 + G_0(z) C(z)} r(t) + \dfrac{H_0(z)}{1 + G_0(z) C(z)} e(t) \\ u(t) = \dfrac{1}{1 + G_0(z) C(z)} r(t) - \dfrac{C(z) H_0(z)}{1 + G_0(z) C(z)} e(t) \end{cases} \tag{3.131}$$

Similarly, define one sensitivity function to reduce the computational complexity.

$$S_0(z) = \dfrac{1}{1 + G_0(z) C(z)} \tag{3.132}$$

Using the above defined sensitivity function, we rewrite equation (3.131) as follows:

$$\begin{cases} y(t) = G_0(z) S_0(z) r(t) + H_0(z) S_0(z) e(t) \\ u(t) = S_0(z) r(t) - C(z) H_0(z) S_0(z) e(t) \end{cases} \tag{3.133}$$

The main problem of closed loop identification is to identify one unknown parameter vector in equation (3.130).

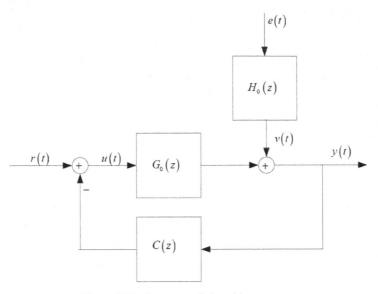

Figure 3.11: Structure of closed loop system

3.5.2   Prediction error identification reviewed

The above considered closed loop system can be divided into two kinds, i.e. parameter identification and non-parameter identification. Here the first case-parameter identification is considered, so firstly we need to parameterize the unknown plant and noise filter, i.e. equation (3.130) is parameterized.

$$
\begin{cases}
y(t,\theta) = \dfrac{G(z,\theta)}{1+G(z,\theta)C(z)}r(t) + \dfrac{H(z,\theta)}{1+G(z,\theta)C(z)}e(t) \\[3mm]
u(t,\theta) = \dfrac{1}{1+G(z,\theta)C(z)}r(t) - \dfrac{C(z)H(z,\theta)}{1+G(z,\theta)C(z)}e(t)
\end{cases}
\tag{3.134}
$$

where in equation (3.134) θ is one unknown parameter vector, which is needed to identify.

The mission of our considered closed loop identification is to estimate that parameter estimator $\hat{\theta}_N$ based on the observed input-output sequence $Z^N = \{y(t), u(t)\}_{t=1}^{N}$. N is the number of observed data.

From above equation (3.134), define one step ahead prediction $y(t, \theta)$ as that

$$
\begin{aligned}
\hat{y}(t,\theta) &= \frac{1+G(z,\theta)C(z)}{H(z,\theta)} \times \frac{G(z,\theta)}{1+G(z,\theta)C(z)}r(t) + \left[1 - \frac{1+G(z,\theta)C(z)}{H(z,\theta)}\right]y(t) \\[3mm]
&= \frac{G(z,\theta)}{H(z,\theta)}r(t) + \frac{H(z,\theta)-1-G(z,\theta)C(z)}{H(z,\theta)}y(t)
\end{aligned}
\tag{3.135}
$$

Combining equation (3.131) and output prediction (3.135), then we have the error function as follows:

$$\varepsilon(t,\theta) = y(t) - \hat{y}(t,\theta) = \frac{1+G(z,\theta)C(z)}{H(z,\theta)}\left[y(t) - \frac{G(z,\theta)}{1+G(z,\theta)C(z)}r(t)\right]$$

(3.136)

Using the above defined error function, we get the following numerical optimization problem to estimate that unknown parameter vector:

$$\hat{\theta}_N = \arg\min_{\theta} V_N(\theta, Z^N) = \arg\min_{\theta} \frac{1}{N}\sum_{t=1}^{N}\varepsilon^2(t,\theta)$$

$$V_N(\theta, Z^N) = \frac{1}{N}\sum_{t=1}^{N}\varepsilon^2(t,\theta)$$

(3.137)

Through using the smooth property with respect to that parameterized prediction error $\varepsilon(t, \theta)$, then that cost function in above optimization process can be reduced to one finding root process, i.e.

$$D(\theta) = \partial_\theta \frac{1}{2}\sum_{t=1}^{N}\varepsilon^2(t,\theta) = -\sum_{t=1}^{N}\left[\partial_\theta\hat{y}(t,\theta)\right]\left[y(t) - \hat{y}(t,\theta)\right]$$

(3.138)

It means that

$$\hat{\theta}_N \in sol\left[D(\theta) = 0\right]$$

(3.139)

where notation ∂_θ means partial derivative operation about θ, and after some tedious mathematical operations, $\partial_\theta\hat{y}(t,\theta)$ is expanded as that

$$\partial_\theta\hat{y}(t,\theta) = \frac{\dfrac{\partial G(z,\theta)}{\partial\theta}H(z,\theta) - G(z,\theta)\dfrac{\partial H(z,\theta)}{\partial\theta}}{H^2(z,\theta)}r(t)$$

$$-\frac{\dfrac{\partial G(z,\theta)}{\partial\theta}C(z)H(z,\theta) - \left(1+G(z,\theta)C(z)\right)\dfrac{\partial H(z,\theta)}{\partial\theta}}{H^2(z,\theta)}y(t) \quad (3.140)$$

The detailed computation process about solving the above parameter optimization problem can be referred to in many papers, for example, our previously published works. But this section is not on the identification algorithm, but on other synthesis identification analysis about our considered closed loop system.

3.5.3 Synthesis identification analysis

Observing Fig. 3.11 again, excitation input {$r(t)$} is a quasi stationary reference signal, and its spectrum is $\phi_r(w)$. Furthermore, through using the condition of that independent and identically distributed while noise with variance λ_0, and that external white noise $e(t)$ has power spectrum λ_0. Furthermore, colored noise $v(t)$ has power spectrum

$$\left| H_0\left(e^{jw}\right) \right|^2 \lambda_0 = H_0\left(e^{jw}\right) H_0^*\left(e^{jw}\right)$$

where notation * is complex conjugate.

From equation (3.133), the uncorrelated property holds between excitation input $\{r(t)\}$ and while noise $\{e(t)\}$, then we obtain some spectrum relations directly, such as

$$\phi_u\left(w\right) = \left|S_0\right|^2 \phi_r\left(w\right) + \left|C\right|^2 \left|S_0\right|^2 \left|H_0\right|^2 \lambda_0 = \phi_u^r\left(w\right) + \phi_u^e\left(w\right) \qquad (3.141)$$

where $\phi_u^r\left(w\right)$ and $\phi_u^e\left(w\right)$ are two part for the excitation input spectrum.

Similarly, we have the following relation about the output spectrum:

$$\phi_y\left(w\right) = \left|G_0\right|^2 \left|S_0\right|^2 \phi_r\left(w\right) + \left|H_0\right|^2 \left|S_0\right|^2 \lambda_0 \qquad (3.142)$$

Similarly, we get another form for the cross spectrum.

$$\begin{cases} \phi_{yu}\left(w\right) = G_0 \left|S_0\right|^2 \phi_r\left(w\right) - C\left|H_0\right|^2 \left|S_0\right|^2 \lambda_0 \\ \phi_{ye}\left(w\right) = -CH_0 S_0 \lambda_0 \end{cases} \qquad (3.143)$$

Observing above three equations simultaneously, it holds

$$\lambda_0 \phi_u\left(w\right) - \left|\phi_{ue}\left(w\right)\right|^2 = \lambda_0 \left|S_0\right|^2 \phi_r\left(w\right) + \lambda_0^2 \left|C\right|^2 \left|S_0\right|^2 \left|H_0\right|^2 - \lambda_0^2 \left|C\right|^2 \left|S_0\right|^2 \left|H_0\right|^2$$

$$= \lambda_0^2 \left|S_0\right|^2 \phi_r\left(w\right) = \lambda_0 \phi_u^r\left(w\right) \qquad (3.144)$$

The advantageous about the above spectral relations concerns on estimating the transfer function $G_0(z)$ with its spectral analysis estimation. More specifically, spectral analysis estimation $\hat{G}\left(e^{jw}\right)$ is set as follows.

$$\hat{G}\left(e^{jw}\right) = \frac{\phi_{yu}\left(w\right)}{\phi_u\left(w\right)} = \frac{G_0 \left|S_0\right|^2 \phi_r\left(w\right) - C\left|H_0\right|^2 \left|S_0\right|^2 \lambda_0}{\left|S_0\right|^2 \phi_r\left(w\right) + \left|C\right|^2 \left|S_0\right|^2 \left|H_0\right|^2 \lambda_0}$$

$$= \frac{G_0\left(e^{jw}\right)\phi_r\left(w\right) - C\left(e^{jw}\right)\left|H_0\left(e^{jw}\right)\right|^2 \lambda_0}{\phi_r\left(w\right) + \left|C\left(e^{jw}\right)\right|^2 \left|H_0\left(e^{jw}\right)\right|^2 \lambda_0}$$

$$= \frac{G_0\left(e^{jw}\right)\phi_r\left(w\right) - C\left(e^{jw}\right)\phi_v\left(w\right)}{\phi_r\left(w\right) + \left|C\left(e^{jw}\right)\right|^2 \phi_v\left(w\right)} \qquad (3.145)$$

Taking the limit operation on both sides of equation (3.145), when N is sufficiently large, that spectral analysis estimation $\hat{G}\left(e^{jw}\right)$ approaches to the following limit:

$$\frac{G_0\left(e^{jw}\right)\phi_r\left(w\right)-C\left(e^{jw}\right)\phi_v\left(w\right)}{\phi_r\left(w\right)+\left|C\left(e^{jw}\right)\right|^2\phi_v\left(w\right)}$$

Consider the cost function (3.135), assume the true parameter is in the considered parameter set, it means one true parameter vector θ_0 exists to satisfy that

$$\begin{cases} G\left(z,\theta_0\right)=G_0\left(z\right), \\ H\left(z,\theta_0\right)=H_0\left(z\right) \end{cases}$$

As the error function (3.136) is not dependent of the unknown parameter explicitly, so to find its detailed relation with the unknown parameter, we rewrite it as another form, i.e.

$$\varepsilon\left(t,\theta\right)=H^{-1}\left(z,\theta\right)\left[y\left(t\right)-G\left(z,\theta\right)u\left(t\right)\right] \tag{3.146}$$

Combining two equations (3.134) and (3.146), we easily obtain the following detailed expression as

$$y\left(t\right)-G\left(z,\theta\right)u\left(t\right)=\frac{G_0\left(z\right)}{1+G_0\left(z\right)C\left(z\right)}r\left(t\right)+\frac{H_0\left(z\right)}{1+G_0\left(z\right)C\left(z\right)}e\left(t\right)$$

$$-\frac{G\left(z,\theta\right)}{1+G_0\left(z\right)C\left(z\right)}r\left(t\right)-\frac{G\left(z,\theta\right)C\left(z\right)H_0\left(z\right)}{1+G_0\left(z\right)C\left(z\right)}e\left(t\right)$$

$$=\frac{H_0\left(z\right)\left(1-G\left(z,\theta\right)C\left(z\right)\right)}{1+G_0\left(z\right)C\left(z\right)}e\left(t\right)+\frac{G_0\left(z\right)-G\left(z,\theta\right)}{1+G_0\left(z\right)C\left(z\right)}r\left(t\right) \tag{3.147}$$

Substituting (3.147) into the prediction error (3.146), we get

$$\varepsilon\left(t,\theta\right)=\frac{1}{1+G_0\left(z\right)C\left(z\right)}r\left(t\right)\times\frac{G_0\left(z\right)-G\left(z,\theta\right)}{H\left(z,\theta\right)}+\frac{H_0\left(z\right)}{H\left(z,\theta\right)}$$

$$\times\frac{1-G\left(z,\theta\right)C\left(z\right)}{1+G_0\left(z\right)C\left(z\right)}e\left(t\right)$$

$$=S_0\left(z\right)r\left(t\right)\times\frac{G_0\left(z\right)-G\left(z,\theta\right)}{H\left(z,\theta\right)}+\frac{H_0\left(z\right)}{H\left(z,\theta\right)}\times\frac{1-G\left(z,\theta\right)C\left(z\right)}{1+G_0\left(z\right)C\left(z\right)}e\left(t\right) \tag{3.148}$$

From above our derivations, another form for that cost function is rewritten as follows:

$$V_N(\theta) = \frac{1}{N}\sum_{t=1}^{N}\varepsilon^2(t,\theta) = \int \frac{\left|G\left(e^{jw},\theta\right)-G_0\left(e^{jw}\right)\right|^2}{\left|H\left(e^{jw},\theta\right)\right|^2}$$

$$\frac{1}{\left|1+G_0\left(e^{jw}\right)C\left(e^{jw}\right)\right|^2}\phi_r(w)\,dw$$

$$+\int\left|\frac{1+G\left(e^{jw},\theta\right)C\left(e^{jw}\right)}{1+G_0\left(e^{jw}\right)C\left(e^{jw}\right)}\right|^2\left|\frac{H_0\left(e^{jw}\right)}{H\left(e^{jw},\theta\right)}\right|^2\lambda_0\,dw \qquad (3.149)$$

It means that we identify the unknown parameter estimations through minimizing the following improved cost function.

$$V_N(\theta) = \int \frac{\left|G\left(e^{jw},\theta\right)-G_0\left(e^{jw}\right)\right|^2}{\left|H\left(e^{jw},\theta\right)\right|^2}\phi_u^r(w)\,dw+\int\left|\frac{1+G\left(e^{jw},\theta\right)C\left(e^{jw}\right)}{1+G_0\left(e^{jw}\right)C\left(e^{jw}\right)}\right|^2$$

$$\frac{1}{\left|H\left(e^{jw},\theta\right)\right|^2}\phi_v(w)\,dw \qquad (3.150)$$

The above equation (3.150) tells us that if the number N of the observed data is sufficiently large, the entire minimum process for the improved cost function $V_N(\theta)$ is guaranteed to be the global minimum, i.e. the following relation holds that

$$G\left(e^{jw},\theta\right) \rightarrow G_0\left(e^{jw},\theta_0\right) = G_0\left(e^{jw}\right)$$

It means that

$$\theta \rightarrow \theta_0$$

The above obtained result is one simplified form for the classical result.

3.5.4   Replace nonlinear controller with linear equivalent controller

Observing Fig. 3.11 and equation (3.140), a feedback controller $C(z)$ exists in the above derivation. During all the designed problems for the closed loop system, for example, system identification, controller design, state filter etc., a detailed form about that feedback controller $C(z)$ must be needed, for example, the PID controller. But in practice, all plants and controllers are all nonlinear forms, so when the feedback controller $C(z)$ is one nonlinear form, then equation (3.140) cannot be solved easily, as we do not know the explicit form for the nonlinear controller.

Now the problem of replacing an unknown nonlinear controller with an equivalent linear controller is studied, and the equivalent property between them is also established.

Observing Fig. 3.11 again, it is emphasized that the feedback controller $f(y(t))$ is not a linear controller, but a nonlinear controller. Then the input for the considered system corresponds to

$$u(t) = r(t) - f\left(y(t)\right) \qquad (3.151)$$

Observing the right and left side for the nonlinear controller $f(y(t))$, and formulating its input signal and output signal as follows:

$$\left\{r(t) - u(t), y(t)\right\}_{t=1}^{N} \qquad (3.152)$$

For the convenience of understanding this main essence, we plot the original Fig. 3.11 again here and modify that feedback controller as one nonlinear feedback controller, then this modified structure is seen in Fig. 3.12.

As the nonlinear system or nonlinear controller is not easily analyzed, and the research on the linear system or linear controller is mature, we adjust that nonlinear feedback system as the commonly used linear feedback system. It corresponds to constructing one linear feedback controller $C(z)$ to replace that original nonlinear feedback controller $f(y(t))$, and this equivalent property holds in the sense of some statistical probability.

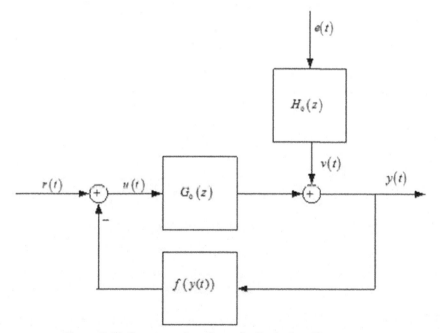

Figure 3.12: Structure of nonlinear feedback closed loop system

More specifically, denote the output of that linear feedback controller $C(z)$ as $C(z)\,y(t)$. Then one measure is used to verify the approximated degree between the linear controller $C(z)$ and nonlinear controller $f(y(t))$. We set this measure as the following error criterion:

$$\lim_{N \to \infty} E\left[C(z)y(t) - f(y(t))\right]^2 = E\left[C(z)\right] \qquad (3.153)$$

Thus the linear feedback controller $C(z)$ is chosen to minimize one constructed error criterion function $E[C(z)]$. The detailed minimization result for that constructed error criterion function is given as the following Theorem 3.5.

Theorem 3.5: Set the feedback controller's input signal as $y(t)$ and its output $f(y(t))$ in the considered nonlinear feedback closed loop system, and $C(z)$ is a linear feedback controller, the condition of the linear feedback controller $C(z)$ to replace the original nonlinear control $f(y(t))$ is

$$\phi_{y,f}(\tau) = \phi_{y,Cy}(\tau), \forall \tau > 0 \qquad (3.154)$$

where $\phi_{y,f}(\tau)$ denotes the power spectral between input signal $y(t)$ and its corresponding output $f(y(t))$ in the above considered nonlinear feedback closed loop system.

Proof: Set $C_1(z)$ is another linear controller, let

$$C_1(z)y(t) = \int_0^t c_1(t-\tau)y(\tau)\,d\tau \qquad (3.155)$$

and similarly we also have

$$C(z)y(t) = \int_0^t c(t-\tau)y(\tau)\,d\tau \qquad (3.156)$$

Define

$$e(t) = c_1(t) - c(t)$$

Then we have

$$E(z)y(t) = C_1(z)y(t) - C(z)y(t) = \int_0^t e(t-\tau)y(\tau)\,d\tau = \int_0^t e(\tau)y(t-\tau)\,d\tau \qquad (3.157)$$

Using the defined error criterion to get

$$E\left[C_1(z)\right] - E\left[C_2(z)\right] = \lim_{N \to \infty} \int_0^T \frac{1}{T}\left[C_1(z)y(t) - f(y(t))\right]^2$$
$$-\left[C(z)y(t) - f(y(t))\right]^2 dt$$
$$= \lim_{N \to \infty} \int_0^T \frac{1}{T}\left[E(z)y(t)\right]^2 + 2E(z)y(t)\left(C(z)y(t) - f(y(t))\right)dt \qquad (3.158)$$

As $C(z)$ minimizes the defined error criterion $E[C(z)]$ on the condition that $E[C_1(z)] \geq E[C(z)]$ for all linear time invariant controller $C_1(z)$. The right hand side is nonnegative for all linear feedback controllers $C_1(z)$ on the condition of that linear part in $E(z) y(t)$ approach to zero, i.e.

$$\lim_{N \to \infty} \int_0^T \frac{1}{T} E(z) y(t) \big(C(z) y(t) - f(y(t)) \big) dt = 0$$
$$\forall E(z) \tag{3.159}$$

Substituting $E(z) y(t)$ and interchanging the order of integration gives

$$0 = \lim_{N \to \infty} \frac{1}{T} \int_0^T \int_0^t e(\tau) y(t-\tau) \big(C(z) y(t) - f(y(t)) \big) d\tau \, dt$$
$$= \frac{1}{T} \int_0^T \left[\int_\tau^T y(t-\tau) \big(C(z) y(t) - f(y(t)) \big) dt \right] d\tau \tag{3.160}$$

Since $e(t)$ is an arbitrary impulse response, the coefficient of $e(t)$ is zero. Then interchanging the order of integration, the above optimality condition becomes

$$0 = \frac{1}{T} \int_\tau^T y(t-\tau) \big(C(z) y(t) - f(y(t)) \big) dt \tag{3.161}$$

i.e.

$$0 = \phi_{y,Cy-f}(\tau) = \phi_{y,Cy}(\tau) - \phi_{y,f}(\tau), \forall \tau \geq 0 \tag{3.162}$$

Doing the same operation on equation (3.162) to get

$$\phi_{y,Cy}(\tau) = \phi_{y,f}(\tau), \forall \tau \geq 0$$

Then the proof of the Theorem 3.5 is completed.

To construct the explicit form for that linear feedback controller $C(z)$, we expand equation (3.154) to its reduced form.

$$Ey(t) C(z) y(t) = Ey(t) f(y(t))$$
$$\updownarrow$$
$$Ey(t) \big(C(z) y(t) - f(y(t)) \big) = 0 \tag{3.163}$$

The above mathematical operations are some basic expectation operations. Observing Fig. 3.12 again, we have

$$C(z) y(t) = f(y(t)) = r(t) - u(t) \tag{3.164}$$

To determine the linear controller $C(z)$, post multiplication of equation (3.164) by $y^T(t)$ and taking the cross correlation operation, we have

$$E\big[r(t) - u(t) \big] y^T(t) = C(z) E\big[y(t) y^T(t) \big]$$
$$\updownarrow$$
$$C(z) R_y(\tau) = R_{ry}(\tau) - R_{uy}(\tau)$$
$$\updownarrow$$
$$C(z) = \frac{R_{ry}(\tau) - R_{uy}(\tau)}{R_y(\tau)} \tag{3.165}$$

where $R_y(\tau)$, $R_{ry}(\tau)$, and $R_{uy}(\tau)$ are given as

$$R_y(\tau) = E\left[y(t)y^T(t)\right], R_{ry}(\tau) = E\left[r(t)y^T(t)\right],$$
$$R_{uy}(\tau) = E\left[r(t)y^T(t)\right]$$

From equation (3.165), it holds that

$$C(z) = \left[R_{ry}(\tau) - R_{uy}(\tau)\right]R_y^{-1}(\tau) \tag{3.166}$$

In case of no excitation input, i.e. $r(t) = 0$, then linear controller $C(z)$ is simplified as

$$C(z) = R_{uy}(\tau)R_y^{-1}(\tau) \tag{3.167}$$

Comment: Generally, Theorem 3.5 gives one condition about the linear feedback controller during the process of replacing the nonlinear feedback controller. Equation (3.166) or (3.167) are two detailed forms about the constructed linear controllers.

3.5.5   Conclusion

A new stealth identification strategy is proposed to identify the closed loop linear time invariant system. The stealth identification modifies the closed loop feedback mechanism so that the unknown controller does not affect the excitation signal. Stealth identification strategy makes the classical prediction error method possible for identifying the closed loop system with an unknown controller. One convergence condition to ensure the global minimum in the process of stealth identification is given.

3.6   Summary

Closed loop system identification is one classical identification problem during the research of system identification theory. As the detailed process of system identification is to use the measured input-output data to identify the unknown model parameter, it is similar to the idea of data driven. This chapter formulates some of our new recent studies for closed loop system identification, such as stealth identification, performance analysis, and identification for minimum variance control strategy.

References

[1] Forssel, U. and Ljung, L. 1999. Closed loop identification revisited, Automatica, 35(7): 1215-1241.

[2] Ljung, L. 1999. System Identification: Theory for the User. Prentice Hall Press, Upper Saddle River, New Jersey, USA.

[3] Pintelon, R. and Schoukens, J. 2001. System Identification: A Frequency Domain Approach. New York: IEEE Press.

[4] Augero, J.C. 2011. A virtual closed loop method for closed loop identification, Automatica, 47(8): 1626-1637.

[5] Forssell, U. and Ljung, L. 2000. Some results on optimal experiment design, Automatica, 36(5): 749-756.

[6] Leskers, M. 2007. Closed loop identification of multivariable process with part of the inputs controlled, International Journal of Control, 80(10): 1552-1561.

[7] Hjalmarssion, H. 2005. From experiment design to closed loop control, Automatica, 41(3): 393-438.

[8] Hjalmarssion, H. 2008. Closed loop experiment design for linear time invariant dynamical systems via LMI, Automatica, 44(3): 623-636.

[9] Bombois, X. 2006. Least costly identification experiment for control, Automatica, 42(10): 1651-1662.

[10] Hildebrand, R. 2003. Identification for control: Optimal input design with respect to a worst case gap cost function, SIAM Journal of Control Optimization, 41(5): 1586-1608.

[11] Gevers, M. 2006. Identification of multi input systems: Variance analysis and input design issues, Automatica, 42(41): 559-572.

[12] Gevers, M. 2009. Identification and information matrix: How to get just sufficiently rich, IEEE Transactions on Automatic Control, 54(12): 2828-2840.

[13] Goodin, G.C. 2002. Bias issues in closed loop identification with application to adaptive control, Communications in Information and Systems, 2(4): 349-370.

[14] Welsh, J.S. 2002. Finite sample properties of indirect nonparametric closed loop identification, IEEE Transactions on Automatic Control, 47(8): 1277-1291.

[15] Douma, S.G. 2008. Validity of the standard cross correlation test for model structure validation, Automatica, 44(4): 1285-1294.

[16] Wang, J. and Wang, Y. 2017a. Model structure validation for closed loop system identification, International Journal of Modelling, Identification and Control, 27(4): 323-331.

[17] Wang, J. and Wang, Y. 2017b. Further results on model structure validation for closed loop system identification, Advances in Wireless Communications and Networks, 30(5): 57-66.

[18] Bombois, X. and Anderson, B.D.O. 2005. Quantification of frequency domain error bounds with guaranteed confidence level in prediction error identification, Systems & Control Letters, 54(5): 471-482.

CHAPTER 4

Data Driven Model Validation for Closed Loop System

4.1   Introduction

Generally, two strategies are used to design the controller in a closed loop, i.e. the model based design and the direct data driven design. The primary step of the model based design is to construct the plant model and apply this mathematical model in the process of designing a controller. Conversely in the direct data driven method, the modeling process is not needed and the controller is directly designed by using the input-output data. As the first model based design strategy is more applied, we do much research on system identification to identify the plant model. The whole theory of system identification can be divided into four categories, i.e. experiment design, model structure selection, model parameter identification, and model structure validation test. Further, more research is concerned with the first three categories. To the authors' knowledge, there is very little study on model structure validation under closed loop conditions. Our research on model validation for a closed loop system are generalized as follows.

(1) The problem of the model structure validation for closed loop system identification. Two probabilistic model uncertainties and an optimum input filter are derived from some statistical properties of the parameter estimation. The probabilistic bounds and optimum input filter are based on an asymptotic normal distribution of the parameter estimator and its covariance matrix, which was estimated from sampled data. The uncertainties bound of the model parameter and cross-correlation function are constructed in the probability sense by using the inner product form of the asymptotic covariance matrix. Further, the input filter is derived from the point of optimization. We modify the sign perturbed sums (SPS) method to construct non-asymptotic confidence regions under a finite number of data points, where some modifications are studied for closed loop systems.

(2) After closed loop system identification is reviewed, the asymptotic analysis and finite sample analysis for closed loop system identification are studied

respectively, corresponding to the infinite data and finite data. More specifically, within the framework of infinite data, the cost function is modified to its simplified form, and one optimal feedback controller is obtained based on our derivations. The simplified cost function and optimal feedback controller are beneficial for practical applications. Furthermore, the asymptotic variance of that optimal feedback controller is also yielded from the point of asymptotic analysis. In the case of finite data, finite sample properties are constructed for closed loop system identification, then one difference between the sampled identification criterion and its corresponding expected criterion is derived as an explicit form, which can bound one guaranteed interval for the sampled identification criterion.

4.2 Model structure validation for closed loop system identification

There are three common identification methods in closed loop identification, i.e. the direct approach, the indirect approach, and the joint input-output approach, where the feedback is neglected in the direct approach and the plant model is identified directly using the input-output data. For the indirect approach, the feedback effect is considered and the input-output from the whole closed loop condition is used to identify the plant model. The joint input-output approach is very similar to the indirect approach. In [1], three methods are presented to identify closed loop system. In [2], research on the system identification theory is introduced in the time domain. Similarly, the frequency domain system identification is given in [3]. A new virtual closed loop method for closed loop identification is proposed in [4]. In [5], one projection algorithm is proposed based on the prediction error recursive method. In [6], when many inputs exist in closed loop, can the closed loop be identified with parts of the inputs controlled? The relationship between closed loop identification and closed loop control is obtained in [7]. In [8], the linear matrix inequality is used to describe the problem of optimal input design in a closed loop. Further, the least cost identification experiment problem is analyzed in [9]. The power spectral of the input signal is considered to be an objective function and the accuracy of the parameter estimations is the constraint [10]. In [11], the H infinity norm from robust control is introduced to be the objective function in the optimal input design problem. Based on the H infinity norm, the uncertainty between the identified model and the nominal model is measured and the optimal input is chosen by minimizing this uncertainty [12]. The selection of the optimal input can also be determined from the point of the asymptotic behavior of the parameter estimation [13]. The Persistent excitation input in closed loop is analyzed and we obtain some conditions about how to obtain persistent excitation [14]. Reference [15] considers how to apply closed loop identification into adaptive control so that bias and covariance terms are isolated separately. All the above results hold when the number of the observed signals will converge to infinity [16].

There are few references for model structure validation test these days. Only in references [2] and [15], model structure validation in open loop identification has been presented and the standard cross correlation test is proposed to test the confidence interval of the cross variance matrix between the prediction error and input from the probabilistic sense. Because of the simple structure of an open loop, the process of deriving the covariance matrix is very easy. For model structure validation in engineering, the more effective strategy is to do one similar experiment again. After exciting the formal system with a group of new inputs, we compare if the actual output is consistent with the identified output. Although this test is simple, we could not analyze the accuracy and credibility of the identified model. So to reflect the identification accuracy, here we first apply the statistical probability framework to derive the variance matrix of the unknown parameters. This variance matrix is decomposed into one inter product form which is used to construct one uncertainty bound about the unknown parameter estimation. This uncertainty bound is called the confidence interval and it constitutes the guaranteed confidence region test with respect to the model parameter estimation under closed loop conditions. Further, we analyze the cross correlation function between the prediction error and input, then one hypothesis testing problem is considered to replace the formal model structure validation testing problem. We construct a probability distribution of the cross-correlation function to guarantee that the original hypothesis testing problem is un-falsified. As each input is obtained by a white noise through a shaping filter [17], to ensure the closeness between the closed loop output predictor and actual output, we consider how to select this shaping filter. Generally, from the above description, to solve the problem of model structure validation for closed loop identification, we test if some conditions of the three categories, (i.e. the confidence interval of parameter estimation, the confidence probability level of one cross correlation function between prediction error and input, the selection of shaping filter) can achieve the expected requirements respectively.

4.2.1   Problem description

Consider the following actual closed loop system with output feedback (see Fig. 4.1). Where $G_0(q)$ is the true plant model; $H_0(q)$ is the noise filter, they are all linear time invariant transfer functions. $K(q)$ is a stable linear time invariant controller, here we assume this controller is priori known. The excited signal $r(t)$ and external disturbance $e(t)$ are assumed to be uncorrelated, $e(t)$ is a white noise with zero mean value and variance λ_0. $v(t)$ is a colored noise which can be obtained by passing white noise $e(t)$ through the noise filter $H_0(q)$. $u(t)$ and $y(t)$ are the input-output signals corresponding to plant model $G_0(q)$.

Rewriting excited signal $r(t)$ as white noise $w(t)$ passing through shaping filter $R(q)$. $R(q)$ is the power spectrum factor with stable non-minimal phase of excited signal $r(t)$. q is the delay operator, it means that $qu(t) = u(t + 1)$.

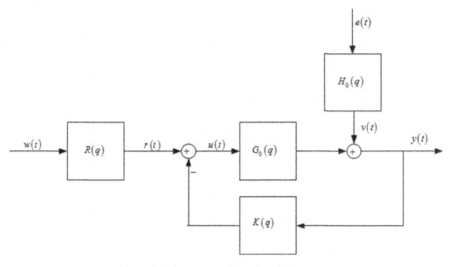

Figure 4.1: Structure of the closed loop system

As $r(t) = R(q) w(t)$, the power spectrum density of excited signal is given as [18]

$$\phi_r(\omega) = R(q) R^*(q) \phi_w(\omega) = |R(q)|^2 \qquad (4.1)$$

In closed loop system structure, through some computations, we derive the transfer function form.

$$y(t) = G_0(q) R(q) w(t) - G_0(q) K(q) y(t) + H_0(q) e(t)$$

Continuing to do some computations and we get

$$y(t) = \frac{G_0(q) R(q)}{1 + G_0(q) K(q)} w(t) + \frac{H_0(q)}{1 + G_0(q) K(q)} e(t)$$

$$u(t) = \frac{R(q)}{1 + G_0(q) K(q)} w(t) - \frac{K(q) H_0(q)}{1 + G_0(q) K(q)} e(t) \qquad (4.2)$$

To simplify the analysis process, define the sensitivity function as

$$S_0(q) = \frac{1}{1 + G_0(q) K(q)}$$

The output of closed loop system can be written as

$$y(t) = G_0(q) R(q) S_0(q) w(t) + H_0(q) S_0(q) e(t)$$

Since our goal is not emphasized on the identification method, but on the model structure validation, only the simple direct approach is used to describe the basic ideas.

4.2.2   Confidence region test of model parameter

Introduce the unknown parameter vector in a closed loop system, the parameterized form given by

$$y(t,\theta) = \frac{G(q,\theta)R(q)}{1+G(q,\theta)K(q)}w(t) + \frac{H(q,\theta)}{1+G(q,\theta)K(q)}e(t) \tag{4.3}$$

where θ denotes the unknown parameter vector, it exists in the parameterized plant model $G(q, \theta)$ and noise model $H(q, \theta)$ respectively. The goal of closed loop identification is to identify the unknown parameter vector $\hat{\theta}_N$ from one given input-output data set $Z^N = \{y(t), u(t)\}_{t=1}^{N}$, where N denotes the number of total observed data.

According to (4.3), the prediction of $y(t, \theta)$ can be calculated as the one step ahead prediction.

$$\hat{y}(t,\theta) = \frac{1+G(q,\theta)K(q)}{H(q,\theta)} \times \frac{G(q,\theta)R(q)}{1+G(q,\theta)K(q)}w(t) + \left[1 - \frac{1+G(q,\theta)K(q)}{H(q,\theta)}\right]y(t)$$

$$= \frac{G(q,\theta)R(q)}{H(q,\theta)}w(t) + \frac{H(q,\theta)-1-G(q,\theta)K(q)}{H(q,\theta)}y(t) \tag{4.4}$$

Computing the one step ahead prediction error or residual, now it becomes

$$\varepsilon(t,\theta) = y(t) - \hat{y}(t,\theta) = \frac{1+G(q,\theta)K(q)}{H(q,\theta)}\left[y(t) - \frac{G(q,\theta)R(q)}{1+G(q,\theta)K(q)}w(t)\right] \tag{4.5}$$

In the standard prediction error algorithm, when using the input-output data $Z^N = \{y(t), u(t)\}_{t=1}^{N}$ with the data number N, the parameter vector is identified by

$$\hat{\theta}_N = \arg\min_{\theta} V_N\left(\theta, Z^N\right) = \arg\min_{\theta} \frac{1}{N}\sum_{t=1}^{N}\varepsilon^2(t,\theta) \tag{4.6}$$

Defining the asymptotic limit parameter estimate θ^* as

$$\theta^* = \arg\min_{\theta} \lim_{N\to\infty} E\left\{V_N\left(\theta, Z^N\right)\right\}$$

where E denotes the expectation operator. In the common identification process, assume that there always exists one true parameter vector θ_0 such that

$$G(q,\theta_0) = G_0(q), \, H(q,\theta_0) = H_0(q)$$

This assumption shows that the identified model is contained in the considered model set. Based on some results from reference [2], we get the asymptotic matrix of the parameter estimate.

$$\operatorname{cov} \hat{\theta}_N = \lambda_0 \langle \varphi, \varphi \rangle^{-1} \tag{4.7}$$

where $\langle \varphi, \varphi \rangle$ denotes some inter product operator, φ is the negative gradient of the predictor error, i.e. it can be computed from

$$\varphi(t,\theta) = -\frac{\partial \varepsilon(t,\theta)}{\partial \theta} = \frac{\partial \hat{y}(t,\theta)}{\partial \theta}$$

Next we give the calculation process of the negative gradient of the predictor error under closed loop condition below. As (4.7) is a basic formula in studying asymptotic analysis, we combine (4.3) and (4.4) to get

$$\hat{y}(t,\theta) = \frac{G(q,\theta)R(q)}{H(q,\theta)} w(t) + \frac{H(q,\theta)-1-G(q,\theta)K(q)}{1+G(q,\theta)K(q)} R(q) w(t)$$
$$+ \frac{H(q,\theta)-1-G(q,\theta)K(q)}{1+G(q,\theta)K(q)} e(t)$$
$$= \frac{G(q,\theta)R(q)}{1+G(q,\theta)K(q)} w(t) + \frac{H(q,\theta)-1-G(q,\theta)K(q)}{1+G(q,\theta)K(q)} e(t) \tag{4.8}$$

Substituting (4.8) into (4.5) and computing the partial derivative operations with respect to unknown parameter vector θ, and then we have

$$\frac{\partial \varepsilon(t,\theta)}{\partial \theta} = \frac{G'(\theta)KH(\theta)-\left[1+G(\theta)K\right]H'(\theta)}{H(\theta)} \times \frac{1}{1+G(\theta)K} e(t)$$
$$- \frac{1}{H(\theta)} \times \frac{G'(\theta)R\left[1+G(\theta)K\right]-G(\theta)RG'(\theta)K}{1+G(\theta)K} w(t)$$
$$= \left[\frac{G'(\theta)K}{H(\theta)\left[1+G(\theta)K\right]} - \frac{H'(\theta)}{H(\theta)}\right] e(t) - \left[\frac{G'(\theta)R}{H(\theta)} - \frac{G(\theta)RG'(\theta)K}{H(\theta)\left[1+G(\theta)K\right]}\right] w(t) \tag{4.9}$$

where $G'(\theta)$ denotes the partial derivative operation with respect to θ, and the delay operator q are all ignored to simplify the derivations.

Using the uncorrelated assumption between white noise $e(t)$ and $w(t)$, i.e. it holds that

$$Ee(t)w^T(t) = 0$$

Putting ahead one $H(\theta)$ in (4.9), and rewriting it as

$$\frac{\partial \varepsilon(t,\theta)}{\partial \theta} = \frac{1}{H(\theta)} \times \left(\left[G'(\theta)KS(\theta)-H'(\theta)\right]e(t)\right.$$
$$\left. -\left[G'(\theta)R-G(\theta)RG'(\theta)S(\theta)K\right]w(t)\right) \tag{4.10}$$

where we use the parameterized sensitivity function.

$$S(\theta) = \frac{1}{1 + G(\theta)K}$$

As the following equality holds.

$$R - G(\theta)RKS(\theta) = \frac{R}{1 + G(\theta)K} = RS(\theta)$$

Rewriting (4.10) as the following matrix form:

$$\frac{\partial \varepsilon(t, \theta)}{\partial \theta} = \frac{1}{H(\theta)} \times \begin{bmatrix} G'(\theta) & H'(\theta) \end{bmatrix} \begin{bmatrix} KS(\theta)H(\theta) & RS(\theta) \\ -1 & 0 \end{bmatrix} \begin{bmatrix} e(t) \\ w(t) \end{bmatrix} \quad (4.11)$$

According to (4.7), the asymptotic covariance matrix is that

$$P_\theta = \operatorname{cov} \hat{\theta}_N = \lambda_0 \left[E\varphi(t, \theta_0) \varphi^T(t, \theta_0) \right]^{-1} = \langle \phi, \phi \rangle^{-1}$$

$$\phi = \frac{1}{\sqrt{\lambda_0}\, H_0} \begin{bmatrix} G'(\theta_0) & H'(\theta_0) \end{bmatrix} \begin{bmatrix} \sqrt{\lambda_0}\, KS_0 H_0 & RS_0 \\ -\sqrt{\lambda_0} & 0 \end{bmatrix} \quad (4.12)$$

where $G'(\theta)$ and $H'(\theta)$ are given respectively as

$$G'(\theta) = \frac{\partial G(\theta)}{\partial \theta}, \quad H'(\theta) = \frac{\partial H(\theta)}{\partial \theta}$$

On basis of (4.12), we have the asymptotic result.

$$\hat{\theta}_N \xrightarrow{N \to \infty} \theta_0$$

It shows that the parameter estimator $\hat{\theta}_N$ will converge to its limit θ_0, and further $\hat{\theta}_N$ will asymptotically converge ($N \to \infty$) to normally distributed random variable with mean θ_0 and variance P_θ.

$$\sqrt{N}\left(\hat{\theta}_N - \theta_0\right) \to \mathbb{N}(0, P_\theta), \text{ as } N \to \infty$$

This asymptotic result can be rewritten in a quadratic form, and then we get one λ^2 distribution.

$$N\left(\hat{\theta}_N - \theta_0\right)^T P_\theta^{-1}\left(\hat{\theta}_N - \theta_0\right) \xrightarrow{N \to \infty} \lambda_n^2 \quad (4.13)$$

where n is the number of degrees of freedom in the λ^2 distribution, being equal to the dimension of the parameter vector. Equation (4.13) implies that the random variable $\hat{\theta}_N$ satisfies one uncertainty bound.

$$\hat{\theta}_N \in D(\alpha, \theta_0); D(\alpha, \theta_0) = \left\{ \theta_0 / N (\theta - \theta_0)^T P_\theta^{-1} (\theta - \theta_0) \le \lambda_{n,\alpha}^2 \right\} \qquad (4.14)$$

with $\lambda_{n,\alpha}^2$ corresponding to a probability level α in λ_n^2 distribution, but now in order to quantity the uncertainty on θ_0 rather than on $\hat{\theta}_N$. For every realization of $\hat{\theta}_N$, it holds that

$$\hat{\theta}_N \in D(\alpha, \theta_0) \Leftrightarrow \theta_0 \in D(\alpha, \hat{\theta}_N)$$

It signifies that

$$\theta_0 \in D(\alpha, \hat{\theta}_N) \text{ with probability } \alpha$$

$$D(\alpha, \hat{\theta}_N) = \left\{ \theta / N (\hat{\theta}_N - \theta)^T P_\theta^{-1} (\hat{\theta}_N - \theta) \le \lambda_{n,\alpha}^2 \right\} \qquad (4.15)$$

Equations (4.14) and (4.15) give the confidence intervals of the unknown parameter estimator under closed loop condition. The probability level of the event $\hat{\theta}_N \in D(\alpha, \theta_0)$ holds is at least α.

4.2.3 Confidence region test of cross correlation function

As the cross correlation function between input excitation signal and predictor error satisfies one probability level in the probability sense, we compute the specific expression of the predictor error. Based on equation (4.5), we have

$$\varepsilon(t, \theta) = \frac{1 + G(\theta) K}{H(\theta)} \left[\frac{G_0 R}{1 + G_0 K} w(t) + \frac{H_0}{1 + G_0 K} e(t) - \frac{G(\theta) R}{1 + G(\theta) K} w(t) \right] \qquad (4.16)$$

Inserting the parameter estimator $\hat{\theta}_N$ into the predictor error $\varepsilon(t, \theta)$, we get

$$\varepsilon(t, \hat{\theta}_N) = \frac{1 + G(\hat{\theta}_N) K}{H(\hat{\theta}_N)} \left[\frac{G_0 R}{1 + G_0 K} w(t) - \frac{G(\theta^*) R}{1 + G(\theta^*) K} w(t) + \frac{G(\theta^*) R}{1 + G(\theta^*) K} w(t) \right.$$

$$\left. - \frac{G(\hat{\theta}_N) R}{1 + G(\hat{\theta}_N) K} w(t) + \frac{H_0}{1 + G_0 K} e(t) \right]$$

$$= \underbrace{\frac{1 + G(\hat{\theta}_N) K}{H(\hat{\theta}_N)} \left[\frac{G_0 R}{1 + G_0 K} - \frac{G(\theta^*) R}{1 + G(\theta^*) K} \right] w(t)}_{A_1(t, G_0, \theta^*)}$$

$$+ \underbrace{\frac{1 + G(\hat{\theta}_N) K}{H(\hat{\theta}_N)} \left[\frac{G(\theta^*) R}{1 + G(\theta^*) K} - \frac{G(\hat{\theta}_N) R}{1 + G(\hat{\theta}_N) K} \right] w(t)}_{A_2(t, \hat{\theta}_N, \theta^*)}$$

$$+\frac{1+G\left(\hat{\theta}_N\right)K}{H\left(\hat{\theta}_N\right)}\underbrace{\frac{H_0}{1+G_0K}e(t)}_{A_3\left(t,\hat{\theta}_N,G_0,H_0\right)}$$

(4.17)

Remark:

(a) The first term $A_1(t, G_0, H_0, \theta^*)$ of equation (4.17) is the residual signal part induced by the asymptotic bias of estimated model $G(\theta^*)$.

(b) The second term $A_2\left(t,\hat{\theta}_N,\theta^*\right)$ is also the residual signal part but induced by the variance error of the parameter estimator $G\left(\hat{\theta}_N\right)$.

(c) The third term $A_3\left(t,\hat{\theta}_N,G_0,H_0\right)$ is the effect of measurement noise $e(t)$, and represents the un-modeling errors in the estimated noise model.

Then from reference [19], the model structure validation problem is formulated into one hypothesis testing problem.

$$\gamma_0 : A_1\left(t,G_0,H_0,\theta^*\right)=0$$

(4.18)

Under the condition that the input signal is persistent excitation, then condition of the above hypothesis testing holds is that

$$\frac{G\left(\theta^*\right)}{1+G\left(\theta^*\right)K}=\frac{G_0}{1+G_0K}\Rightarrow\theta^*=\theta_0$$

when hypothesis testing problem γ_0 holds, then the residual signal $\varepsilon\left(t,\hat{\theta}_N\right)$ contains only two terms.

$$\varepsilon_1\left(t,\hat{\theta}_N\right)=A_2\left(t,\hat{\theta}_N,\theta^*\right)+A_3\left(t,\hat{\theta}_N,G_0,H_0\right)$$

(4.19)

Computing the sample cross correlation function between residual and input

$$\hat{R}_{\varepsilon_1 w}\left(\tau\right)=\frac{1}{N}\sum_{t=1}^{N}\varepsilon_1\left(t,\hat{\theta}_N\right)w\left(t-\tau\right),\ \tau=0\cdots n-1$$

where n is a user's choice, being the total number of the parameter vector. Defining the vector as

$$\hat{R}_{\varepsilon_1 w}=\left[\hat{R}_{\varepsilon_1 w}\left(0\right)\ \hat{R}_{\varepsilon_1 w}\left(1\right)\ \cdots\ \hat{R}_{\varepsilon_1 w}\left(n-1\right)\right]^T$$

Reformulating above equation, we have

$$\hat{R}_{\varepsilon_1 w}=\frac{1}{N}\begin{bmatrix} w(1) & w(2) & \cdots & w(N) \\ 0 & w(1) & \cdots & w(N-1) \\ \vdots & \vdots & \vdots & \vdots \\ 0 & 0 & \cdots & w(N-n+1) \end{bmatrix}\begin{bmatrix} \varepsilon_1\left(1,\hat{\theta}_N\right) \\ \varepsilon_1\left(2,\hat{\theta}_N\right) \\ \vdots \\ \varepsilon_1\left(N,\hat{\theta}_N\right) \end{bmatrix}$$

using the assumed condition that $e(t)$ is uncorrelated with $w(t)$, the above equation can be simplified

$$
\hat{R}_{\varepsilon_1 w} =
\begin{bmatrix}
\hat{R}_{\varepsilon_1 w}(0) \\
\hat{R}_{\varepsilon_1 w}(1) \\
\vdots \\
\hat{R}_{\varepsilon_1 w}(n-1)
\end{bmatrix}
= \frac{A_4}{N}
\begin{bmatrix}
w(1) & w(2) & \cdots & w(N) \\
0 & w(1) & \cdots & w(N-1) \\
\vdots & \vdots & \vdots & \vdots \\
0 & 0 & \cdots & w(N-n+1)
\end{bmatrix}
$$

$$
\begin{bmatrix}
w(1) \\
w(2) \\
\vdots \\
w(N)
\end{bmatrix}
= \frac{A_4}{N}
\begin{bmatrix}
N \\
0 \\
\vdots \\
0
\end{bmatrix}
= A_4
\begin{bmatrix}
1 \\
0 \\
\vdots \\
0
\end{bmatrix}
$$

$$
A_4 = \frac{1+G(\hat{\theta}_N)K}{H(\hat{\theta}_N)}
\left[\frac{G(\theta^*)R}{1+G(\theta^*)K} - \frac{G(\hat{\theta}_N)R}{1+G(\hat{\theta}_N)K} \right] \tag{4.20}
$$

From equation (4.20), we see that the cross correlation function between predictor error and input satisfies.

$$
\hat{R}_{\varepsilon_1 w}(0) = A_4, \quad \hat{R}_{\varepsilon_1 w}(\tau) = 0, \quad \tau = 1 \cdots n-1
$$

to compute the mean and covariance of cross correlation $\hat{R}_{\varepsilon_1 w}(0) = A_4$, by observing A_4, we have

$$
E\left[\hat{R}_{\varepsilon_1 w}(0) \right] = EA_4 = 0
$$

Applying Taylor series expansion to the difference in A_4, we get

$$
\frac{G(\theta^*)R}{1+G(\theta^*)K} - \frac{G(\hat{\theta}_N)R}{1+G(\hat{\theta}_N)K} = -\frac{1}{\left[1+G(\theta^*)K\right]} \Delta G \tag{4.21}
$$

According to equations (4.20) and (4.21), the variance value of cross correlation function $\hat{R}_{\varepsilon_1 w}(0)$ is given as

$$
\text{cov}\, \hat{R}_{\varepsilon_1 w}(0) = E\left[\hat{R}_{\varepsilon_1 w}(0) \right]^2 = \frac{1}{H_0 \left[1+G_0 K\right]^2} E\Delta G \Delta G^T = \frac{1}{H_0 \left[1+G_0 K\right]^2} \text{cov}\, G \tag{4.22}
$$

where $\text{cov}\, G$ denotes the variance of the estimator plant $G(\hat{\theta}_N)$. We expand $\text{cov}\, G$ to get

$$
\text{cov}\, G = G_0' P_\theta G_0' = G_0' \langle \phi, \phi \rangle^{-1} G_0'
$$

Based on the above asymptotic analysis, we see the cross correction function $\hat{R}_{\varepsilon_1 w}(0)$ will converge to one asymptotic Gaussian distribution with zero mean.

$$\hat{R}_{\varepsilon_1 w}(0) \rightarrow \mathbb{N}\left(0, \frac{1}{N}\text{cov}\,\hat{R}_{\varepsilon_1 w}(0)\right) = \mathbb{N}\left(0, \frac{G_0' \langle \phi, \phi \rangle^{-1} G_0'}{N H_0 \left[1 + G_0 K\right]^2}\right) \tag{4.23}$$

The sufficient condition that the hypothesis testing holds is that for every realization $\hat{R}_{\varepsilon_1 w}$ of random variable $\hat{R}_{\varepsilon_1 w}(0)$, we have

$$\gamma_0 \text{ holds} \Rightarrow \hat{R}_{\varepsilon_1 w}^T(0)\left[\text{cov}\,\hat{R}_{\varepsilon_1 w}(0)\right]^{-1}\hat{R}_{\varepsilon_1 w}(0) \in \lambda^2(n) \tag{4.24}$$

The hypothesis is un-falsified if

$$\hat{R}_{\varepsilon_1 w}^T(0)\left[\text{cov}\,\hat{R}_{\varepsilon_1 w}(0)\right]^{-1}\hat{R}_{\varepsilon_1 w}(0) \leq c_\lambda(\alpha, n) \tag{4.25}$$

where $c_\lambda(\alpha, n)$ is the $(1 - \alpha)$ probability level of the λ^2 distribution with degrees of freedom n.

$$x \in \lambda^2(n) \Rightarrow pr\left(x \leq c_\lambda(\alpha, n)\right) = \alpha \tag{4.26}$$

In (4.24) and (4.26), the confidence interval about the cross correction function between predictor error and input signal under closed loop condition is given.

4.2.4   Shaping filter test

To characterize the specific expression of the shaping filter $R(q)$, we combine the actual output system and its parameterized form.

$$\begin{cases} y(t) = \dfrac{G_0 R}{1 + G_0 K}w(t) + \dfrac{H_0}{1 + G_0 K}e(t) \\[4mm] \hat{y}(t, \theta) = \dfrac{G(\theta) R}{1 + G(\theta) K}w(t) + \dfrac{H(\theta)}{1 + G(\theta) K}e(t) \end{cases}$$

Computing the prediction error at parameter estimator $\hat{\theta}_N$.

$$y(t) - \hat{y}(t, \hat{\theta}_N) = \left[\frac{G_0}{1 + G_0 K} - \frac{G(\hat{\theta}_N)}{1 + G(\hat{\theta}_N)K}\right]Rw(t) + \left[\frac{H_0}{1 + G_0 K} - \frac{H(\hat{\theta}_N)}{1 + G(\hat{\theta}_N)K}\right]e(t) \tag{4.27}$$

Similarly applying the Taylor series expansion to above two equations, we obtain

$$\frac{G_0}{1 + G_0 K} - \frac{G(\hat{\theta}_N)}{1 + G(\hat{\theta}_N)K} \approx -\frac{1}{\left[1 + G_0 K\right]^2}\Delta G$$

$$\frac{H_0}{1+G_0 K} - \frac{H(\hat{\theta}_N)}{1+G(\hat{\theta}_N)K} \approx -\frac{1}{1+G_0 K}\Delta H - \frac{H_0 K}{[1+G_0 K]^2}\Delta G$$

$$\Delta G = G(\hat{\theta}_N) - G_0, \quad \Delta H = H(\hat{\theta}_N) - H_0 \qquad (4.28)$$

Substituting these two Taylor series expansions into the prediction error, we get

$$y(t) - \hat{y}(t, \hat{\theta}_N) = \frac{1}{[1+G_0 K]^2}\Delta G R w(t) + + \left[\frac{H_0 K}{[1+G_0 K]^2}\Delta G + \frac{1}{1+G_0 K}\Delta H\right]e(t)$$

$$(4.29)$$

Using the Parseval theorem, we have

$$E\left\{\left\|y(t) - \hat{y}(t, \hat{\theta}_N)\right\|^2\right\} = \frac{1}{2\pi}\int_{-\pi}^{\pi}\frac{R^2}{[1+G_0 K]^4}E\left(\Delta G \Delta G^T\right)$$

$$+\frac{\lambda_0 H_0^2 K}{[1+G_0 K]^4}E\left(\Delta G \Delta G^T\right) + \frac{\lambda_0}{[1+G_0 K]^2}E\left(\Delta H \Delta H^T\right)$$

$$+\frac{H_0 K}{[1+G_0 K]^3}E\left(\Delta G \Delta H^T\right) \qquad (4.30)$$

Computing above each variance matrix respectively, we formulate (4.12) and do some derivations to get

$$\begin{bmatrix} \text{cov}\, G(\hat{\theta}_N) & E\left(\Delta G \Delta H^T\right) \\ E\left(\Delta H \Delta G^T\right) & \text{cov}\, H(\hat{\theta}_N) \end{bmatrix} = \frac{n}{N}\frac{\lambda_0 H_0^2}{[1+G_0 K]^2}$$

$$(4.31)$$

$$\times \begin{bmatrix} R^2 S_0^2 + \lambda_0 K^2 S_0^2 H_0^2 & -\lambda_0 K H_0 S_0 \\ -\lambda_0 K H_0 S_0 & \lambda_0 \end{bmatrix}^{-1}$$

The inverse matrix is solved by

$$\frac{\begin{bmatrix} \lambda_0 & \lambda_0 K H_0 S_0 \\ \lambda_0 K H_0 S_0 & R^2 S_0^2 + \lambda_0 K^2 S_0^2 H_0^2 \end{bmatrix}}{\lambda_0^2 R^2 S_0^2}$$

where in the process of solving the inverse matrix, one identity is used.

$$\lambda_0\left(R^2 S_0^2 + \lambda_0 K^2 S_0^2 H_0^2\right) - \lambda_0^2 S_0^2 K^2 H_0^2 = \lambda_0^2 R^2 S_0^2$$

So the following variances are obtained.

$$\mathrm{cov}\, G\left(\hat{\theta}_N\right) = \frac{n}{N}\frac{H_0^2}{R^2}, \quad E\left(\Delta G \Delta H^T\right) = \frac{n}{N}\frac{H_0^3 KS_0}{R^2}; \mathrm{cov}\, H\left(\hat{\theta}_N\right)$$

$$= \frac{n}{N}\left[\frac{H_0^2}{\lambda_0} + \frac{K^2 S_0^2 H_0^2}{R^2}\right] \tag{4.32}$$

Substituting (4.32) into (4.30) and do some computations, we get

$$E\left\{\|y(t) - \hat{y}(t)\|^2\right\} = \frac{n}{2N\pi}\int_{-\pi}^{\pi} \frac{\lambda_0 H_0^4 KS_0^4 + \lambda_0 H_0^2 KS_0^3 + H_0^4 K^3 S_0^4}{R^2}\,dq \tag{4.33}$$

As the input signal $u(t)$ corresponding to plant is that

$$u(t) = \frac{R}{1 + G_0 K}w(t) - \frac{KH_0}{1 + G_0 K}e(t)$$

So the power spectral density of input signal $u(t)$ can be computed as

$$\phi_u\left(\omega\right) = R^2 S_0^2 + \lambda_0 K^2 H_0^2 S_0^2 \tag{4.34}$$

The selection of shaping filter R can be determined by solving one optimization problem.

$$\begin{cases} \min_{R^2} \dfrac{1}{2\pi}\int \dfrac{a}{R^2(q)}dq \\[2mm] \text{subject to } \int R^2 S_0^2 \le L - \lambda_0 K^2 H_0^2 S_0^2 \\[2mm] a = \dfrac{n}{N}KH_0^2 S_0^3\left(\lambda_0 H_0^2 S_0 + \lambda_0 + H_0^2 KS_0\right) \end{cases} \tag{4.35}$$

In this optimization problem, L is a positive constant being the user's choice and it can be used to limit the size of the power spectral corresponding to the input signal.

Here we only give the result, i.e. the optimization variable R^2 satisfies that

$$\phi_r = R^2\left(q\right) = \mu\sqrt{a} = \mu\sqrt{\frac{n}{N}KH_0^2 S_0^3\left(\lambda_0 H_0^2 S_0 + \lambda_0 + H_0^2 KS_0\right)} \tag{4.36}$$

where constant μ satisfies the equation

$$\mu = \frac{\dfrac{L - \lambda_0 K^2 H_0^2 S_0^2}{|S_0|^2}}{\sqrt{\dfrac{n}{N}KH_0^2 S_0^3\left(\lambda_0 H_0^2 S_0 + \lambda_0 + H_0^2 KS_0\right)}} \tag{4.37}$$

After solving this shaping filter, it is then used to devise the power spectral of the external excitation input.

4.2.5 Conclusion

In this section we analyze the model structure validation problem for a closed loop condition from three aspects:

(1) Confidence region of model parameter.
(2) Confidence region of cross correlation function.
(3) Selection of shaping filter. Two uncertainty bounds about model parameters and cross correlation function are constructed in probability sense by using asymptotic variance analysis. Similarly, the shaping filter of the excitation input is determined by solving one minimization problem.

4.3 Non-asymptotic confidence regions in closed loop model validation

Using some results from our previous results, in this section we continue to study the problem of model structure validation for closed loop system identification. So to reflect the identification accuracy, here we first apply the statistical probability framework to derive the variance matrix of the unknown parameters. This variance matrix is decomposed into one inter product form which is used to construct one uncertainty bound about the unknown parameter estimation. This uncertainty bound is called the confidence region and it constitutes the guaranteed confidence region with respect to the model parameter estimation under closed loop conditions. A statistical probability framework needs infinite data points, but in practice, we only have a finite number of data points and limited statistical knowledge about the noise. So to relax this strict condition on the number of data points, we introduce Sign-Perturbed Sums (SPS) to construct a non-asymptotic confidence region. As the original SPS method is suited for the linear regression model, it can not be appropriate for our closed loop system. This section introduces and analyzes the modified SPS method for the closed loop system. This confidence region constructed by the modified SPS method contains the true parameters with a user chosen exact probability for finite data points. To the best of our knowledge, the asymptotic confidence region or non-asymptotic confidence region corresponds to only the linear regression model, not our closed loop system. To achieve our ultimate goal of closed loop model validation, we do much work to extend the existing theories and modify them to be suitable for a closed loop system.

Both confidence regions (4.14) and (4.15) are guaranteed only asymptotically when the number of measured data points tends to infinity, i.e. $N \to \infty$. But in practice, a finite number of measured data points and limited statistical knowledge about the noise are given, this fact strongly restricts the above asymptotic confidence region. In this section, we use the Sign-Perturbed Sums method to construct a non-asymptotic confidence region under relaxed statistical assumption on noise. SPS method is proposed for linear regression model in [17], and its advantage is that the confidence region obtained by the SPS method has exact confidence probability, i.e. it contains the true parameter with a user-chosen exact probability as the number of measured data points is finite. Now the SPS process is introduced into model structure validation for closed loop system, i.e. we extend the linear regression model in [19], [20] to closed loop system and explain how SPS can be applied to construct guaranteed finite sample confidence region for closed loop system.

For the sake of completeness, from equation (4.6), we rewrite it as that

$$\hat{\theta}_N = \arg\min_{\theta} V_N\left(\theta, Z^N\right) = \arg\min_{\theta} \frac{1}{N}\sum_{t=1}^{N} \varepsilon^2\left(t,\theta\right) = \arg\min_{\theta} \frac{1}{N}\sum_{t=1}^{N}\left[y(t)-\hat{y}(t,\theta)\right]^2$$

and the negative gradient of the predictor error

$$\varphi\left(t,\theta\right) = -\frac{\partial \varepsilon\left(t,\theta\right)}{\partial \theta} = \frac{\partial \hat{y}\left(t,\theta\right)}{\partial \theta}$$

Using the necessary condition on (4.6) to obtain one normal equation

$$\frac{1}{N}\sum_{t=1}^{N}\varphi\left(t,\theta\right)\varepsilon\left(t,\theta\right) = \frac{1}{N}\sum_{t=1}^{N}\varphi\left(t,\theta\right)\left[y(t)-\hat{y}(t,\theta)\right] = 0 \qquad (4.38)$$

Due to the parameter estimate $\hat{\theta}_N$ is the solution to the normal equation, it means that

$$\frac{1}{N}\sum_{t=1}^{N}\varphi\left(t,\hat{\theta}_N\right)\varepsilon\left(t,\hat{\theta}_N\right) = \frac{1}{N}\sum_{t=1}^{N}\varphi\left(t,\hat{\theta}_N\right)\left[y(t)-\hat{y}\left(t,\hat{\theta}_N\right)\right] = 0 \qquad (4.39)$$

The SPS method builds the confidence region by using the perturb equation and exploits the information in the data as much as possible, while assuming minimal prior statistical knowledge on the noise [19]. In the whole SPS method, $(m-1)$ sign-perturbed sums are introduced.

$$H_i\left(\theta\right) = \sum_{t=1}^{N}\varphi\left(t,\theta\right)\alpha_{it}\varepsilon\left(t,\theta\right) = \sum_{t=1}^{N}\varphi\left(t,\theta\right)\alpha_{it}\left[y(t)-\hat{y}(t,\theta)\right], i = 1,2\cdots m-1$$

$$(4.40)$$

where $\{\alpha_{it}\}$ are random signs, i.e. independent and identically distributed random variables which takes on the values ± 1 with equal probabilities $\frac{1}{2}$. When no sign perturbations are used, the reference sum is defined.

$$H_0\left(\theta\right) = \sum_{t=1}^{N}\varphi\left(t,\theta\right)\varepsilon\left(t,\theta\right) \qquad (4.41)$$

Then the SPS method used to construct non-asymptotic confidence region is dependent on the following sums:

$$\begin{cases} S_0(\theta) = \Phi^{-\frac{1}{2}}(\theta) \sum_{t=1}^{N} \varphi(t,\theta)\varepsilon(t,\theta) \\ S_i(\theta) = \Phi^{-\frac{1}{2}}(\theta) \sum_{t=1}^{N} \varphi(t,\theta)\alpha_{it}\varepsilon(t,\theta) \end{cases} \qquad (4.42)$$

where $i = \{1, 2 \cdots m - 1\}$ and matrices $\Phi(\theta)$ are perturbed covariance estimates.

$$\begin{cases} \Phi(\theta) = \sum_{t=1}^{N} \varphi(t,\theta)\varphi^T(t,\theta) \\ \Phi^{\frac{1}{2}}(\theta)\Phi^{\frac{1}{2}}(\theta)^T = \Phi(\theta) \end{cases}$$

The above sums (4.42) coming from [17] are used to construct a non-asymptotic confidence region for only the linear regression model, not our closed loop system. To ensure that the SPS method is suitable for the closed loop system, some modifications must be added to achieve this goal. The one step ahead prediction is modified, after considering random signs.

$$\hat{y}(t,\theta,\alpha_i) = G(q,\theta)u(t,\theta,\alpha_i) + H(q,\theta)(\alpha_{it}e(q,\theta)) \qquad (4.43)$$

Then using the feedback structure, input signal on plant is changed as

$$u(t,\theta,\alpha_i) = r(t) - K(q)\hat{y}(t,\theta,\alpha_i) \qquad (4.44)$$

Computing the one step ahead prediction error or residual, with the random signs.

$$\varepsilon(t,\theta,\alpha_i) = y(t) - \hat{y}(t,\theta,\alpha_i) \qquad (4.45)$$

Comparing (4.43), (4.44), (4.45) and their corresponding expressions, the difference is that here random signs α_{it} are immerged into their forms respectively, i.e. their relations are listed as

$$\begin{cases} \varepsilon(t,\theta) \rightarrow \varepsilon(t,0,\alpha_i) \\ \hat{y}(t,\theta) \rightarrow \hat{y}(t,\theta,\alpha_i) \\ u(t,\theta) \rightarrow u(t,\theta,\alpha_i) \end{cases}$$

where α_i denotes the vector $(\alpha_{i1}, \alpha_{i2} \cdots \alpha_{im})$, and α_{it} is defined as above.

Based on these three modifications about $\varepsilon(t, \theta, \alpha_i)$, $\hat{y}(t,\theta,\alpha_i)$, $u(t, \theta, \alpha_i)$, the reference sum and $(m - 1)$ sign-perturbed sums can be evaluated for closed loop system.

$$\begin{cases} S_{0c}\left(\theta\right) = \Phi^{-\frac{1}{2}}\left(\theta,\alpha_i\right)\sum_{t=1}^{N}\varphi\left(t,\theta,\alpha_i\right)\varepsilon\left(t,\theta,\alpha_i\right) \\[2mm] S_{ic}\left(\theta\right) = \Phi^{-\frac{1}{2}}\left(\theta,\alpha_i\right)\sum_{t=1}^{N}\varphi\left(t,\theta,\alpha_i\right)\alpha_{it}\varepsilon\left(t,\theta,\alpha_i\right) \end{cases}$$

$$\varphi\left(t,\theta,\alpha_i\right) = -\frac{\partial\varepsilon\left(t,\theta,\alpha_i\right)}{\partial\theta} = \frac{\partial\hat{y}\left(t,\theta,\alpha_i\right)}{\partial\theta} \qquad\qquad (4.46)$$

Also the perturbed covariance estimate is

$$\Phi\left(\theta,\alpha_i\right) = \sum_{t=1}^{N}\varphi\left(t,\theta,\alpha_i\right)\varphi^{T}\left(t,\theta,\alpha_i\right) \qquad\qquad (4.47)$$

Combining modifications (4.43)-(4.47), the non-asymptotic confidence region is constructed by the modified SPS method, where the main procedures are formulated as

1. Given two integer parameters $m > q > 0$, such that $p = 1 - \frac{q}{m}$, and $p\in(0,\,1)$ is a confidence probability.
2. Generate $N(m-1)$ independent and identically distributed random signs $\{\alpha_{it}\}$ with

$$P\left(\alpha_{it} = 1\right) = P\left(\alpha_{it} = -1\right) = \frac{1}{2}$$

For $i = \{1, 2 \cdots m - 1\}$ and $i = \{1, 2 \cdots N\}$
3. For the given θ, compute the prediction outputs, inputs and errors.

$$\begin{cases} \hat{y}\left(t,\theta,\alpha_i\right) = G\left(q,\theta\right)u\left(t,\theta,\alpha_i\right) + H\left(q,\theta\right)\left(\alpha_{it}e\left(q,\theta\right)\right) \\[1mm] u\left(t,\theta,\alpha_i\right) = r\left(t\right) - K\left(q\right)\hat{y}\left(t,\theta,\alpha_i\right) \\[1mm] \varepsilon\left(t,\theta,\alpha_i\right) = y\left(t\right) - \hat{y}\left(t,\theta,\alpha_i\right) \end{cases}$$

4. Calculate the gradient of $\varepsilon(t,\,\theta,\,\alpha_i)$ and the perturbed covariance estimate.

$$\begin{cases} \varphi\left(t,\theta,\alpha_i\right) = -\dfrac{\partial\varepsilon\left(t,\theta,\alpha_i\right)}{\partial\theta} = \dfrac{\partial\hat{y}\left(t,\theta,\alpha_i\right)}{\partial\theta} \\[3mm] \Phi\left(\theta,\alpha_i\right) = \displaystyle\sum_{t=1}^{N}\varphi\left(t,\theta,\alpha_i\right)\varphi^{T}\left(t,\theta,\alpha_i\right) \end{cases}$$

5. Calculate the factor $\Phi^{-\frac{1}{2}}\left(\theta,\alpha_i\right)$ such that

$$\Phi^{-\frac{1}{2}}\left(\theta,\alpha_i\right)\Phi^{-\frac{1}{2}}\left(\theta,\alpha_i\right)^{T} = \Phi\left(\theta,\alpha_i\right)$$

6. Compute the reference sum and $(m - 1)$ sign-perturbed sums and compute the rank $R(\theta)$ of $\left\| S_{0c}(\theta) \right\|^2$ as the smallest in the ordering of the variables $\left\| S_{ic}(\theta) \right\|^2$

7. Construct non-asymptotic confidence region as

$$\hat{\Theta}_N = \left\{ \theta : R(\theta) \leq m - q \right\} \tag{4.48}$$

Remark: One important advantage of the SPS method is that the confidence region constructed by (4.48) in a closed loop system, has the exact confidence probability for any finite data points. The confidence probability of the constructed confidence region is exact p, that is

$$P\left\{ \theta \in \hat{\Theta}_N \right\} = 1 - \frac{q}{m} = p$$

It means that this non-asymptotic confidence region has exact confidence probability as the number of data points is finite.

4.4   Further results on model structure validation

To reflect the identification accuracy, we first apply the statistical probability framework to derive the variance matrix of the unknown parameters. This variance matrix is decomposed into one inter product form which is used to construct one uncertainty bound about the unknown parameter estimation. This uncertainty bound is called the confidence interval and it constitutes the guaranteed confidence region test with respect to the model parameter estimation under closed loop condition. Furthermore, our aim is to find if the identified model is any good at describing the measured data and being a basis for latter controller design. To validate the effectivity of the identified model, we construct one bound corresponding to the model error. This bound is proposed by an inequality condition, which connects the algebraic nature between the model error, the input signal, and the model validation test quantity. Based on the bound and the proposed inequality condition, the problem of model structure validation for the closed loop system is reformulated to verify if the absolute value of model error satisfies the above inequality. Since the closed loop system is considered here, some priori properties of the closed loop system are provided to simplify the bound.

4.4.1   One bound on model error

A new technique for estimating bias and variance contributions to the model error is suggested in this section. The model structure validation procedure gives a direct measurement of the model error of a closed loop system, so the problem of model structure validation for closed loop system identification can be changed to test whether the model error obeys one inequality condition. Our problem is to figure out if the identified model $G\left(\hat{\theta}_N\right)$ is any good at describing the measured data and to give a statement on how far away the model might be from a true description.

As it is very useful to consider the prediction error or model residual, we could write model residual as follows based on equation (4.5).

$$\varepsilon(t,\theta) = \frac{1+G(\theta)K}{H(\theta)}\left[\frac{G_0 R}{1+G_0 K}w(t) + \frac{H_0}{1+G_0 K}e(t) - \frac{G(\theta)R}{1+G(\theta)K}w(t)\right] \quad (4.49)$$

Inserting the parameter estimator $\hat{\theta}_N$ into the predictor error $\varepsilon(t, \theta)$, we get

$$\varepsilon\left(t,\hat{\theta}_N\right) = \frac{1+G\left(\hat{\theta}_N\right)K}{H\left(\hat{\theta}_N\right)}\left[\frac{G_0 R}{1+G_0 K}w(t) - \frac{G\left(\theta^*\right)R}{1+G\left(\theta^*\right)K}w(t) + \frac{G\left(\theta^*\right)R}{1+G\left(\theta^*\right)K}w(t)\right.$$

$$\left. - \frac{G\left(\hat{\theta}_N\right)R}{1+G\left(\hat{\theta}_N\right)K}w(t) + \frac{H_0}{1+G_0 K}e(t)\right]$$

$$= \underbrace{\frac{1+G\left(\hat{\theta}_N\right)K}{H\left(\hat{\theta}_N\right)}\left[\frac{G_0 R}{1+G_0 K} - \frac{G\left(\theta^*\right)R}{1+G\left(\theta^*\right)K}\right]w(t)}_{A_1\left(t,G_0,\theta^*\right)}$$

$$+ \underbrace{\frac{1+G\left(\hat{\theta}_N\right)K}{H\left(\hat{\theta}_N\right)}\left[\frac{G\left(\theta^*\right)R}{1+G\left(\theta^*\right)K} - \frac{G\left(\hat{\theta}_N\right)R}{1+G\left(\hat{\theta}_N\right)K}\right]w(t)}_{A_2\left(t,\hat{\theta}_N,\theta^*\right)}$$

$$+ \underbrace{\frac{1+G\left(\hat{\theta}_N\right)K}{H\left(\hat{\theta}_N\right)}\frac{H_0}{1+G_0 K}e(t)}_{A_3\left(t,\hat{\theta}_N,G_0,H_0\right)}$$

$$(4.50)$$

Remark:

(a) The first term $A_1(t, G_0, H_0, \theta^*)$ of equation (4.50) is the residual signal part induced by the asymptotic bias of estimated model $G(\theta^*)$.

(b) The second term $A_2\left(t,\hat{\theta}_N,\theta^*\right)$ is also the residual signal part but induced by the variance error of the parameter estimator $G\left(\hat{\theta}_N\right)$.

(c) The third term $A_3\left(t,\hat{\theta}_N,G_0,H_0\right)$ is the effect of measurement of noise $e(t)$, and represents the un-modeling errors in the estimated noise model.

Then the model structure validation problem is formulated into one hypothesis testing problem.

$$\gamma_0 : A_1\left(t, G_0, H_0, \theta^*\right) = 0 \tag{4.51}$$

Under the condition that the input signal is persistent excitation, then the condition of the above hypothesis testing holds is that

$$\frac{G\left(\theta^*\right)}{1 + G\left(\theta^*\right)K} = \frac{G_0}{1 + G_0 K} \Rightarrow \theta^* = \theta_0$$

when hypothesis testing problem γ_0 holds, then the residual signal $\varepsilon\left(t, \hat{\theta}_N\right)$ contains only two terms.

$$\varepsilon_1\left(t, \hat{\theta}_N\right) = A_2\left(t, \hat{\theta}_N, \theta^*\right) + A_3\left(t, \hat{\theta}_N, G_0, H_0\right) \tag{4.52}$$

To give a detailed expression on model residual $\varepsilon_1\left(t, \hat{\theta}_N\right)$, we continue to rewrite it

$$\varepsilon_1\left(t, \hat{\theta}_N\right) = A_2\left(t, \hat{\theta}_N, \theta^*\right) + A_3\left(t, \hat{\theta}_N, G_0, H_0\right)$$

$$= \frac{1 + G\left(\hat{\theta}_N\right)K}{H\left(\hat{\theta}_N\right)} \left[\frac{G\left(\theta^*\right)}{1 + G\left(\theta^*\right)K} - \frac{G\left(\hat{\theta}_N\right)}{1 + G\left(\hat{\theta}_N\right)K}\right] r(t) + \frac{1 + G\left(\hat{\theta}_N\right)K}{H\left(\hat{\theta}_N\right)} \frac{H_0}{1 + G_0 K} e(t)$$

$$= \underbrace{\frac{1 + G\left(\hat{\theta}_N\right)K}{H\left(\hat{\theta}_N\right)} \frac{1}{1 + G\left(\theta^*\right)K}}_{L_1(q)} \tilde{G}r(t) + \underbrace{\frac{1 + G\left(\hat{\theta}_N\right)K}{H\left(\hat{\theta}_N\right)} \frac{H_0}{1 + G_0 K}}_{L_2(q)} e(t)$$

$$= L_1(q)\tilde{G}r(t) + L_2(q)e(t) \tag{4.53}$$

In the derivation process of equation (4.53), we use two known equalities.

$$\begin{cases} r(t) = R(q)w(t) \\ \tilde{G}(q) = G\left(\theta^*\right) - G\left(\hat{\theta}_N\right) \end{cases} \tag{4.54}$$

By observing equation (4.53), now the problem is, what can be said about model error $\tilde{G}(q)$ based on the information in $Z^N = \left\{y(t), u(t)\right\}_{t=1}^{N}$? Ideally we want to give one bound on $\left|\tilde{G}\left(e^{iw}\right)\right|$ valid for each frequency w. Before giving our main technical result of one bound on model error, we define some preliminaries as the basis for our next derivation process.

Then input-output data record corresponding to the whole closed loop system is that

$$Z^N = \{y(t), r(t)\}_{t=1}^{N}$$

The periodogram of the input sequence $\{r(t), t = 1, 2 \cdots N\}$ is defined by

$$\left|R_N(w)\right|^2 = \frac{1}{N}\left|\sum_{t=1}^{N} r(t)e^{-iwt}\right|^2 \tag{4.55}$$

Let regression vector $\varphi(t)$ be that

$$\varphi(t) = \begin{bmatrix} r(t) & r(t-1) & \cdots & r(t-M+1) \end{bmatrix}^T \tag{4.56}$$

where M is the delay time in regression vector, and its autocorrelation function is defined as

$$R_N = \frac{1}{N}\sum_{t=1}^{N}\varphi(t)\varphi^T(t) \tag{4.57}$$

A scalar measurement of the correlation between past input and model residual is constructed.

$$\tilde{\xi}_N^M = \frac{1}{N}\left\|\sum_{t=1}^{N}\varphi(t)\varepsilon_1(t)\right\|_{R_N^{-1}}^2 \tag{4.58}$$

Here $\|.\|$ denotes the common Euclidian norm, such as $\|v\|_R = v^* R v$

The procedure to obtain one bound on model error lies in the following theorem.

Theorem 4.1: Assume that the input-output data record $Z^N = \{y(t), r(t)\}_{t=1}^{N}$ is subject to equation (4.20), let $G(\hat{\theta}_N)$ be an identified model, and construct measure $\tilde{\xi}_N^M$ from (4.55)-(4.58). Assume that there exists a δ, such that $R_N > \delta I$, then the model error $\tilde{G}(q)$ satisfies the following inequality:

$$\left[\frac{1}{2\pi}\int_{-\pi}^{\pi}\left|L_1(e^{iw})\right|^2\left|\tilde{G}(e^{iw})\right|^2\left|R_N(e^{iw})\right|^2 dw\right]^{\frac{1}{2}} \le (1+\eta)\left[\frac{1}{N}\tilde{\xi}_N^M\right]^{\frac{1}{2}}$$

$$+ (1+\eta)x_N + (2+\eta)C_r\sum_{k=M}^{\infty}|\rho_k| \tag{4.59}$$

Here

$$x_N = \left\|\frac{1}{N}\sum_{t=1}^{N}L_2(q)e(t)\varphi(t)\right\|_{R_N^{-1}};$$

ρ_k is the impulse response of $L_1(q)\tilde{G}(q)$; $|R_N(w)|^2$ is the periodogram of the input sequence;

$$C_r = \max_{1 \le t \le N} |r(t)|; \quad \eta = \frac{C_r M}{\sqrt{N}\delta};$$

By the appropriate choice of feedback controller $K(q)$, the size of model error over arbitrarily small frequency intervals can be arbitrarily small. From the above equation (4.59), we see that if one identified model $G(\hat{\theta}_N)$ can be used for the next control process, then the absolute value of model error must satisfy this inequality condition.

Before giving an explicit proof about the inequality condition (4.59), one important lemma from [2] is used here.

Lemma 4.1: Let

$$w(t) = N(q)u(t) = \sum_{k=0}^{\infty} n_k u(t-k)$$

$$\varphi(t) = \begin{bmatrix} u(t) & u(t-1) & \cdots & u(t-M+1) \end{bmatrix}^T$$

$$R_N = \frac{1}{N}\sum_{t=1}^{N} \varphi(t)\varphi^T(t); U_N(w) = \frac{1}{\sqrt{N}}\sum_{t=1}^{N} u(t)e^{-iwt}$$

Assume that

$$|u(t)| \le C_u \quad t = 1, 2 \cdots N; R_N > \delta I$$

Let

$$\tilde{\rho}_M = \sum_{k=M}^{\infty} |n_k|$$

and define that

$$\beta^2 = \left\| \frac{1}{N}\sum_{t=1}^{N} w(t)\varphi_1(t) \right\|_{R_N^{-1}}^2; \quad B^2 = \frac{1}{2\pi}\int_{-\pi}^{\pi} |N(e^{iw})|^2 |U_N(w)|^2 \, dw$$

Then for $\eta = \dfrac{C_r M}{\sqrt{N}\delta}$, we have that

$$B \le (1+\eta)\beta + (2+\eta)C_u \tilde{\rho}_M$$

Based on this lemma, now we give a short proof about Theorem 1.

Proof: As

$$\tilde{G}(q) = G(\theta^*) - G(\hat{\theta}_N)$$

and

$$\varepsilon_1\left(t,\hat{\theta}_N\right) = L_1\left(q\right)\tilde{G}r\left(t\right) + L_2\left(q\right)e\left(t\right)$$

Thus

$$\left[\frac{1}{N}\check{\xi}_N^M\right]^{\frac{1}{2}} = \left\|\frac{1}{N}\sum_{t=1}^{N}\varphi(t)\varepsilon_1\left(t\right)\right\|_{R_N^{-1}}$$

$$= \left\|\frac{1}{N}\sum_{t=1}^{N}\left(L_1\left(q\right)\tilde{G}\left(q\right)r\left(t\right) + L_2\left(q\right)e\left(t\right)\right)\varphi(t)\right\|_{R_N^{-1}} \geq \gamma - x_N \quad (4.60)$$

where

$$\gamma = \left\|\frac{1}{N}\sum_{t=1}^{N}L_1\left(q\right)\tilde{G}\left(q\right)r\left(t\right)\varphi(t)\right\|_{R_N^{-1}} \quad (4.61)$$

Then

$$\gamma \leq \left[\frac{1}{N}\check{\xi}_N^M\right]^{\frac{1}{2}} + x_N \quad (4.62)$$

Using lemma, we see that

$$\left[\frac{1}{2\pi}\int_{-\pi}^{\pi}\left|L_1\left(e^{iw}\right)\right|^2\left|\tilde{G}\left(e^{iw}\right)\right|^2\left|R_N\left(e^{iw}\right)\right|^2 dw\right]^{\frac{1}{2}} \leq \left(1+\eta\right)\gamma + \left(2+\eta\right)C_r\tilde{\rho}_M \quad (4.63)$$

Substituting (4.62) into (4.63), then inequality condition (4.59) is proved.

As a closed loop system is considered here, then some properties of the closed loop system can be used to simplify that inequality condition using the assumption that $e(t)$ is uncorrelated with $w(t)$.

$$Er\left(t\right)e\left(t\right) = ER\left(q\right)w\left(t\right)e\left(t\right) = 0$$

We compute the first term in $\left[\frac{1}{N}\check{\xi}_N^M\right]^{\frac{1}{2}}$ as follows:

$$\left\|\frac{1}{N}\sum_{t=1}^{N}L_2\left(q\right)e\left(t\right)\varphi(t)\right\|_{R_N^{-1}} = L_2\left(q\right)r_{re}^{T}R_N^{-1}r_{re}L_2\left(q\right)$$

$$= L_2\left(q\right)\left[r_{re}\left(0\right) \quad r_{re}\left(1\right) \quad \cdots \quad r_{re}\left(M-1\right)\right]R_N^{-1}\begin{bmatrix} r_{re}\left(0\right) \\ r_{re}\left(1\right) \\ \vdots \\ r_{re}\left(M-1\right) \end{bmatrix}L_2\left(q\right) = 0 \quad (4.64)$$

Then the second term in $\left[\frac{1}{N}\xi_N^M\right]^{\frac{1}{2}}$ is also computed.

$$\left\|\frac{1}{N}\sum_{t=1}^{N}L_1(q)\tilde{G}(q)r(t)\varphi(t)\right\|_{R_N^{-1}} = L_1(q)\tilde{G}(q)$$

$$\left[r_r(0) \quad r_r(1) \quad \cdots \quad r_r(M-1)\right]\times R_N^{-1}\begin{bmatrix} r_r(0) \\ r_r(1) \\ \vdots \\ r_r(M-1) \end{bmatrix}L_1(q)\tilde{G}(q)$$

$$= \left|L_1(q)\tilde{G}(q)\right|^2\left[\left|R(q)\right|^2 \quad 0 \quad \cdots \quad 0\right]R_N^{-1}\begin{bmatrix} |R(q)|^2 \\ 0 \\ \vdots \\ 0 \end{bmatrix}$$

$$= \left|L_1(q)\tilde{G}(q)\right|^2\left|R(q)\right|^4\left[1 \quad 0 \quad \cdots \quad 0\right]R_N^{-1}\begin{bmatrix} 1 \\ 0 \\ \vdots \\ 0 \end{bmatrix} \tag{4.65}$$

where

$$\begin{cases} r_{re}(\tau) = \dfrac{1}{\sqrt{N}}\sum_{t=1}^{N}r(t)e(t-\tau) \\ r_r(\tau) = \dfrac{1}{\sqrt{N}}\sum_{t=1}^{N}r(t)r(t-\tau) \end{cases}$$

Before substituting equation (4.64) and (4.65) into (4.59), we should obtain the (1,1) element of matrix R_N. Observing the explicit structure of regression vector $\varphi(t)$ in matrix R_N, we have that

$$\varphi(t) = R(q)\left[w(t) \quad w(t-1) \quad \cdots \quad w(t-M+1)\right]^T$$

and

$$R_N = \frac{1}{N}\sum_{t=1}^{N}\varphi(t)\varphi^T(t) = \left|R(q)\right|^2\frac{1}{N}N = \left|R(q)\right|^2 \tag{4.66}$$

The above derivation process corresponding to matrix R_N is very easy, for example we only give one equality.

$$\varphi(1)\varphi^T(1) = \left|R(q)\right|^2 \begin{bmatrix} w(1) \\ w(0) \\ \vdots \\ w(-M+1) \end{bmatrix} \times \begin{bmatrix} w(1) & w(0) & \cdots & w(-M+1) \end{bmatrix}^T = \left|R(q)\right|^2 I$$

So for all $t = 1, 2 \cdots N$, we have that

$$\varphi(t)\varphi^T(t) = \left|R(q)\right|^2 I, \quad t = 1, 2 \cdots N \tag{4.67}$$

substituting (4.66) into (4.65), we get that

$$\left\| \frac{1}{N} \sum_{t=1}^{N} L_1(q)\tilde{G}(q)r(t)\varphi(t) \right\|_{R_N^{-1}} = \left|L_1(q)\tilde{G}(q)\right|^2 \left|R(q)\right|^2 \tag{4.68}$$

combining equation (4.64) and (4.68), the inequality condition in Theorem 4.1 can be simplified as follows:

$$\left[\frac{1}{2\pi} \int_{-\pi}^{\pi} \left|L_1(e^{iw})\right|^2 \left|\tilde{G}(e^{iw})\right|^2 \left|R_N(e^{iw})\right|^2 dw \right]^{\frac{1}{2}}$$

$$\leq (1+\eta)\left|L_1(e^{iw})\right|\left|\tilde{G}(e^{iw})\right|\left|R(e^{iw})\right| + (2+\eta)C_r \sum_{k=M}^{\infty} \left|\rho_k\right| \tag{4.69}$$

where variable q is changed to variable e^{iw}.

From equation (4.69), the identification model $G(\hat{\theta}_N)$ is useful on the condition that the model error $\tilde{G}(q)$ must obey one bound, which is described by inequality condition (4.69). So the problem of model structure validation for closed loop system identification is reformulated to verify if the absolute value of model error satisfies the above inequality. This bound tells us that if one appropriate feedback controller is chosen, the model error will converge to zero, so this bound can guide us to design controllers with two-degrees of freedom.

4.5   Finite sample properties for closed loop identification

From our previously published work about stealth identification for a closed loop system structure, the identification problem for a closed loop system with an unknown controller and nonlinear controller is tackled, and the new prediction error and inverse covariance matrix are found to be independent of the unknown or nonlinear controller. In this chapter, we do not concentrate on the identification algorithm for the closed loop system again but turn to identification accuracy analysis for the closed loop system. As there are two kinds of accuracy analysis, i.e. asymptotic analysis and finite sample analysis, so we consider the identification accuracy analysis for the closed loop system from the point of asymptotic analysis and finite sample analysis respectively. More specifically, after reviewing some existed results about closed loop system identification, the cost function, used to identify the unknown parameter vector in the case of a parameterized plant, is reformulated to its reduced form. The spectral estimation for an unknown plant is obtained as a nonparametric estimation, which is widely used in flutter identification. Moreover, based on our derived reduced cost function, one optimal feedback controller is yielded to be dependent on the identified plant. These reduced cost functions and optimal feedback controllers are beneficial for later practical applications, such as closed loop model validation, optimal input design, etc. Then the asymptotical analysis is applied to derive the variance for the optimal feedback controller, and the derivation of the reduced cost function and the optimal feedback controller are full of some asymptotic results. On the contrary, finite sample properties for the closed loop system are also studied to quantify the difference between the identification criterion and the expected criterion. Furthermore, the VC dimension, coming from machine learning, is used to bound this difference.

4.5.1 Closed loop system identification

Some different forms exist for closed loop system structure in practice, here consider one typical closed loop system with one linear time invariant feedback controller in Fig. 4.2 where in Fig. 4.2, $r(t)$ is the external excitation input, chosen by the designer, $P_0(z)$ is the true or nominal plant, $H_0(z)$ is one noise filter, $C(z)$ is the feedback controller, whether it is unknown or nonlinear, our proposed stealth

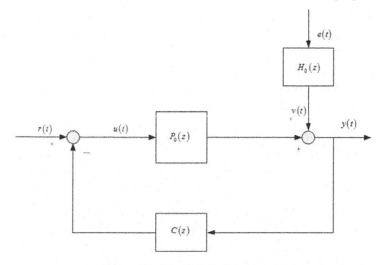

Figure 4.2: One typical closed loop system

identification strategy can be used to design it. $u(t)$ and $y(t)$ are input and output for plant $P_0(z)$. $e(t)$ is a white noise with zero mean white Gaussian and mutually uncorrelated, $v(t)$ is the filtered noise by passing white noise $e(t)$ through that noise filter $H_0(z)$, z^{-1} is the backward shift operator.

Observing Fig. 4.2 again, the process of identifying the plant $P_0(z)$ and noise filter $H_0(z)$ corresponds to closed loop system identification, and we have $v(t) = H_0(z) e(t)$. From Fig. 4.1, we have the following equities easily:

$$\begin{cases} y(t) = P_0(z) u(t) + H_0(z) e(t) \\ u(t) = r(t) - C(z) y(t) \end{cases} \tag{4.70}$$

After simple computations, we have

$$\begin{cases} y(t) = \dfrac{P_0(z)}{1 + P_0(z) C(z)} r(t) + \dfrac{H_0(z)}{1 + P_0(z) C(z)} e(t) \\ u(t) = \dfrac{1}{1 + P_0(z) C(z)} r(t) - \dfrac{C(z) H_0(z)}{1 + P_0(z) C(z)} e(t) \end{cases} \tag{4.71}$$

To simplify notation, one sensitivity function $S_0(z)$ is defined as

$$S_0(z) = \dfrac{1}{1 + P_0(z) C(z)} \tag{4.72}$$

Then equation (4.71) can be simplified as

$$\begin{cases} y(t) = P_0(z) S_0(z) r(t) + H_0(z) S_0(z) e(t) \\ u(t) = S_0(z) r(t) - C(z) H_0(z) S_0(z) e(t) \end{cases} \tag{4.73}$$

Here $(P_0(z), H_0(z))$ are unknown in the closed loop system and are regarded as true values or nominated values. The closed loop identification is to use input-output variables $\{r(t), y(t)\}$ to estimate $(P_0(z), H_0(z))$. To show $(P_0(z), H_0(z))$ are unknown, one unknown parameter vector θ is introduced to parametrize the closed loop system, i.e. $(P(z, \theta), H(z, \theta))$. It means closed loop system is parametrized by one unknown parameter vector θ, then the goal of closed loop identification is to estimate that unknown parameter vector θ.

Substituting the parametrized form $(P(z, \theta), H(z, \theta))$ into equation (4.71), the parametrized relations are given as

$$\begin{cases} y(t) = \dfrac{P(z, \theta)}{1 + P(z, \theta) C(z)} r(t) + \dfrac{H(z, \theta)}{1 + P(z, \theta) C(z)} e(t) \\ u(t) = \dfrac{1}{1 + P(z, \theta) C(z)} r(t) - \dfrac{C(z) H(z, \theta)}{1 + P(z, \theta) C(z)} e(t) \end{cases} \tag{4.74}$$

Combining equation (4.71) and (4.74), $\{u(t), y(t)\}_{t=1}^{N}$ are input-output variables for plant $P_0(z)$, and $\{u(t,\theta), y(t,\theta)\}_{t=1}^{N}$ are parametrized input-output variables for plant $P(z, \theta)$. To identify that unknown parameter vector θ in $(P(z, \theta), H(z, \theta))$, classical prediction error strategy is applied here, where N denotes the number of input-output variables.

$$
\hat{y}(t,\theta) = \frac{1+P(z,\theta)C(z)}{H(z,\theta)} \frac{P(z,\theta)}{1+P(z,\theta)C(z)} r(t) + \left[1 - \frac{1+P(z,\theta)C(z)}{H(z,\theta)} \right] y(t)
$$

$$
= \frac{P(z,\theta)}{H(z,\theta)} r(t) + \frac{H(z,\theta)-1-P(z,\theta)C(z)}{H(z,\theta)} y(t) \tag{4.75}
$$

Define the residual $\varepsilon(t, \theta)$ as that

$$
\varepsilon(t,\theta) = y(t) - \hat{y}(t,\theta) = \frac{1+P(z,\theta)C(z)}{H(z,\theta)} \left[y(t) - \frac{P(z,\theta)}{1+P(z,\theta)C(z)} r(t) \right] \tag{4.76}
$$

Based on the measured input-output variables $Z_N = \{r(t), y(t)\}_{t=1}^{N}$, that unknown parameter vector θ is given as the following numerical optimization problem:

$$
\hat{\theta}_N = \arg\min_{\theta} V_N(t,\theta,Z_N) = \arg\min_{\theta} \frac{1}{N} \sum_{t=1}^{N} \varepsilon^2(t,\theta) \tag{4.77}
$$

where $\hat{\theta}_N$ denoted the parameter estimation based on measured input-output $Z_N = \{r(t), y(t)\}_{t=1}^{N}$. Substituting residual $\varepsilon(t, \theta)$ into numerical optimization (4.77), then lots of optimization methods can be applied to minimize that cost function or identification criterion $\frac{1}{N} \sum_{t=1}^{N} \varepsilon^2(t,\theta)$, for example, least squares method, Newton method, and gradient method, etc.

4.5.2 Asymptotic analysis

As it does not concern the optimal input design and identification algorithm for the closed loop system here, but on finite sample analysis of closed loop system identification. Before starting our new results about finite sample analysis, we firstly give our derived asymptotic analysis result, as some results in this section will be used in further study. From some knowledge about statistical signal processing, assume external excitation input $r(t)$ be quasi-stationary with power spectrum $\phi_r(w)$, where w is one spectrum variable. The variance for white noise $e(t)$ is λ_0, then power spectrum $\phi_v(w)$ for the filter noise $v(t)$ is $\phi_v(w) = H_0(e^{jw}) H_0^*(e^{jw}) \lambda_0$, where notation * means complex conjugate.

4.5.2.1 Asymptotic analysis for cost function

Using the uncorrelated condition between $r(t)$ and $e(t)$, the output spectrum and input spectrum in equation (4.74) are yielded respectively as

$$\begin{cases} \phi_y(w) = |P_0|^2 |S_0|^2 \phi_r(w) + |H_0|^2 |S_0|^2 \lambda_0 \\ \phi_u(w) = |S_0|^2 \phi_r(w) + |C|^2 |H_0|^2 |S_0|^2 \lambda_0 = \phi_u^r(w) + \phi_u^e(w) \end{cases} \qquad (4.78)$$

where in equation (4.78), variance z is neglected, $\phi_u^r(w)$ and $\phi_u^e(w)$ correspond to components, coming from the external excitation input and external noise respectively.

Similarly, the cross spectrums are given as

$$\begin{cases} \phi_{yu}(w) = P_0 |S_0|^2 \phi_r(w) - C|H_0|^2 |S_0|^2 \lambda_0 \\ \phi_{ue}(w) = -CH_0 S_0 \lambda_0 \end{cases} \qquad (4.79)$$

Observing equation (4.78) and (4.79), we see that

$$\lambda_0 \phi_u(w) - |\phi_{ue}(w)|^2 = \lambda_0 |S_0|^2 \phi_r(w) + |C|^2 |H_0|^2 |S_0|^2 \lambda_0^2 - |C|^2 |H_0|^2 |S_0|^2 \lambda_0^2$$

$$= \lambda_0 |S_0|^2 \phi_r(w) = \lambda_0 \phi_u^r(w) \qquad (4.80)$$

The above spectrum relations benefit the spectral estimation of the transfer function $P_0(z)$. From a practical point of view, the spectral estimate is defined as

$$\hat{P}\left(e^{jw}\right) = \frac{\phi_{yu}(w)}{\phi_u(w)} = \frac{P_0 |S_0|^2 \phi_r(w) - C|H_0|^2 |S_0|^2 \lambda_0}{|S_0|^2 \phi_r(w) + |C|^2 |H_0|^2 |S_0|^2 \lambda_0}$$

$$= \frac{P_0\left(e^{jw}\right)\phi_r(w) - C\left(e^{jw}\right)\left|H_0\left(e^{jw}\right)\right|^2 \lambda_0}{\phi_r(w) + \left|C\left(e^{jw}\right)\right|^2 \left|H_0\left(e^{jw}\right)\right|^2 \lambda_0} = \frac{P_0\left(e^{jw}\right)\phi_r(w) - C\left(e^{jw}\right)\phi_v(w)}{\phi_r(w) + \left|C\left(e^{jw}\right)\right|^2 \phi_v(w)}$$

$$(4.81)$$

When the number of measured data approaches to infinity ($N \to \infty$), the above relations hold, i.e.

$$\begin{cases} P_0\left(e^{jw}\right)\phi_r(w) - C\left(e^{jw}\right)\phi_v(w) \to \phi_{yu}(w) \\ \phi_r(w) + \left|C\left(e^{jw}\right)\right|^2 \phi_v(w) \to \phi_u(w) \end{cases} \qquad (4.82)$$

From the asymptotic theory, we see that if N tends to infinity, spectral estimate $\hat{P}\left(e^{jw}\right)$ converges to expression $\dfrac{P_0\left(e^{jw}\right)\phi_r(w) - C\left(e^{jw}\right)\phi_v(w)}{\phi_r(w) + \left|C\left(e^{jw}\right)\right|^2 \phi_v(w)}$, but it is one biased estimate.

As residual $\varepsilon(t, \theta)$ is implicit with unknown parameter vector θ, we rewrite it as that

$$\varepsilon(t,\theta) = H^{-1}(z,\theta)\left[y(t) - P(z,\theta)u(t)\right] \tag{4.83}$$

Further suppose one true parameter vector θ_0 exist and satisfy

$$P(z,\theta_0) = P_0(z); H(z,\theta_0) = H_0(z)$$

After substituting equation (4.71) into (4.83), it holds that

$$y(t) - P(z\quad)u(t) = \frac{P_0(z)}{+P_0(z)C(z)}r(t)\quad\frac{H_0(z)}{+P_0(z)C(z)}e(t)$$
$$-\frac{P(z,\theta)}{1+P_0(z)C(z)}r(t) - \frac{P(z,\theta)C(z)H(z)}{1+P_0(z)C(z)}e()$$
$$\frac{P(z)\,P(z\quad)}{P_0(z)C(z)}r(t)\quad\frac{H(z)\left[\quad P(z\quad)C(z)\right]}{P_0(z)C(z)}e(t) \tag{4.84}$$

Then residual $\varepsilon(t, \theta)$ is rewritten again as

$$\varepsilon(t,\theta) = \frac{P_0(z) - P(z,\theta)}{H(z,\theta)}\frac{1}{1+P_0(z)C(z)}r(t) + \frac{H_0(z)}{H(z,\theta)}\frac{\left[1 - P(z,\theta)C(z)\right]}{1+P_0(z)C(z)}e(t)$$
$$= \frac{P_0(z) - P(z,\theta)}{H(z,\theta)}S_0(z)r(t) + \frac{H_0(z)}{H(z,\theta)}\frac{\left[1 - P(z,\theta)C(z)\right]}{1+P_0(z)C(z)}e(t) \tag{4.85}$$

On the condition of this above new residual, then the cost function $V_N(\theta)$ in equation (4.77) is also rewritten as that

$$V_N(\theta) = \frac{1}{N}\sum_{t=1}^{N}\varepsilon^2(t,\theta) = \int_{-\pi}^{\pi}\frac{\left|P_0(e^{jw}) - P(e^{jw},\theta)\right|^2}{H(e^{jw},\theta)}\phi_u^r(w)\,dw$$

$$+\int_{-\pi}^{\pi}\frac{1}{\left|H(e^{jw},\theta)\right|^2}\left|\frac{1+P(e^{jw},\theta)C(e^{jw})}{1+P_0(e^{jw})C(e^{jw})}\right|^2\phi_v(w)\,dw \tag{4.86}$$

where in equation (4.86), Parseval's relations is used. From equation (4.86), the minimum value is yielded in case of $P(e^{jw}, \theta) \to P_0(e^{jw})$ or $\hat{\theta} \to \theta_0$, but they hold when N tends to infinity. This condition corresponds to the asymptotic analysis.

4.5.2.2 Asymptotic analysis for optimal feedback controller

In above section, the asymptotic theory is applied to rewrite the cost function $V_N(\theta)$, based on our derived cost function (4.86), we seek its optimal feedback controller $C(z)$.

Adding one weight matrix $N(w)$ into the derived cost function $V_N(\theta)$, we denote the cost function $V_N(\theta)$ as $V(\phi_u(w), \phi_{ue}(w))$, i.e.

$$
V\left(\phi_u(w), \phi_{ue}(w)\right)
$$

$$
= \int_{-\pi}^{\pi} \left[\frac{\lambda_0 N_{12}(w) - 2\,\mathrm{Re}\left[N_{12}(w)\phi_{ue}(-w)\right]}{\lambda_0 \phi_u(w) - |\phi_{ue}(w)|^2} + \frac{N_{22}(w)\phi_u(w)}{\lambda_0 \phi_u(w) - |\phi_{ue}(w)|^2} \right] \phi_v(w)\,dw \tag{4.87}
$$

where the weight matrix $N(w)$ is one 2×2 matrix, i.e.

$$
N(w) = \begin{bmatrix} N_{11}(w) & N_{12}(w) \\ N_{21}(w) & N_{22}(w) \end{bmatrix}
$$

Rewrite the weight matrix $N(w)$ as one special form, i.e.

$$
N(w) = S(w) \begin{bmatrix} \left|M\left(e^{jw}\right)\right|^2 & M\left(e^{jw}\right) \\ M\left(e^{-jw}\right) & 1 \end{bmatrix} \tag{4.88}
$$

Substituting the above chosen weight matrix into cost function, it holds that

$$
\frac{\lambda_0 S(w)\left|M\left(e^{jw}\right)\right|^2 - 2\left[S(w)M\left(e^{jw}\right)\phi_{ue}(w)\right]}{\lambda_0 \phi_u(w) - |\phi_{ue}(w)|^2} + \frac{S(w)\phi_u(w)}{\lambda_0 \phi_u(w) - |\phi_{ue}(w)|^2} \phi_v(w)
$$

$$
= S(w) \frac{\lambda_0 \left|M\left(e^{jw}\right)\right|^2 - 2\left[M\left(e^{jw}\right)\phi_{ue}(w)\right] + \phi_u(w)}{\lambda_0 \phi_u(w) - |\phi_{ue}(w)|^2} \phi_v(w)
$$

$$
= \frac{S(w)\phi_v(w)}{\lambda_0} \frac{\lambda_0^2 \left|M\left(e^{jw}\right)\right|^2 - 2\lambda_0 M\left(e^{jw}\right)\phi_{ue}(w) + \phi_{ue}^2(w)}{\lambda_0 \phi_u(w) - |\phi_{ue}(w)|^2} \phi_v(w)
$$

$$
+ \frac{\lambda_0 \phi_u(w) - \phi_{ue}^2(w)}{\lambda_0 \phi_u(w) - |\phi_{ue}(w)|^2} \phi_v(w)
$$

$$
= \frac{S(w)\phi_v(w)}{\lambda_0} \left[1 + \frac{\left|\lambda_0^2 M\left(e^{jw}\right) - \phi_{ue}(w)\right|^2}{\lambda_0 \phi_u(w) - |\phi_{ue}(w)|^2} \right] \tag{4.89}
$$

Then the cost function is that

$$
\min_{\phi_u(w), \phi_{ue}(w)} \int_{-\pi}^{\pi} \frac{S(w)\phi_v(w)}{\lambda_0} \left[1 + \frac{\left|\lambda_0^2 M\left(e^{jw}\right) - \phi_{ue}(w)\right|^2}{\lambda_0 \phi_u(w) - |\phi_{ue}(w)|^2} \right] dw \tag{4.90}
$$

Applying our existed relations in equation (4.90)

$$\begin{cases} \phi_{ue}(w) = -CH_0 S_0 \lambda_0 \\ \phi_u(w) = |S_0|^2 \phi_r(w) + |C|^2 |H_0|^2 |S_0|^2 \lambda_0 = \phi_u^r(w) + \phi_u^e(w) \end{cases}$$

i.e.

$$\frac{\left|\lambda_0^2 M\left(e^{jw}\right) - \phi_{ue}(w)\right|^2}{\lambda_0 \phi_u(w) - |\phi_{ue}(w)|^2} = \frac{\lambda_0^2 M\left(e^{jw}\right) + CH_0 S_0 \lambda_0}{\lambda_0 S_0 \phi_r(w)} \tag{4.91}$$

Substituting equation (4.91) into (4.90), the cost function is reduced to

$$\min_{\phi_u(w), \phi_{ue}(w)} \int_{-\pi}^{\pi} \frac{\lambda_0^2 M\left(e^{jw}\right) + CH_0 S_0 \lambda_0}{\lambda_0 S_0 \phi_r(w)} \, dw \tag{4.92}$$

It is easy to see that the optimal feedback controller $C(z)$ must satisfy that

$$C(z) H_0(z) S_0(z) = -M(z)$$

i.e.

$$\frac{C(z) H_0(z)}{1 + P_0(z) C(z)} = -M(z)$$

$$\Updownarrow$$

$$C(z) H_0(z) = -M(z)\left[1 + P_0(z) C(z)\right]$$

It yields that

$$C(z) = -\frac{M(z)}{P_0(z) M(z) + H_0(z)} \tag{4.93}$$

where in equation (4.93), $M(z)$ is given, and $(P_0(z), H_0(z))$ can be replaced by their parametrized form $(P(z, \theta), H(z, \theta))$, then optimal feedback controller $C(z)$ is chosen as

$$C(z) = -\frac{M(z)}{P(z, \theta) M(z) + H(z, \theta)} \tag{4.94}$$

During the above mathematical derivations, asymptotic theory is used as well, for example, Parseval's relation and some spectral relations.

Observing equation (4.94) again, optimal feedback controller $C(z)$ is dependent of the parametrized form $(P(z, \theta), H(z, \theta))$, so we need to analyze this closed dependence through the variance expression for our derived optimal feedback controller.

Consider two disturbances or perturbs exist in parametrized forms($P(z, \theta)$, $H(z, \theta)$), without loss of generality set $(P(z, \theta), H(z, \theta)) = (P_0(z), H_0(z)) + (\Delta P, \Delta H)$, i.e.

$$\left(P(z,\theta), H(z,\theta) \right) = \left(P_0(z), H_0(z) \right) + \left(\Delta P, \Delta H \right)$$

$$\begin{cases} \Delta P = P(z,\theta) - P_0(z) \\ \Delta H = H(z,\theta) - H_0(z) \end{cases}$$

$$\hspace{10cm} (4.95)$$

Substituting these two perturbs into optimal feedback controller, we have

$$C(z) = \frac{M(z)}{\left[P_0(z) + \Delta P \right] M(z) + \left[H_0(z) + \Delta H \right]}$$

$$= \frac{M(z)}{P_0(z) M(z) + \Delta P M(z) + \left[H_0(z) + \Delta H \right]}$$

$$= M(z) \left[\frac{1}{P_0(z) M(z) + H_0(z)} - \frac{1}{\left[P_0(z) M(z) + H_0(z) \right]^2} \left[\Delta P M(z) + \Delta H \right] \right]$$

$$= \frac{M(z)}{P_0(z) M(z) + H_0(z)} - \frac{M(z)}{\left[P_0(z) M(z) + H_0(z) \right]^2} \left[\Delta P M(z) + \Delta H \right]$$

$$= C_0(z) - \frac{M(z)}{\left[P_0(z) M(z) + H_0(z) \right]^2} \left[\Delta P M(z) + \Delta H \right] \hspace{2cm} (4.96)$$

where for the latter analysis, $C_0(z)$ means the nominate controller.

Define the perturb, existing in optimal feedback controller as

$$\Delta C = C(z) - C_0(z) = -\frac{M(z)}{\left[P_0(z) M(z) + H_0(z) \right]^2} \begin{bmatrix} M(z) & 1 \end{bmatrix} \begin{bmatrix} \Delta P \\ \Delta H \end{bmatrix}$$

$$= -\begin{bmatrix} C_0^2(z) & \dfrac{C_0(z)}{P_0(z) M(z) + H_0(z)} \end{bmatrix} \begin{bmatrix} \Delta P \\ \Delta H \end{bmatrix} \hspace{2cm} (4.97)$$

where the following relations are used:

$$C_0(z) - \frac{M(z)}{\left[P_0(z) M(z) + H_0(z) \right]^2} \left[\Delta P M(z) + \Delta H \right]$$

$$= C_0(z) - \frac{M^2(z)}{\left[P_0(z) M(z) + H_0(z) \right]^2} \Delta P - \frac{M(z)}{\left[P_0(z) M(z) + H_0(z) \right]^2} \Delta H$$

$$= C_0(z) - C_0^2(z) \Delta P - \frac{C_0(z)}{P_0(z) M(z) + H_0(z)} \Delta H$$

Based on the derived perturb ΔC, variance of the optimal feedback controller $C(z)$ is given as

$$P_C = E\left[C(z) - C_0(z)\right]^2 = \left[C_0^2(z) \quad \frac{C_0(z)}{P_0(z)M(z) + H_0(z)}\right]\left[\begin{array}{c} \Delta P \\ \Delta H \end{array}\right]\left[\Delta P \quad \Delta H\right]$$

$$\left[\begin{array}{c} C_0^2(z) \\ C_0(z) \\ \hline P_0(z)M(z) + H_0(z) \end{array}\right]$$

$$= \left[C_0^2(z) \quad \frac{C_0(z)}{P_0(z)M(z) + H_0(z)}\right] Cov_{(P,H)} \left[\begin{array}{c} C_0^2(z) \\ C_0(z) \\ \hline P_0(z)M(z) + H_0(z) \end{array}\right] \quad (4.98)$$

where $Cov_{(P,H)}$ is variance for parametrized forms $(P(z, \theta), H(z, \theta))$.

Substituting our existed result in equation (4.98), then variance P_C is yielded as that

$$P_C = \left[C_0^2(z) \quad \frac{C_0(z)}{P_0(z)M(z) + H_0(z)}\right]\left[\begin{array}{cc} \dfrac{n}{N} \dfrac{|H_0|^2 \lambda_0}{|S_0|\phi_r(w)} & 0 \\ 0 & \dfrac{n}{N}|H_0|^2 \end{array}\right]$$

$$\left[\begin{array}{c} C_0^2(z) \\ C_0(z) \\ \hline P_0(z)M(z) + H_0(z) \end{array}\right] \quad (4.99)$$

where n is the number of parameter vector θ.

Further observing equation (4.97) again, the biased error is that

$$E[\Delta C] = -C_0^2(z) E[\Delta P] + \frac{C_0(z)}{P_0(z)M(z) + H_0(z)} E[\Delta H] \quad (4.100)$$

So if all parametrized forms $(P(z, \theta), H(z, \theta))$ are unbiased estimates, i.e. $E[\Delta P] = E[\Delta H] = 0$, then our derived optimal feedback controller $C(z)$ is also one unbiased estimate, i.e.

$$E\left[C(z)\right] = C_0(z), N \to \infty$$

Generally, the asymptotic theory is applied to analyze that cost function, optimal feedback controller and its corresponding variance.

4.5.3   Finite sample analysis

During the above mathematical derivations, some results hold based on the infinite data, i.e. $N \to \infty$, for example, that asymptotic variance (4.99) for the optimal feedback controller $C(z)$. We regard those results in section 3 as asymptotic analysis, but in practice, that finite data is an ideal case, and the number of measured data is truly finite. Then within this framework of finite sample data, how can we use the sampled identification criterion to replace its corresponding limit or expected criterion? This section is to measure the difference between the sampled identification criterion and the expected criterion. Based on this explicit difference, the number of measured data is obtained to guarantee the difference is not a greater one given a positive scalar. This necessary number of sampled data corresponds to sample complexity in a closed loop structure.

For the sake of completeness, consider that residual $\varepsilon(t, \theta)$ (4.76) again.

$$\varepsilon\left(t,\theta\right) = y\left(t\right) - \hat{y}\left(t,\theta\right) = \frac{1 + P\left(z,\theta\right)C\left(z\right)}{H\left(z,\theta\right)}\left[y\left(t\right) - \frac{P\left(z,\theta\right)}{1 + P\left(z,\theta\right)C\left(z\right)}r\left(t\right)\right]$$

Substituting that optimal feedback controller $C(z)$ (4.94) into above residual, it holds that

$$P\left(z,\theta\right)C\left(z\right) = -\frac{P\left(z,\theta\right)M\left(z\right)}{P\left(z,\theta\right)M\left(z\right) + H\left(z,\theta\right)}$$

$$1 + P\left(z,\theta\right)C\left(z\right) = 1 - \frac{P\left(z,\theta\right)M\left(z\right)}{P\left(z,\theta\right)M\left(z\right) + H\left(z,\theta\right)} = \frac{H\left(z,\theta\right)}{P\left(z,\theta\right)M\left(z\right) + H\left(z,\theta\right)}$$

$$\frac{P\left(z,\theta\right)}{1 + P\left(z,\theta\right)C\left(z\right)} = \frac{P\left(z,\theta\right)\left[P\left(z,\theta\right)M\left(z\right) + H\left(z,\theta\right)\right]}{H\left(z,\theta\right)} \tag{4.101}$$

Combining above equities, we get

$$\varepsilon\left(t,\theta\right) = \frac{1}{P\left(z,\theta\right)M\left(z\right) + H\left(z,\theta\right)}\left[y\left(t\right) - \frac{P\left(z,\theta\right)\left[P\left(z,\theta\right)M\left(z\right) + H\left(z,\theta\right)\right]}{H\left(z,\theta\right)}r\left(t\right)\right]$$

$$= \frac{1}{P\left(z,\theta\right)M\left(z\right) + H\left(z,\theta\right)}y\left(t\right) - \frac{P\left(z,\theta\right)}{H\left(z,\theta\right)}r\left(t\right) \tag{4.102}$$

Observing that identification criterion (4.77) again, as only N terms exist in this identification criterion, we call it the empirical value or sampled value of identification criterion, and denote it as

$$V_N\left(\theta\right) = \frac{1}{N}\sum_{t=1}^{N}\varepsilon^2\left(t,\theta\right) \tag{4.103}$$

To measure this sampled identification criterion, we need one ideal value to compare, i.e. one expected criterion, which is defined as follows:

$$V(\theta) = \frac{1}{N} \sum_{t=1}^{N} E\left[\varepsilon^2(t,\theta)\right] \tag{4.104}$$

where expectation operation $E[\varepsilon^2(t, \theta)]$ is introduced in the above equation (4.104).

To compare the difference between $V_N(\theta)$ and $V(\theta)$, i.e. $\left|V_N(\theta) - V(\theta)\right|$, the square term is needed. From equation (4.70), we have

$$\begin{cases} y^2(t) = P_0^2(z)S_0^2(z)r^2(t) + H_0^2(z)S_0^2(z)e^2(t) + 2P_0(z)H_0(z)S_0^2(z)r(t)e(t) \\ y(t)r(t) = P_0(z)S_0(z)r^2(t) + H_0(z)S_0(z)r(t)e(t) \end{cases} \tag{4.105}$$

Using equation (4.105), we compute the square term $\varepsilon^2(t, \theta)$ as follows:

$$\varepsilon^2(t,\theta) = \left[\frac{1}{P(z,\theta)M(z)+H(z,\theta)}\right]^2 \left[\begin{array}{c} P_0^2(z)S_0^2(z)r^2(t)+H_0^2(z)S_0^2(z)e^2(t) \\ +2P_0(z)H_0(z)S_0^2(z)r(t)e(t) \end{array}\right]$$

$$+\left[\frac{P(z,\theta)}{H(z,\theta)}\right]^2 r^2(t) - \frac{2P(z,\theta)}{\left[P(z,\theta)M(z)+H(z,\theta)\right]H(z,\theta)}$$

$$\left[P_0(z)S_0(z)r^2(t)+H_0(z)S_0(z)r(t)e(t)\right]$$

$$=\left[\begin{array}{c} \dfrac{P_0^2(z)S_0^2(z)}{\left[P(z,\theta)M(z)+H(z,\theta)\right]^2}+\left[\dfrac{P(z,\theta)}{H(z,\theta)}\right]^2 \\ -\dfrac{2P(z,\theta)P_0(z)S_0(z)}{\left[P(z,\theta)M(z)+H(z,\theta)\right]^2 H(z,\theta)} \end{array}\right] r^2(t)$$

$$+\left[\frac{2P_0(z)H_0(z)S_0^2(z)}{\left[P(z,\theta)M(z)+H(z,\theta)\right]^2}-\frac{2P(z,\theta)H_0(z)S_0(z)}{\left[P(z,\theta)M(z)+H(z,\theta)\right]H(z,\theta)}\right]r(t)e(t)$$

$$+\frac{H_0^2(z)S_0^2(z)}{\left[P(z,\theta)M(z)+H(z,\theta)\right]^2}e^2(t) \tag{4.106}$$

Taking the expectation operation $E\varepsilon^2(t, \theta)$ on both sides of equation (4.106), it yields the difference between $\left|V_N(\theta) - V(\theta)\right|$.

$$\frac{1}{N}\sum_{t=1}^{N}\varepsilon^2(t,\theta) - \frac{1}{N}\sum_{t=1}^{N}E\varepsilon^2(t,\theta)$$

$$=\frac{1}{N}\sum_{t=1}^{N}\sum_{k=0}^{\infty}\sum_{l=0}^{\infty}\left[\frac{H_0^2(z)S_0^2(z)}{\left[P(z,\theta)M(z)+H(z,\theta)\right]^2}\right]_{k,l}$$

$$\left(e(t-k)e(t-l)-E\left[e(t-k)e(t-l)\right]\right)$$

$$+\frac{1}{N}\sum_{t=1}^{N}\sum_{k=0}^{\infty}\sum_{l=0}^{\infty}\begin{bmatrix}\dfrac{2P_0(z)H_0(z)S_0^2(z)}{\left[P(z,\theta)M(z)+H(z,\theta)\right]^2}\\[4mm]-\dfrac{2P(z,\theta)H_0(z)S_0(z)}{\left[P(z,\theta)M(z)+H(z,\theta)\right]H(z,\theta)}\end{bmatrix}_{k,l}$$

$$\left(r(t-k)e(t-l)-E\left[r(t-k)e(t-l)\right]\right)$$

$$=\frac{1}{N}\sum_{t=1}^{N}\sum_{k=0}^{\infty}\sum_{l=0}^{\infty}\left[N_1(z)\right]_{k,l,\theta}\left(e(t-k)e(t-l)-\lambda_0\right)$$

$$+\frac{1}{N}\sum_{t=1}^{N}\sum_{k=0}^{\infty}\sum_{l=0}^{\infty}\left[N_2(z)\right]_{k,l,\theta}\left(r(t-k)e(t-l)-E\left[r(t-k)e(t-l)\right]\right) \quad (4.107)$$

where in equation (4.107), we use the following known equities:

$$r(t-k)r(t-l)=E\left(r(t-k)r(t-l)\right)$$

$$E\left(e(t-k)e(t-l)\right)=\begin{cases}\lambda_0 & k=l\\0 & k\neq l\end{cases}$$

$$E\left(r(t-k)e(t-l)\right)=0$$

To give a bound for the difference (4.107), set

$$n_1=\sup_{\theta}\left[N_1(z)\right]_{k,l,\theta}$$

$$n_2=\sup_{\theta}\left[N_2(z)\right]_{k,l,\theta}$$

Then equation (4.107) is simplified as that

$$\frac{1}{N}\sum_{t=1}^{N}\varepsilon^2(t,\theta)-\frac{1}{N}\sum_{t=1}^{N}E\varepsilon^2(t,\theta)\leq\frac{1}{N}\sum_{t=1}^{N}\sum_{k=0}^{\infty}\sum_{l=0}^{\infty}n_1\left(e(t-k)e(t-l)-\lambda_0\right)$$

$$+\frac{1}{N}\sum_{t=1}^{N}\sum_{k=0}^{\infty}\sum_{l=0}^{\infty}n_2\left(r(t-k)e(t-l)-E\left[r(t-k)e(t-l)\right]\right)$$

From the statistical signal processing, the bound for the first subtract is given as

$$\frac{1}{N}\sum_{k=0}^{\infty}\sum_{l=0}^{\infty}n_1\left(e(t-k)e(t-l)-\lambda_0\right)\leq\sum_{k=0}^{\infty}\sum_{l=0}^{\infty}n_1(k+l+1)\varepsilon_0\leq\varepsilon_1 \quad (4.108)$$

where ε_0 and ε_1 are two positive scalars, similarly the similar result holds for the second subtract

$$\frac{1}{N}\sum_{t=1}^{N}\sum_{k=0}^{\infty}\sum_{l=0}^{\infty}n_2\left(r(t-k)e(t-l)-E\left[r(t-k)e(t-l)\right]\right)\le\varepsilon_2 \qquad (4.109)$$

where ε_2 is also one positive scalar.

Observing equation (4.108) and (4.109), and set $\varepsilon=\varepsilon_1+\varepsilon_2$, then we have

$$\left|\frac{1}{N}\sum_{t=1}^{N}\varepsilon^2(t,\theta)-\frac{1}{N}\sum_{t=1}^{N}E\varepsilon^2(t,\theta)\right|\le\varepsilon=\varepsilon_1+\varepsilon_2 \qquad (4.110)$$

Equation (4.110) means after chosen one suitable sample complexity N, the difference $\left|\dfrac{1}{N}\displaystyle\sum_{t=1}^{N}\varepsilon^2(t,\theta)-\dfrac{1}{N}\displaystyle\sum_{t=1}^{N}E\varepsilon^2(t,\theta)\right|$ will at least be one given positive scalar ε. If ε is small enough, then N will be increased, i.e. N is increased with ε reducing.

4.5.4 Conclusion

In this section, we study the asymptotic analysis and finite sample analysis for closed loop system identification, which correspond to infinite data and finite data. Firstly, in the case of infinite data, the cost function and optimal feedback controller are obtained by our mathematical derivations in order to simplify the identification process for a closed loop system, then the variance for the optimal feedback controller is derived too. Secondly, for the finite sample data, the bound for the difference between the sampled identification criterion and the expected criterion is constructed.

4.6 Summary

Model validation is an important process during the system identification theory, whose goal is to testify whether the identified model is appropriated or not. Our studying of model validation is around the points of asymptotic and non-asymptotic analysis, which correspond to the infinite data and finite data. More specifically, the asymptotic confidence region and non-asymptotic confidence region are constructed for the parameter estimation, and finite sample analysis between the true identification cost and its sampled cost is derived to show the number of iterations, while guaranteeing the given identification accuracy.

References

[1] Forssel, U. and Ljung, L. 1999. Closed loop identification revisited, Automatica, 35(7): 1215-1241.

[2]   Ljung, L. 1999. System Identification: Theory for the User. Prentice Hall, Upper Saddle River, New Jersey, USA.

[3]   Pintelon, R. and Schoukens, J. 2001. System Identification: A Frequency Domain Approach. New York: IEEE Press.

[4]   Augero, J.C. 2011. A virtual closed loop method for closed loop identification, Automatica, 47(8): 1626-1637.

[5]   Forssell, U. and Ljung, L. 2000. Some results on optimal experiment design, Automatica, 36(5): 749-756.

[6]   Leskers, M. 2007. Closed loop identification of multivariable process with part of the inputs controlled, International Journal of Control, 80(10): 1552-1561.

[7]   Hjalmarssion, H. 2005. From experiment design to closed loop control, Automatica, 41(3): 393-438.

[8]   Hjalmarssion, H. 2008. Closed loop experiment design for linear time invariant dynamical systems via LMI, Automatica, 44(3): 623-636.

[9]   Bombois, X. 2006. Least costly identification experiment for control, Automatica, 42(10): 1651-1662.

[10] Hildebrand, R. 2003. Identification for control: Optimal input design with respect to a worst case gap cost function, SIAM Journal of Control Optimization, 41(5): 1586-1608.

[11] Gevers, M. 2006. Identification of multi input systems: Variance analysis and input design issues, Automatica, 42(41): 559-572.

[12] Gevers, M. 2009. Identification and information matrix: How to get just sufficiently rich, IEEE Transactions on Automatic control, 54(12): 2828-2840.

[13] Goodin, G.C. 2002. Bias issues in closed loop identification with application to adaptive control, Communications in Information and Systems, 2(4): 349-370.

[14] Welsh, J.S. 2002. Finite sample properties of indirect nonparametric closed loop identification, IEEE Transactions on Automatic control, 47(8): 1277-1291.

[15] Douma, S.G. 2008. Validity of the standard cross correlation test for model structure validation, Automatica, 44(4): 1285-1294.

[16] Mary, X.A. 2015. Modelling and control of MIMO gasifier system during coal quality variations, International Journal of Modelling, Identification and Control, 23(2): 131-139.

[17] Taeib, A. 2015. Tuning optimal PID controller, International Journal of Modelling, Identification and Control, 23(2): 140-147.

[18] Yuzhu, G. 2014. Identification of nonlinear systems with non-persistent excitation using an iterative forward orthogonal least squares regression algorithm, International Journal of Modelling, Identification and Control, 23(1): 1-7.

[19] Tarhouni, M. 2014. Support kernels regression for NARMA system identification, International Journal of Modelling, Identification and Control, 22(2): 139-149.

[20] Csorji, B. 2015. Sign perturbed sums: A new system identification approach for constructing exact non-asymptotic confidence region in linear regression models, IEEE Transactions on Signal Processing, 69(1): 169-181.

[21] Kolumhan, S. and Vajk, I. 2015. Perturb data sets methods for hypothesis testing and structure of corresponding confidence sets, Automatica, 51(1): 326-331.

[22] Kieffer, M. and Walter, E. 2014. Guaranteed characterization of exact non-asymptotic confidence regions as defined by LSCR and SPS, Automatica, 50(2): 507-512.

CHAPTER 5

Data Driven Identification for Nonlinear System

5.1 Introduction

In this chapter, firstly, the nonlinear system is modeled by the one polynomial nonlinear state space model. Two polynomial functions of input or state are added in the common state space model and these two polynomial functions correspond to the nonlinear factors. This polynomial nonlinear state space model can be represented by the special nonlinear closed feedback system. To identify every system matrix in the polynomial nonlinear state space model, firstly, each system matrix is vectorized as an unknown parameter vector and then two parallel distribution algorithms are applied to identify this unknown parameter vector in the unconstrained or constrained conditions respectively. When some state equation equalities are deemed as the constrained conditions, the new optimization variables are the state instants and the unknown parameter vector, consisting of all system matrices. These complex optimization variables are solved by the parallel distribution algorithm and the whole process of the parallel distribution algorithm is given explicitly. Generally, the main contributions consist of two folds: one is to apply that polynomial function to represent the nonlinear factor, existing in the nonlinear state space model, and the other is to propose the parallel distribution algorithm to identify the unknown parameter vector, while analyzing its property. Finally, the simulation example is used to prove the efficiency of this parallel distribution algorithm.

Secondly, we probe the recursive identification of piecewise affine Hammerstein models directly by using input-output data. To explain the identification process of a parametric piecewise affine nonlinear function, we prove that the inverse function corresponding to the given piecewise affine nonlinear function is also an equivalent piecewise affine form. Based on this equivalent property, during our detailed identification process concerning the piecewise affine function and linear dynamical system, three recursive least squares methods are proposed to identify those unknown parameters under the probabilistic description or bounded property on noise. Based on complex mathematical derivation, the inverse function of a given piecewise affine nonlinear function is also an equivalent piecewise affine form. As the least squares method is suited under one condition that the considered noise may be a zero mean random signal, a projection algorithm with a dead zone in the presence of bounded noise can enhance the robustness of the parameter update equation. To the best knowledge of the authors, this is the first attempt at identifying piecewise affine Hammerstein models, which combine a piecewise affine function and a linear dynamical system. In the presence of bounded noise, the modified recursive least squares methods are efficient in identifying two kinds of unknown parameters, so the common set membership method can be replaced by our proposed methods.

5.2   Parallel distributed estimation for polynomial nonlinear state space models

The essence of system identification belongs to the research field of adaptive control theory, it means that the one mathematical model is constructed for our considered plant through the observed input-output data sequence so that the constructed mathematical model is used as the basis for the next adaptive control. The categories of system identification can be divided into two kinds: linear system identification and nonlinear system identification. For linear system identification, as one linear relation exists between the input-output data, this special linear relation can be expanded into the linear regression form, which simplifies the whole identification process. But this linear regression form is an ideal one, nonlinear form exists in our real life systems and phenomena widely, such as nature and engineering. The cause of the commonly used linear form is that the existing theory of linear system identification can be directly applied. Considering nonlinear system identification, the nonlinear system is always linearized in one given range. To apply the identification method and asymptotic analysis to nonlinear system identification, some contributions to nonlinear system identification have been developed in recent years.

Theoretical research on linear system identification has matured, for example, the detailed statistical identification methods and the asymptotic or convergence properties are analyzed from the point of the time domain in [1], where some stochastic probability inequalities, corresponding to the guaranteed interval estimations, are proposed. Furthermore, these various statistical identification methods are the theoretical basis for our new identification strategies. Reference [2] studies the identification methods in the frequency domain and points out how to avoid the aliasing effect in carrying out the Fourier transformation. The recursive identification methods on line are proposed in [3], where the computational complexity and real time algorithms are also given for the linear system. At present, the identification of nonlinear system is still under investigation, as there is no universal identification method that can be applied to any form of nonlinear systems. So far some special nonlinear systems are studied, such as the Wiener system, Hammerstein system, and the linear parameter varying system, etc. The instrumental variable method is studied for the linear parameter varying system [4], and the norm regularization identification is proposed in identifying the switched linear regression model [5].

Reference [6] gives the state space identification for a nonlinear system with dual linear blocks. For two special classical nonlinear systems—the Wiener system and the Hammerstein system, the frequency weighted maximum likelihood identification is given for identifying the Wiener system, and a real-time robust set estimation is given for the Hammerstein system. Based on these two special cases, many researchers start to study other identification strategies and the identifiability of different series combinations [7]. Since white noise is an ideal case, it does not exist in engineering, in addition, deriving statistical properties of noise is often very difficult in practice, as it is usually not possible to measure the external noise directly. The use of the orthonormal basis functions is proposed to be a systematic construction mechanism of stable time domain kernels for impulse response estimation of linear time invariant systems [8], further two proposed weights of the orthonormal basis functions are regarded as decay terms, guaranteeing stability in the associated hypothesis space. A Bayesian nonparametric approach is proposed to estimate multi-input multi-output linear parameter varying models under the general noise model structure [9], where this Bayesian nonparametric approach is based on the estimation of the one step ahead predictor. Because linear parameter varying models are special nonlinear systems, efficient maximum likelihood estimation is considered based on conditional probability density and total probability theorem [10]. In [11], adopting maximum entropy arguments, a new L2 penalty is derived and the Hankel matrix is exploited to achieve the maximum entropy vector kernels for multi-input multi-output system identification. Alternatively, a new kernel based regulation approach that combines ideas from machine learning, statistics, and dynamical systems, has been introduced recently [12], and further extended in [13], where a new sparse multiple kernel based regularization is provided to alleviate the latter computational cost.

Moreover, the prediction error method framework is well suited to identify a large variety of noise and plant models, see [14] for an overview. Since linear parameter varying models cover a large number of processes and noise representations where the Box-Jenkins model is the most general form, the prediction error identification of Box-Jenkins models leads to a nonlinear optimization problem [15], which is solved by a scaled gradient projection method. In the parametric case, the structural dependency is generally characterized by using a pre-specified set of basis functions [16]. The nonparametric methods for the identification of linear parameter varying models offer attractive alternative approaches to capture the underlying dependencies directly from the measured data without any parameterization [17]. A Bayesian nonparametric approach based on the estimation of the one step ahead predictor is proposed to estimate the linear parameter varying models [18], and some predictors associated with the input and output signals are modeled as the stable impulse response models. The problem of controlling the linear parameter varying model is considered in [19], and the control strategy can be applied in a wafer stage. During the system identification process, many identification methods are proposed to identify these unknown parameters, for example, the classical least squares method, instrumental variable method, maximum likelihood estimation method, prediction error method, Bayesian method, etc [20].

It is well known that the dual of the quadratic programming problem is an unconstrained optimization problem [21]. After deriving the prediction of output value by the prediction error method and substituting it with the one considered cost function, we take the derivative of the cost function concerning input value to obtain one optimal input [22]. But the problem of deriving the prediction of output value is dependent on external noise, which is always assumed to be independent and identically distributed white noise. As white noise is an ideal case, it does not exist in engineering, and in addition, deriving statistical properties of noise is often very difficult in practice as it is usually not possible to measure noise directly [23]. Although the model predictive control is a robust type of control in most reported applications, some new and very promising results allow one to think that this control technique will experience greater expansion within this community in the coming years [24]. The problem of applying the interval predictor model to the model predictive control is studied in [25], wherein robust model predictive control, the obtained interval predictor model is used in one min–max optimization problem, and furthermore the max operation is a piecewise affine function form. Generally, to the best of our knowledge, it is urgent to carry out in-depth research on this nonlinear system identification.

As the nonlinear system can be approximated by using the linear combination of some basic functions, for example, the classical Volterra series, the kernel function from the support vector machine theory, orthonormal basis functions, etc. To keep the similarity between linear system identification and nonlinear system identification, some trade-off forms for the nonlinear system are proposed to retain the linear form. It means that these trade-off forms are nonlinear systems, as they show the properties of the linear system, they can be expanded as linear state space models, such as the linear time vary system, linear piecewise system, the linear parameter varying system, etc. In our opinion, although these trade-off forms are similar to the linear system, they characterize some properties of nonlinear systems. The existed theories and methods from linear system identification can be adapted to identify these trade-off forms directly, and these trade-off forms can be suited to express the nonlinear phenomenon in practice. In classical control theory, the transfer function form is used to describe the system or model, but on the contrary in the modern control theory state space form is applied more widely. So in this chapter, the considered model is also the commonly used state space model with two polynomial functions, which correspond to input and state respectively. These two polynomial functions are added to show the effects of nonlinearity. This polynomial nonlinear state space model as first proposed in [24] and the interior point algorithm is applied to identify each matrix in this polynomial nonlinear state space model.

Here, after combining all unknown matrices as one unknown parameter vector, we study the problem of how to identify all matrices in this polynomial nonlinear state space model deeply. Firstly, the feasibility of the polynomial nonlinear state space model is illustrated to show that it can be used to describe many other nonlinear systems, such as the bilinear system, the affine system, and the block structure nonlinear system. More specifically, the nonlinear feedback system and block structure nonlinear system can be transformed to our considered polynomial nonlinear state space model through some algebraic manipulations, and their detailed transformation processes are given. To identify each matrix in this polynomial nonlinear state space model, we extract one unknown parameter vector, coming from each system matrix, and apply two novel parallel distributed identification strategies to estimate this unknown parameter vector with unconstraint and constraint conditions respectively. In particular, when considering the state equation as equality constraints, we need to merge all the instantaneous state values of each state and the unknown parameter vector into one new decision vector. For the problem of identifying this complex decision vector, the whole process of our applied parallel distributed identification strategy and its corresponding updated iteration are also derived.

5.2.1 Polynomial nonlinear state space model

As multivariable system contains multi-input and multi-output, so the state space model from the modern control theory is used to describe the multivariable system. The system is modeled as a block box, and the only prior information about this system is the input-output data sequence, then the goal of system identification is to identify all system matrices from this state space model.

Set the dimensions of the input and output signal as n_u and n_y respectively, and consider the following discrete time nonlinear state space model:

$$\begin{cases} x(t+1) = f\big(x(t), u(t)\big) \\ y(t) = g\big(x(t), u(t)\big) \end{cases} \tag{5.1}$$

where in equation (5.1) $u(t) \in R^{n_u}$ is input signal and $y(t) \in R^{n_y}$ is output signal, $t = [1 \cdots N]$ denotes the discrete time instant, N is the total number of discrete time instant, $x(t) \in R^n$ is system state and n is the model order. Roughly speaking, the first equation in (5.1) is the state equation, it describes the evolutionary process of the state, and the second equation in (5.2) is the observed equation, it means that the output signal is a nonlinear function about the state and input. f and g are two nonlinear functions.

As the first step for system identification is to choose the considered model structure, so these two nonlinear functions f and g in equation (5.1) can be expanded by some basic functions. After choosing one certain of polynomial basis function, the obtained polynomial nonlinear state space model is that

$$\begin{cases} x(t+1) = Ax(t) + Bu(t) + E\xi\big(x(t), u(t)\big) \\ y(t) = Cx(t) + Du(t) + F\eta\big(x(t), u(t)\big) \end{cases}$$
$$(5.2)$$

where in equation (5.2) two nonlinear terms $E\xi$ and $F\eta$ are added in the polynomial nonlinear state space model on the basis of the classical linear state space model. The system matrices of two linear terms $x(t)$ and $u(t)$ are $A \in R^{n \times n}$ and $B \in R^{n \times n_u}$ respectively in the first state equation, similarly the system matrices corresponding to linear terms $x(t)$ and $u(t)$ are $C \in R^{n_y \times n}$ and $D \in R^{n_y \times n_u}$ respectively in the second observed equation. Vectors $\xi(t) \in R^{n_\xi}$ and $\eta(t) \in R^{n_\eta}$ include the nonlinear function forms with their order 2 to d, which are related with $x(t)$ and $u(t)$. Then the system matrices for the nonlinear function terms are $E \in R^{n \times n_\xi}$ and $F \in R^{n_y \times n_\eta}$ respectively. The cause of adding these two nonlinear function terms is to express the practical nonlinear factor. If chosen approximated polynomial basis functions, then these two nonlinear function terms can be expanded as the following combination about $x(t)$ and $u(t)$:

$$x_1^{\alpha_1} x_2^{\alpha_2} \cdots x_n^{\alpha_n} u_1^{\beta_1} u_2^{\beta_2} \cdots u_{n_u}^{\beta_{n_u}}$$
$$(5.3)$$

where $\alpha_1 \cdots \alpha_n, \beta_1 \cdots \beta_{n_u}$ denote the orders of their corresponding state and input, and the nonlinear degree of freedom d satisfies that

$$2 \le \sum_{i=1}^{n} \alpha_i + \sum_{j=1}^{n_u} \beta_j \le d$$

From equation (5.2), we see that if polynomial (5.3) is added to those n state equations and n_y observed equations, then one considered polynomial nonlinear state space model (5.2) is obtained. The advantage of this model is that it can describe numerous nonlinear systems, such as the bilinear system, the affine system, and the block-structured nonlinear system. However, on the contrary, the disadvantage of this model is that in using a full-parameterized system, if the system has a high degree d of nonlinearity, the number of parameter vectors will be greatly increased, then the curse of dimensionality will be caused. So to alleviate this curse of dimensionality, we can apply model reduction or similar transformation to this polynomial nonlinear state space model (5.2) to obtain its minimal state space realization before the identification process.

The main goal of identifying this polynomial nonlinear state space model is to identify all system matrices by using the input-output data sequence. All system matrices are formulated as follows:

$$\big(A, B, E, C, D, F\big)$$

Vectorize these six system matrices and concentrate on one unknown parameter vector $\theta \in R^{n_\theta}$, which contains all system parameters.

$$\theta^T = vec\left[A, B, E, C, D, F\right] \tag{5.4}$$

Our goal here is to apply the parallel distributed identification strategy to solve one cost function, whose decision variable is the above constructed unknown parameter vector $\theta \in R^{n_\theta}$.

5.2.2   Transformation of nonlinear feedback system

There are detailed transformation processes between three common nonlinear systems and our considered polynomial nonlinear state space model [25]. To verify that the polynomial nonlinear state space model can describe many other nonlinear systems, this section gives the transformation process between nonlinear feedback system and polynomial state space model to illustrate the equivalence between them.

Consider the following structure of nonlinear feedback system as Fig. 5.1.

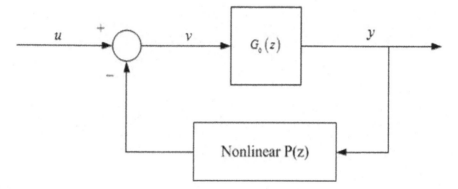

Figure 5.1: The structure of nonlinear feedback system

Where in Fig. 5.1, $G_0(z)$ is the system plant, and feedback controller $P(z)$ is a nonlinear form, the structure of that linear plant $G_0(z)$ can be described as follows:

$$\begin{cases} x_0\left(t+1\right) = A_0 x_0\left(t\right) + B_0 v\left(t\right) \\ y\left(t\right) = C_0 x_0\left(t\right) + D_0 v\left(t\right) \end{cases} \tag{5.5}$$

where the input signal corresponding to linear plant $G_0(z)$ is that.

$$v\left(t\right) = u\left(t\right) - P\left(y\left(t\right)\right) = u\left(t\right) - \sum_{i=1}^{r} P_i y^i\left(t\right) \tag{5.6}$$

Substituting equation (5.6) into (5.5), it holds that

$$\begin{cases} x_0(t+1) = A_0 x_0(t) + B_0 \left(u(t) - \sum_{i=1}^{r} P_i y^i(t) \right) \\ y(t) = C_0 x_0(t) + D_0 \left(u(t) - \sum_{i=1}^{r} P_i y^i(t) \right) \end{cases} \tag{5.7}$$

For the sake of brevity, assume $D_0 = 0$, then equation (5.7) is simplified as that

$$\begin{cases} x_0(t+1) = A_0 x_0(t) + B_0 \left(u(t) - \sum_{i=1}^{r} P_i y^i(t) \right) \\ y(t) = C_0 x_0(t) \end{cases} \tag{5.8}$$

Define the following matrices as that

$$A = A_0 - P_1 B_0 C_0, B = B_0, C = C_0, D = 0, F = 0$$

$$E = -B_0 \begin{bmatrix} P_2 C_0^2 & \cdots & P_r C_0^r \end{bmatrix}^T,$$

$$\xi(t) = \begin{bmatrix} x_0^2(t) & \cdots & x_0^r(t) \end{bmatrix}^T, \eta(t) = 0$$

Then using the above defined matrices, equation (8) can be transformed as one polynomial nonlinear state space model (5.2). When $D_0 \neq 0$, the transformation process is similar to the above.

5.2.3 Transformation of block structure nonlinear system

Consider one block structure nonlinear system with four linear dynamic block structures $G_{11}, G_{12}, G_{21}, G_{22}$ and one static nonlinear system as Fig. 5.2.

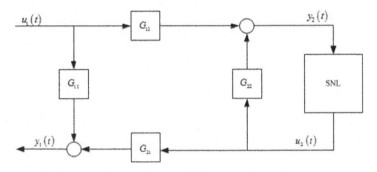

Figure 5.2: The structure of one block structure nonlinear system

Where in Fig. 5.2 $G_{11}(s)$, $G_{12}(s)$, $G_{21}(s)$, $G_{22}(s)$ are four linear time invariant frequent response functions with single input and single output, s is a time delay operation. $u_1(t)$ and $y_1(t)$ are the input-output data sequence at time instant t for the whole block structure. $u_2(t)$ and $y_2(t)$ are the partial input-output data sequence at time instant t for the unknown nonlinear block structure. SNL means the static nonlinear block and it can be expanded as polynomial form.

Observing Fig. 5.2 again, we see that this block structure nonlinear system can be used to describe many kinds of nonlinear block system. For example, when $G_{11} = G_{22} = 0$ and $G_{12} = 1$, it is the Wiener system, and if $G_{11} = G_{22} = 0$, then it corresponds to the Wiener–Hammerstein system. When $G_{21} = 1$ and $G_{11} = G_{22} = 0$, it is the Hammerstein system. Also the above mentioned nonlinear feedback system is a special case of this block structure nonlinear system in case of $G_{22} \neq 0$. The existence of that static nonlinearity is similar to the nonlinear damping structure in mechanical systems.

From Fig. 5.2, in case of no interior signal and partial model, $u_1(t)$ and $u_2(t)$ can be regarded as the input signal, further $y_1(t)$ and $y_2(t)$ the observed signal. Then we obtain the following transfer function as that

$$\begin{cases} y_2(t) = G_{12}(z)u_1(t) + G_{22}(z)u_2(t) \\ y_1(t) = G_{11}(z)u_1(t) + G_{21}(z)u_2(t) \end{cases} \tag{5.9}$$

The two nonlinear least squares methods can be applied to identify those four unknown frequent response functions in equation (5.9), here we only use the state space model to replace equation (5.9).

$$\begin{cases} x(t+1) = A_1 x(t) + \begin{pmatrix} B_1 & B_2 \end{pmatrix} \begin{pmatrix} u_1(t) \\ u_2(t) \end{pmatrix} \\ \begin{pmatrix} y_1(t) \\ y_2(t) \end{pmatrix} = \begin{pmatrix} C_1 \\ C_2 \end{pmatrix} x(t) + \begin{pmatrix} D_1 & D_2 \\ D_3 & D_4 \end{pmatrix} \begin{pmatrix} u_1(t) \\ u_2(t) \end{pmatrix} \end{cases} \tag{5.10}$$

where that static nonlinear part is defined as the following polynomial form:

$$u_2(t) = \sum_{p=1}^{d} a_p y_2^p(t) \tag{5.11}$$

where in equation (5.10) $u_1(t) \in R^{n_{u1}}$ is input signal, $y_1(t) \in R^{n_{y1}}$ is observed signal, $u_2(t) \in R^{n_{u2}}$ is the unknown nonlinear output, $y_2(t) \in R^{n_{y2}}$ is the unknown nonlinear input, $t = [1 \cdots N]$ denotes the discrete time instant, N is the total number of discrete time instant, $x(t) \in R^n$ is system state and n is the model order. The system matrices in state equation are formulated as that

$$A_1 \in R^{n \times n}, B_1 \in R^{n \times n_{u1}}, B_2 \in R^{n \times n_{u2}}$$

Similarly, the system matrices in the observed equation are formulated as that

$$C_1 \in R^{n_{y1} \times n}, C_2 \in R^{n_{y2} \times n}, D_1 \in R^{n_{y1} \times n_{u1}}, D_2 \in R^{n_{y1} \times n_{u2}}, D_3 \in R^{n_{y2} \times n_{u1}}, D_4 \in R^{n_{y2} \times n_{u2}}$$

The merit of transforming the transfer function (5.9) into the state space model (5.10) is that the relations between each variable can be shown explicitly and the equivalence between them holds. Due to some knowledge from the modern control theory, the observed equation (5.10) can be rewritten as

$$
\begin{pmatrix} y_1(t) \\ y_2(t) \end{pmatrix} = \left(\begin{pmatrix} C_1 \\ C_2 \end{pmatrix} (sI - A_1)^{-1} \begin{pmatrix} B_1 & B_2 \end{pmatrix} + \begin{pmatrix} D_1 & D_2 \\ D_3 & D_4 \end{pmatrix} \right) \begin{pmatrix} u_1(t) \\ u_2(t) \end{pmatrix}
\tag{5.12}
$$

Formulating the above equation, we have

$$
\begin{pmatrix} y_1(t) \\ y_2(t) \end{pmatrix} = \begin{pmatrix} C_1(sI - A_1)^{-1} B_1 + D_1 & C_1(sI - A_1)^{-1} B_2 + D_2 \\ C_2(sI - A_1)^{-1} B_1 + D_3 & C_2(sI - A_1)^{-1} B_2 + D_4 \end{pmatrix} \begin{pmatrix} u_1(t) \\ u_2(t) \end{pmatrix}
\tag{5.13}
$$

Comparing equation (5.12) and (5.13), the relations between four frequent response functions and their system matrices are given as follows:

$$
\begin{cases}
G_{11}(s) = C_1(sI - A_1)^{-1} B_1 + D_1 \\
G_{12}(s) = C_1(sI - A_1)^{-1} B_2 + D_2 \\
G_{21}(s) = C_2(sI - A_1)^{-1} B_1 + D_3 \\
G_{22}(s) = C_2(sI - A_1)^{-1} B_2 + D_4
\end{cases}
\tag{5.14}
$$

From equation (5.14), we see that if all system matrices in equation (5.2) can be identified, then the four frequent response functions can be obtained easily through simple but tedious calculation. As the static nonlinear part is not shown explicitly in equation (5.10), we rewrite it as one condensed form.

$$
\begin{aligned}
x(t+1) &= A_1 x(t) + B_1 u_1(t) + B_2 u_2(t) \\
y_1(t) &= C_1 x(t) + D_1 u_1(t) + D_2 u_2(t) \\
y_2(t) &= C_2 x(t) + D_3 u_1(t) \\
u_2(t) &= \sum_{p=1}^{d} a_p y_2^p(t) = \sum_{p=1}^{d} a_p (C_2 x(t) + D_3 u_1(t))^p
\end{aligned}
\tag{5.15}
$$

For convenience, assume $D_4 = 0$ in the above condensed form. If $D_4 \neq 0$, the next derivation is similar. As in the block structure nonlinear system, the observed data sequences are $\{u_1(t), y_1(t)\}_{t=1}^{N}$, $u_2(t)$ and $y_2(t)$ are unknown variables. To eliminate these unknown variables $u_2(t)$ and $y_2(t)$, we formulate equation (5.15) again as

$$
\begin{cases}
x(t+1) = A_1 x(t) + B_1 u_1(t) + B_2 \sum_{p=1}^{d} a_p (C_2 x(t) + D_3 u_1(t))^p \\
y_1(t) = C_1 x(t) + D_1 u_1(t) + D_2 \sum_{p=1}^{d} a_p (C_2 x(t) + D_3 u_1(t))^p
\end{cases}
\tag{5.16}
$$

where in equation (5.16), define the following matrices as

$$A = A_1 + B_2 a_1 C_2, B = B_1 + B_2 a_1 D_3$$
$$C = C_1 + D_2 a_1 C_2, D = D_1 + D_2 a_1 D_3$$

Formulating equation (5.16) again, then we have

$$\begin{cases} x(t+1) = Ax(t) + Bu_1(t) + E\xi(x(t), u_1(t)) \\ y_1(t) = Cx(t) + Du_1(t) + F\xi(x(t), u_1(t)) \end{cases} \qquad (5.17)$$

Then those two last nonlinear terms $E\xi$ and $F\xi$ are given as

$$E\xi(x(t), u_1(t)) = B_2 (a_2 \quad \cdots \quad a_d) \times \begin{pmatrix} (C_2 x(t) + D_3 u_1(t))^2 \\ \vdots \\ (C_2 x(t) + D_3 u_1(t))^P \end{pmatrix}$$

$$F\xi(x(t), u_1(t)) = D_2 (a_2 \quad \cdots \quad a_d) \times \begin{pmatrix} (C_2 x(t) + D_3 u_1(t))^2 \\ \vdots \\ (C_2 x(t) + D_3 u_1(t))^P \end{pmatrix} \qquad (5.18)$$

From equation (5.17) and the above detailed derivations, the block structure nonlinear system can be also transformed into our considered polynomial state space model.

Vectorized these five system matrices and concentrate on one unknown parameter vector $\theta \in R^{n_\theta}$, which contains all system parameters.

$$\theta^T = vec[A, B, E, C, D, F] \qquad (5.19)$$

Then the identification of this block structure nonlinear system is similar to our original goal in section 5.2.2

5.2.4　Parallel distributed identification for polynomial nonlinear state space models

As our goal is to apply the parallel distributed identification strategy to solve one cost function, whose decision variable is the above constructed unknown parameter vector $\theta \in R^{n_\theta}$, we collect the observed data sequence for the whole nonlinear system.

$$y_m = \begin{bmatrix} y_m(1) & y_m(2) & \cdots & y_m(N) \end{bmatrix}^T$$

and its parameterized model output is that

$$y(\theta) = \begin{bmatrix} y(1,\theta) & y(2,\theta) & \cdots & y(N,\theta) \end{bmatrix}^T$$

The unknown parameter vector θ can be estimated by solving one least squares cost function $V(\theta)$.

$$\hat{\theta} = \arg\min_{\theta} V(\theta) = \sum_{t=1}^{N} \varepsilon^T(t,\theta)\varepsilon(t,\theta) = \varepsilon^T(\theta)\varepsilon(\theta)$$
$$\varepsilon(t,\theta) = y(t,\theta) - y_m(t),$$
$$\varepsilon(\theta) = y(\theta) - y_m \qquad (5.20)$$

where in equation (5.20) $\varepsilon(t, \theta) \in R^n$ is the model error, some iterative forms can be applied to solve the unknown parameter vector θ, such as gradient method, Newton method, etc. The common property of these iterations is formulated as

$$\theta^{(i+1)} = \theta^{(i)} + \Delta\theta^{(i)} \qquad (5.21)$$

where in equation (5.21) $\theta^{(i+1)}$ is the $i+1$th iteration, $\theta^{(i)}$ is the ith iteration, and $\Delta\theta^{(i)}$ is one correct term. Further, to avoid the computation of the Hessian matrix, we always use the gradient matrix to express $J^{(i)}$.

$$\begin{cases} \left(J^{(i)T} J^{(i)} \right) \Delta\theta^{(i)} = -J^{(i)T} \varepsilon^{(i)} \\ J^{(i)} = \dfrac{\partial\varepsilon}{\partial\theta}\big|_{\theta^{(i)}} \end{cases} \qquad (5.22)$$

Then from equation (5.22), the correct term $\Delta\theta^{(i)}$ is chosen as that

$$\Delta\theta^{(i)} = -\left(J^{(i)T} J^{(i)} \right)^{-1} J^{(i)T} \varepsilon^{(i)} \qquad (5.23)$$

Since the matrix inverse operation occurs in the calculation process of (5.23), in order to ensure that the inverse operation holds, the product of the gradient matrix $(J^{(i)T} J^{(i)})$ is required to be one nonsingular matrix. If it does not hold, then one other constant matrix can be added to satisfy this requirement.

From equation (5.23), the Hessian matrix of the model error can be avoided, but its gradient matrix is needed, so here we give the detailed derivations about the computation of the gradient matrix.

From the definition of the model error in equation (5.9), we have

$$\frac{\partial\varepsilon}{\partial\theta} = \frac{\partial y}{\partial\theta}$$

as a straight forward calculation shows that

$$J_A(t) = \frac{\partial y(t,\theta)}{\partial A} = \left(C + F\frac{\partial \eta(t)}{\partial x(t)}\right)\frac{\partial x(t)}{\partial A}; J_B(t) = \frac{\partial y(t,\theta)}{\partial B} = \left(C + F\frac{\partial \eta(t)}{\partial x(t)}\right)\frac{\partial x(t)}{\partial B}$$

$$J_C(t) = \frac{\partial y(t,\theta)}{\partial C} = x(t); J_D(t) = \frac{\partial y(t,\theta)}{\partial D} = u(t)$$

$$J_E(t) = \frac{\partial y(t,\theta)}{\partial E} = \left(C + F\frac{\partial \eta(t)}{\partial x(t)}\right)\frac{\partial x(t)}{\partial E}; J_F(t) = \frac{\partial y(t,\theta)}{\partial F} = \eta(t) \tag{5.24}$$

Taking the partial derivative with respect to $x(t)$ of nonlinear function $\eta(t)$, it holds that

$$\frac{\partial \eta(t)}{\partial x(t)} = \left[\frac{\partial \eta(t)}{\partial x_1(t)} \quad \cdots \quad \frac{\partial \eta(t)}{\partial x_n(t)}\right]^T$$

Making use of the property of the state equation from polynomial nonlinear system, we have that

$$\frac{\partial x(t+1)}{\partial A} = x(t) + \left(A + E\frac{\partial \xi(t)}{\partial x(t)}\right)\frac{\partial x(t)}{\partial A}; \frac{\partial x(t+1)}{\partial B} = u(t) + \left(A + E\frac{\partial \xi(t)}{\partial x(t)}\right)\frac{\partial x(t)}{\partial B}$$

$$\frac{\partial x(t+1)}{\partial E} = \xi(t) + \left(A + E\frac{\partial \xi(t)}{\partial x(t)}\right)\frac{\partial x(t)}{\partial E} \tag{5.25}$$

Equation (5.25) is one recursive form for each gradient computation.

Taking the partial derivative with respect to $x(t)$ of nonlinear function $\xi(t)$, it holds that

$$\frac{\partial \xi(t)}{\partial x(t)} = \left[\frac{\partial \xi(t)}{\partial x_1(t)} \quad \cdots \quad \frac{\partial \xi(t)}{\partial x_n(t)}\right]^T$$

After combining equation (5.24) and (5.25), the gradient matrix of the correct term $\Delta\theta^{(i)}$ is obtained.

In this section, the parallel distributed identification strategy is applied to identify the unknown parameter vector θ, so that the computational complexity of solving equation (5.23) is decreased.

5.2.4.1 Unconstrained parallel distributed identification strategy

To apply the parallel distributed identification strategy into that minimization problem (5.20), firstly we rewrite the minimization problem (5.20) as that

$$\hat{\theta} = \arg\min_{\theta} \sum_{i=1}^{N} f_i(\theta)$$

$$f_i(\theta) = \varepsilon^T(i,\theta)\varepsilon(i,\theta) \tag{5.26}$$

Restricting the selection of unknown parameter vectors θ into a series of bounded polyhedron subsets, i.e. $\theta \in P_i$, $i = 1 \cdots n_\theta$, then $\theta \in P_i$, $i = 1 \cdots n_\theta$ and formulating minimization problem (5.26) into its corresponding equivalent separable problem.

$$\min_{\theta_i} \sum_{i=1}^{N} f_i(\theta_i)$$

$$\text{subject to } \theta_i = \theta \quad i = 1 \cdots n_\theta; \theta_i \in P_i, i = 1 \cdots n_\theta \qquad (5.27)$$

where in equation (5.27), $\theta_i \in P_i$, $i = 1 \cdots n_\theta$ is additional virtual variables, applying recursive relation of Lagrange multiplier method [3] on equation (5.16) to get

$$p_i(t+1) = p_i(t) + c(t)\big(\theta(t+1) - \theta_i(t+1)\big) \qquad (5.28)$$

where $c(t)$ is one positive sequence, and it must be satisfy the following condition:

$$\liminf_{t \to \infty} c(t) > 0$$

So $\theta_i(t + 1)$ and $\theta(t + 1)$ can be obtained from the following optimization problem:

$$\min \sum_{i=1}^{N} \left\{ f_i(\theta_i) + p_i(t)(\theta - \theta_i) + \frac{c(t)}{2}\|\theta - \theta_i\|_2^2 \right\}$$

$$\text{subject to } \theta_i \in P_i, i = 1 \cdots n_\theta \qquad (5.29)$$

Applying the iterative relation to minimize the extended Lagrangian function for optimization variable θ.

$$\theta = \frac{\sum\limits_{i=1}^{N} \theta_i}{N} - \frac{\sum\limits_{i=1}^{N} p_i(t)}{Nc(t)} \qquad (5.30)$$

again by minimizing that extended Lagrangian function with respect to θ_i, the iterative relation is

$$\theta_i = \arg\min_{\theta_i \in P_i} \left\{ f_i(\theta_i) - P_i(t)\theta_i + \frac{c(t)}{2}\|\theta - \theta_i\|_2^2 \right\} \qquad (5.31)$$

Repeating the iterative operation on equation (5.31) until it converges to the minimum of the extended Lagrangian function. The above method can also perform parallel distributed processing to a large extent, because the minimization process in (5.31) can be parallelized. In the information flow system, the averaging step of (5.30) can be used for updating θ, and this update is implemented by a single node acquisition algorithm based on some distributed processors.

5.2.4.2 *Constrained parallel distributed identification strategy*

Substituting equation (5.1) into the minimization problem (5.26), some equity constrained conditions are obtained

$$x(t+1) = f\big(x(t),u(t)\big)$$

$$\Downarrow$$

$$\begin{cases} x(1) = f\big(x(0),u(0)\big) \\ x(2) = f\big(x(1),u(1)\big) \\ \vdots \\ x(0) = x(N) = f\big(x(N-1),u(N-1)\big) \end{cases} \tag{5.32}$$

where in equation (5.32) state is deemed as one periodic variable with its period N, but in the system identification theory, only input $u(t)$ and output $y(t)$ are known, and state $x(t)$ at each time instant is unknown. So we merge N state values and that unknown parameter vector as one new unknown parameter vector.

$$v = \begin{bmatrix} x(0) & \cdots & x(N-1) & \theta \end{bmatrix}^T \tag{5.33}$$

Combining the cost function (5.26) and the constrained condition (5.2), one new constrained optimization problem is constructed.

$$\hat{v} = \arg\min_{v} V(v) = \sum_{i=1}^{N} f_i(v)$$

$$\text{subject to } F(v) = 0 \tag{5.34}$$

where in equation (5.34) the constrained condition is that

$$F(v) = \begin{bmatrix} f\big(x(0),u(0)\big) - x(1) \\ f\big(x(1),u(1)\big) - x(2) \\ \vdots \\ f\big(x(N-1),u(N-1)\big) - x(0) \end{bmatrix} = 0$$

Expanding the first equity in the constrained condition as a Taylor series

$$f\big(x(0),u(0)\big) - x(1) = f\big(u(0)\big) + f_x\big(u(0)\big)x(0) + f_{\theta_0}\big(u(0)\big)(\theta - \theta_0) - x(1) = 0 \tag{5.35}$$

where f_x and f_θ are the partial derivatives of nonlinear function f with respect to x and θ, θ_0 is one initial value at the original time instant. We rewrite equation (5.35) as

$$\begin{bmatrix} f_x\big(u(0)\big) & f_{\theta_0}\big(u(0)\big) \end{bmatrix} \begin{bmatrix} x(0) \\ \theta \end{bmatrix} - x(1) = -f\big(u(0)\big) + f_{\theta_0}\big(u(0)\big)\theta_0 \tag{5.36}$$

Formulating the above equation, and we have

$$\left[f_x\left(u(0)\right) \quad -1 \quad 0 \quad \cdots \quad 0 \quad f_{\theta_0}\left(u(0)\right) \right] \times \begin{bmatrix} x(0) \\ x(1) \\ \vdots \\ x(N-1) \\ \theta \end{bmatrix} = -f\left(u(0)\right) + f_{\theta_0}\left(u(0)\right)\theta_0 \tag{5.37}$$

Similarly, the second equity in the constrained condition can also be rewritten as

$$\left[0 \quad f_x\left(u(1)\right) \quad -1 \quad 0 \quad \cdots \quad 0 \quad f_{\theta_0}\left(u(1)\right) \right] \times \begin{bmatrix} x(0) \\ x(1) \\ \vdots \\ x(N-1) \\ \theta \end{bmatrix} = -f\left(u(1)\right) + f_{\theta_0}\left(u(1)\right)\theta_0 \tag{5.38}$$

According to the above formulas, all N Taylor series expansions from the constrained condition are merged into

$$\underbrace{\begin{bmatrix} f_x(0) & -1 & 0 & \cdots & 0 & f_{\theta_0}(0) \\ 0 & f_x(1) & -1 & \cdots & 0 & f_{\theta_0}(1) \\ & \vdots & & & \vdots & \\ -1 & 0 & 0 & \cdots & f_x(N-1) & f_{\theta_0}(N-1) \end{bmatrix}}_{e} \times \underbrace{\begin{bmatrix} x(0) \\ x(1) \\ \vdots \\ x(N-1) \\ \theta \end{bmatrix}}_{v}$$

$$= \underbrace{\begin{bmatrix} -f(0) + f_{\theta_0}(0)\theta_0 \\ -f(1) + f_{\theta_0}(1)\theta_0 \\ \vdots \\ -f(N-1) + f_{\theta_0}(N-1)\theta_0 \end{bmatrix}}_{s} \tag{5.39}$$

For convenience, equation (5.39) is simplified as

$$ev = s \text{ or } e_j v = s_j, \ j = 0,1 \cdots N-1 \tag{5.40}$$

Based on above equation (5.40), the original constrained optimization problem can be simplified as

$$\hat{v} = \arg\min_{v} V(v) = \sum_{i=1}^{N} f_i(v)$$

$$\text{subject to } e_j v = s_j, \ j = 0,1 \cdots N-1; \ v \in P \tag{5.41}$$

Then the parallel distributed strategy is applied to solve the constrained optimization problem. For clarity of presentation, the choice of index j is changed as $j = 1 \cdots N$.

$$\hat{v} = \arg\min_{v} V(v) = \sum_{i=1}^{N} f_i(v)$$

subject to $\quad e_j v = s_j, j = 1 \cdots N; v_i \in P_i, i = 1 \cdots m \qquad (5.42)$

Set e_{ji} as the sub-vector in vector e_j, corresponding to v_i, then for all arbitrary j, set $I(j)$ as one index set in the jth equity, with respect to sub-vector v_i, then we have that

$$I(j) = \{i \backslash e_{ji} \neq 0\}, j = 1 \cdots N$$

Adding one new variable z_{ji}, $i \in I(j)$, then minimization problem (5.42) is changed into

$$\hat{v} = \arg\min_{v} V(v) = \sum_{i=1}^{N} f_i(v_i)$$

subject to $\quad e_j v = z_{ji}, j = 1 \cdots N, i \in I(j); v_i \in P_i, i = 1 \cdots m;$

$$\sum_{i \in I(j)} z_{ji} = s_j, j = 1 \cdots N \qquad (5.43)$$

For each $j = 1, 2 \cdots N$, consider the Lagrange multiplier p_{ji} for equity constraint $e_{ji} v_i = z_{ji}$, $i \in I(j)$, then the recursive relation for this Lagrange multiplier p_{ji} is

$$p_{ji}(t+1) = p_{ji}(t) + c(t)\left(e_{ji} v_i(t+1) - z_{ji}(t+1)\right),$$
$$j = 1, 2 \cdots N, i \in I(j) \qquad (5.44)$$

Further $v_i(t + 1)$ and $z_{ji}(t + 1)$ must minimize the extended Lagrangian function, i.e.

$$\sum_{i=1}^{N} f_i(v_i) + \sum_{j=1}^{N}\sum_{i \in I(j)} p_{ji}(t)\left(e_{ji} v_i - z_{ji}\right) + \frac{c(t)}{2}\sum_{j=1}^{N}\sum_{i \in I(j)} \left(e_{ji} v_i - z_{ji}\right)^2$$

subject to $\quad \sum_{i \in I(j)} z_{ji} = s_j, j = 1, 2 \cdots N, v_i \in P_i, i = 1, 2 \cdots m$

The iterative formulas for vector v_i and z_{ji} are as follows:

$$v_i = \arg\min_{\varsigma_i \in P_i}\left\{ f_i(\varsigma_i) + \sum_{\{i \backslash i \in I(j)\}}\left\{ p_{ji}(t) e_{ji}\varsigma_i + \frac{c(t)}{2}\left(e_{ji}\varsigma_i - z_{ji}\right)\right\}\right\}, \forall i = 1, 2 \cdots m$$

$$\{z_{ji} \backslash i \in I(j)\} = \arg\min_{\left\{\varsigma_{ji} \backslash i \in I(j), \sum_{i \in I(j)}\varsigma_{ji} = s_j\right\}}$$

$$\left\{ -\sum_{i \in I(j)} p_{ji}(t)\zeta_{ji} + \frac{c(t)}{2} \sum_{i \in I(j)} \left(e_{ji}\varsigma_i - \zeta_{ji} \right) \right\} \forall j = 1, 2 \cdots N \qquad (5.45)$$

Separable quadratic cost and equity constraint are contained in the process of minimizing operations with respect to variable $\left\{ \zeta_{ji} \setminus i \in I(j) \right\}$. And the minimization value is taken as

$$\zeta_{ji} = e_{ji}v_i + \frac{p_{ji}(t) - \lambda_j}{c(t)}, j = 1, 2 \cdots N, i \in I(j) \qquad (5.46)$$

where λ_j is another Lagrange multiplier, and it may satisfy the following equity:

$$\sum_{i \in I(j)} \zeta_{ji} = s_j$$

The above satisfactory is equivalent to

$$\lambda_j = \frac{1}{m_j} \sum_{i \in I(j)} p_{ji}(t) + \frac{c(t)}{m_j} \sum_{i \in I(j)} \left(e_{ji}v_i - s_{ji} \right); \forall j = 1, 2 \cdots N \qquad (5.47)$$

m_j is the total number of the index set $I(j)$, i.e.

$$m_j = \left| I(j) \right|$$

From equation (5.46), we see that the optimal value $z_{ji}(t + 1)$ is

$$z_{ji}(t+1) = e_{ji}v_i(t+1) + \frac{p_{ji}(t) - \lambda_j(t+1)}{c(t)}, j = 1, 2 \cdots N, i \in I(j)$$

Through comparing with equation (5.44), we have that

$$p_{ji}(t+1) = \lambda_j(t+1), j = 1, 2 \cdots N, i \in I(j)$$

Applying that updated relation on the Lagrange multiplier, it is that

$$\lambda_j(t+1) = \lambda_j(t) + \frac{c(t)}{m_j} \left[\sum_{i \in I(j)} \left(e_{ji}v_i(t+1) - z_{ji}(t+1) \right) \right] \qquad (5.48)$$

Then the above equation is equivalent to

$$\lambda_j(t+1) = \lambda_j(t) + \frac{c(t)}{m_j} \left(e_j v - s_j \right) \qquad (5.49)$$

After replacing $\lambda_j(t)$ by $p_{ji}(t)$, the updated relation for z_{ji} is

$$z_{ji} = e_{ji}v_i + \frac{\lambda_j(t) - \lambda_j}{c(t)} \qquad (5.50)$$

where λ_j is derived as

$$\lambda_j = \lambda_j(t) + \frac{c(t)}{m_j} \sum_{i \in I(j)} \left(e_{ji} v_i - s_{ji} \right) \tag{5.51}$$

Combining the above two equation to obtain the iterative relation for z_{ji}.

$$z_{ji} = e_{ji} v_i - \frac{1}{m_j} \left(e_j v - s_j \right) \tag{5.52}$$

substituting (5.52) into (5.46) and eliminating variable z_{ji}, the parallel iteration of the minimized extended Lagrangian function is

$$v_i = \underset{\varsigma_i \in P_i}{\arg\min} \left\{ f_i(\varsigma_i) + \sum_{\{i \setminus i \in I(j)\}} \left\{ \begin{array}{c} \lambda_j(t) e_{ji} \varsigma_i \\ + \frac{c(t)}{2} \left(e_{ji} \left(\varsigma_i - v_i \right) + \omega_j \right)^2 \end{array} \right\} \right\} \tag{5.53}$$

Then on the basis of v, ω_j is chosen as

$$\omega_j = \frac{1}{m_j} \left(e_j v - s_j \right), j = 1, 2 \cdots N$$

5.2.5 Simulation example

In this simulation example, one unstable nonlinear system with three order is considered.

$$\begin{bmatrix} x_1(t+1) \\ x_2(t+1) \\ x_3(t+1) \end{bmatrix} = A \begin{bmatrix} x_1(t) \\ x_2(t) \\ x_3(t) \end{bmatrix} + Bu(t) + \begin{bmatrix} 1.2x_1(t)^2 \\ 0.3x_2(t)^2 \\ -0.1x_3(t)^2 \end{bmatrix}$$

$$y(t) = C \begin{bmatrix} x_1(t) \\ x_2(t) \\ x_3(t) \end{bmatrix} \tag{5.54}$$

Each system matrix is defined as follows:

$$A = \begin{bmatrix} 0.35 & 0 & 0.12 \\ 0.24 & -0.05 & 0.05 \\ 0.13 & 0.28 & -0.02 \end{bmatrix}, B = \begin{bmatrix} -0.06 \\ 0.9 \\ 0.1 \end{bmatrix}, C = \begin{bmatrix} 0.3 & 0.08 & -0.3 \end{bmatrix}$$

The unstable system is placed in a closed loop feedback structure, so that the feedback term w is added to the reference signal r in additive form.

$$u(t) = r(t) + w(t)$$

This feedback loop does not guarantee stability for any input signal, but when the input signal is selected as a pseudo-random binary sequence with a root mean square 0.2, the output signal of the entire closed-loop system is guaranteed to be bounded. All elements in the three matrices A, B, C are grouped into an unknown parameter vector.

The results of using the parallel distribution algorithm to solve this unknown parameter vector are shown in Figs 5.3 and 5.4, where 15 parameter estimations are given in these two figures. It can be seen from the two figures that as the number of iterations increases, the parameter estimations will tend to their respective actual values. After using the parallel distributed identification algorithm to iterate with each other, the decreasing curve of the cost function in (5.45) is shown in Fig. 5.5. It can be seen from Fig. 5.4 that after 80 iterations, the cost function has gradually become zero, and the algorithm can be terminated, then the parameter estimations at this iteration can be deemed as the final values.

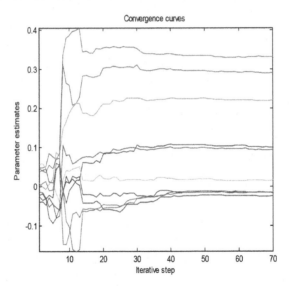

Figure 5.3: The convergence curves for ten parameter estimations

From Fig. 5.3 and Fig. 5.4, all convergence curves, corresponding to each parameter estimation, will approach their true values, after the initial flutter process. It means that with the iterative step increases, the parameter estimation is near its true value.

5.3　Recursive least squares identification for piecewise affine Hammerstein models

Modeling, identification, and prediction are the three main ubiquitous phenomena in our daily lives. Through our ideas and senses, we collect information about

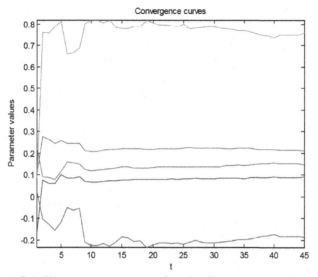

Figure 5.4: The convergence curves for other five parameter estimations

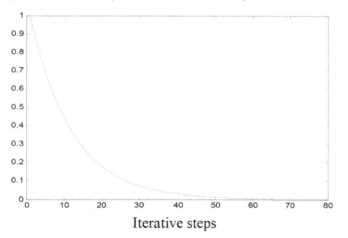

Iterative steps

Figure 5.5: The decreasing curve of the cost function

the world, and then we interpret, predict and respond to actions by using our perceptions. In natural science, lots of experiments or observations guide us to formulate laws of nature, which are used to describe different aspects of the world and let us predict all sorts of things, like planet movements or weather forecasts. Furthermore, in modern technology, modeling and identification can offer us one mathematical description of the physical object. Everywhere and everything around us, there is a need for automatic control mechanisms such as aero-planes, cars, chemical process plants, mobile phones, heating of houses, etc. However, to be able to control a system, one needs to know at least something about how it behaves and reacts to different actions taken on it. Hence we need a model of the system. A system can informally be defined as an entity that interacts with the rest of the world through more or less well defined input and output data. A model is then an approximate description of the system, and an ideal model may be simple, accurate, and general. This approximate description of the system can be constructed by the system identification strategy, as the goal of system identification is to build a mathematical model of a dynamic system based on some initial information about the system and the measurement data collected from the system. The process of system identification consists of designing and conducting the identification experiment to collect the measurement data, selecting the structure of the model and specifying the parameters to be identified, and eventually fitting the model parameters to the obtained data. Finally, the quality of the obtained model is evaluated through a model validation process. Generally, system identification is an iterative process and if the quality of the obtained model is not satisfactory, some or all of the listed phases can be repeated to obtain one satisfied model.

During the system identification procedure, the most common models are linear difference equation descriptions, such as ARX and ARMAX models, as well as linear state space models. When linear models are not sufficient for describing accurately the dynamics of a system, nonlinear identification can be employed. A large number of nonlinear model structures have been constructed to investigate their properties where some real time fast convex algorithms are proposed to identify model parameters. Many tools for identification, as well as for control, and stability analysis, have emerged in recent years. To be able to use these proposed tools, a mathematical model of the system is needed.

As the above piecewise affine models consider only the one piecewise affine function, i.e. there is one kind of unknown parameter to be identified. Let us now turn to one special class of Hammerstein models. These special models form a common class of nonlinear models that consist of a static nonlinearity followed by a linear dynamic system. There are two main special models considered in the nonlinear system identification problem, i.e. Wiener models and Hammerstein models, as many physical phenomena can be reduced to these two special models under some priori known assumption. Identification of Wiener models or Hammerstein models has been recorded widely in the literature. Now we introduce piecewise affine property into the static nonlinearity to establish one piecewise affine Hammerstein models. The goal of our contribution is to identify these two kinds of unknown model parameters.

Assume that nonlinearity is a piecewise affine, inverse function, we need to invert this piecewise affine nonlinear function. As a result, we see that the inverse function of the original piecewise affine nonlinear function is also a piecewise affine nonlinear function. We call this fact the equivalent property. No matter which piecewise affine nonlinear function is used, by using simple but tedious calculations, we construct one regression model to identify the model parameters coming from the piecewise affine Hammerstein models. To be able to identify the unknown parameter vector in the regression model, the basic recursive least squares method is used easily to give a rough parameter estimation. But in this way, the parameter estimation may be biased. So to alleviate this biased estimation, the multi-innovation recursive least squares method is modified by expanding the length of innovation. The advantage of the modified multi-innovation recursive least squares method is that the convergence rate or identification accuracy corresponding to an unknown parameter vector can be greatly increased. As we all know, the least squares method is suited under one condition that the considered noise may be a zero mean random signal that is statistically independent of the input. This condition corresponds to the classical probabilistic system identification problem. To relax this probabilistic description of noise, we investigate one projection algorithm with dead zone in the presence of bounded noise. This bound noise is considered in set membership identification widely. The robustness of the projection algorithm with dead zone can be enhanced by increasing a dead zone in the parameter update equation. Generally speaking, in the whole identification process, the basic recursive least squares and its modified form are applied for the probabilistic noise, and also the projection algorithm with dead zone for bounded noise.

5.3.1 Piecewise affine Hammerstein models

Hammerstein models consist of a static nonlinearity followed by a linear dynamical system, where this static nonlinearity is always expanded by a rational orthonormal basis function. But here the complexity of this static nonlinearity is increased, as the static nonlinearity is a piecewise affine form. This gives an overall piecewise affine model with a special structure, which let us possibly propose an identification algorithm whose worst case complexity is polynomial in the number of observed data.

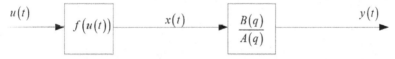

Figure 5.6: Piecewise affine Hammerstein models

Piecewise affine Hammerstein models considered here will be in the form shown in Fig. 5.6, described by the following relations:

$$\begin{cases} y(t) = G(q)x(t) = \dfrac{B(q)}{A(q)}x(t) \\ x(t) = f(u(t)) \end{cases} \tag{5.55}$$

That linear dynamical system $G(q)$ is parameterized as one polynomial form.

$$G(q) = \frac{B(q)}{A(q)} = \frac{b_1 q^{-1} + b_2 q^{-2} + \cdots + b_{n_b} q^{-n_b}}{1 + a_1 q^{-1} + a_2 q^{-2} + \cdots + a_{n_a} q^{-n_a}}$$

n_a, n_b are orders of polynomials $A(q)$ and $B(q)$ respectively, q is the delay operator, $q^{-1}x(t) = x(t-1)$. $y(t)$ is the observed output, $u(t)$ is the observed input, $x(t)$ is the unknown intermediate signal. We assume that $f(u(t))$ is a piecewise affine, inverse function, and parameterize it as

$$x(t) = f\left(u(t)\right) = u(t) - \alpha_0 + \sum_{i=1}^{M} \sigma_i \max\left\{\beta_i u(t) - \alpha_i, 0\right\} \qquad (5.56)$$

where $\sigma_i \in \{-1,1\}$, depending on the sign of the max function, $\{\alpha_0, \alpha_i, \beta_i\}_{i=1}^{M}$ are unknown parameters used to parameterize that piecewise affine, inverse function $f(u(t))$.

The above defined piecewise affine function $f(u(t))$ is similar to the hinging hyper-plane function which is defined as a sum of hinge functions, where each consists of two half hyper-planes, parameterized by unknown parameters $\{\alpha_i, \beta_i\}$. $\sigma_i = \pm 1$ is used to represent both convex and non-convex functions.

Observing (5.55) and (5.56), there are many unknown parameters in piecewise affine Hammerstein models, such as $\left\{\{\alpha_0, \alpha_i, \beta_i\}_{i=1}^{M}, \{a_i\}_{i=1}^{n_a}, \{b_i\}_{i=1}^{n_b}\right\}$, which are needed to identify.

In many other nonlinear system identification references, the inverse function form $u(t) = f^{-1}(x(t))$ is used widely to replace the piecewise affine function $x(t) = f(u(t))$. So during the latter recursive identification process, we encounter a problem, i.e. whether to apply the original piecewise affine function $x(t) = f(u(t))$ or its inverse function $u(t) = f^{-1}(x(t))$. To tackle this problem, we need to derive the inverse function form $u(t) = f^{-1}(x(t))$.

5.3.2   Equivalence between piecewise affine nonlinear functions

In this section, we derive the inverse function $u(t) = f^{-1}(x(t))$ in accordance with piecewise affine function $x(t) = f(u(t))$. As the inverse function $u(t) = f^{-1}(x(t))$ is also a piecewise affine function form, we use any form of these two functions in identification of piecewise affine Hammerstein models. According to piecewise affine nonlinear function $x(t) = f(u(t))$, without loss of generality, assume that $f(u(t))$ is strictly increasing, i.e.

$$0 < \sum_{m=1}^{k} \sigma_m \beta_m < \infty, \ \forall k = 1, 2 \cdots M$$

Due to $\sigma_i \in \{-1,1\}$, assume that

$$\beta_i > 0, \frac{\alpha_1}{\beta_1} < \frac{\alpha_2}{\beta_2} < \cdots < \frac{\alpha_M}{\beta_M}$$

Proposition 5.1: The inverse function $u(t) = f^{-1}(x(t))$ is parameterized as

$$u(t) = f^{-1}\left(x(t)\right) = x(t) + \alpha_0 - \sum_{k=1}^{M} \sigma_k \max\left\{\tilde{\beta}_k x(t) - \tilde{\alpha}_k, 0\right\} \qquad (5.57)$$

where unknown parameters $\tilde{\alpha}_k$ and $\tilde{\beta}_k$ are given as

$$\tilde{\alpha}_k = \frac{\sum_{m=0}^{k-1} \sigma_m \left(\beta_m \alpha_k - \beta_k \alpha_m \right)}{\sum_{m=0}^{k-1} \sigma_m \beta_m \sum_{m=0}^{k} \sigma_m \beta_m}, \tilde{\beta}_k = \frac{\beta_k}{\sum_{m=0}^{k-1} \sigma_m \beta_m \sum_{m=0}^{k} \sigma_m \beta_m}$$

where $\{\alpha_k, \beta_k\}_{k=1}^{M}, \sigma_m$ are from the original piecewise affine function $x(t) = f(u(t))$.

Proof: When $M = 0$, we have

$$x(t) = f\left(u(t)\right) = u(t) - \alpha_0$$

So

$$u(t) = f^{-1}\left(x(t)\right) = x(t) + \alpha_0$$

Now suppose the statement holds for one certain $M - 1$, then we need to prove the statement by induction for certain M. In the following, we discuss two cases respectively.

(i) If $u(t) < \dfrac{\alpha_M}{\beta_M}$, i.e. $\beta_M u(t) - \alpha_M < 0$, then the piecewise affine function can be written as

$$x(t) = f\left(u(t)\right) = u(t) - \alpha_0 + \sum_{i=1}^{M-1} \sigma_i \max\left\{\beta_i u(t) - \alpha_i, 0\right\}$$

$$+ \sigma_M \underbrace{\max\left\{\beta_M u(t) - \alpha_M, 0\right\}}_{0}$$

$$= u(t) - \alpha_0 + \sum_{i=1}^{M-1} \sigma_i \max\left\{\beta_i u(t) - \alpha_i, 0\right\} \tag{5.58}$$

According to the induction assumption, $u(t) = f^{-1}(x(t))$ can be written as in (5.57).

(ii) If $u(t) > \dfrac{\alpha_M}{\beta_M}$, i.e. $\beta_M u(t) - \alpha_M < 0$, then piecewise affine function can be written as

$$x(t) = u(t) - \alpha_0 + \sum_{i=1}^{M-1} \sigma_i \max\left\{\beta_i u(t) - \alpha_i, 0\right\} + \sigma_M \left(\beta_M u(t) - \alpha_M\right) \tag{5.59}$$

Rewriting (5.59) as

$$x(t) - \sigma_M\left(\beta_M u(t) - \alpha_M\right) = u(t) - \alpha_0 + \sum_{i=1}^{M-1} \sigma_i \max\left\{\beta_i u(t) - \alpha_i, 0\right\} \quad (5.60)$$

According to the induction assumption, the right side of above equation (5.60) satisfies the statement for one certain $M-1$. For the sake of compact notation, let $\sigma_0 = \beta_0 = 1$, from (5.60) we get

$$x(t) = u(t) - \alpha_0 + \sum_{i=1}^{M-1} \sigma_i\left(\beta_i u(t) - \alpha_i\right) + \sigma_M\left(\beta_M u(t) - \alpha_M\right)$$

$$= \left[1 + \sum_{i=1}^{M-1} \sigma_i \beta_i + \sigma_M \beta_M\right] u(t) - \left[\alpha_0 + \sum_{i=1}^{M-1} \sigma_i \alpha_i + \sigma_M \alpha_M\right]$$

$$= \sum_{i=0}^{M} \sigma_i \beta_i u(t) - \sum_{i=0}^{M} \sigma_i \alpha_i \quad (5.61)$$

where in (5.61), we abbreviate the max function operation, for one $i \in 1, 2 \cdots M$, if $\beta_i u(t) - \alpha_i < 0$, above equality reduces to the first case for certain $M-1$, then the induction assumption can be applied again. So in second case, we assume

$$\beta_i u(t) - \alpha_i > 0, \forall i = 1, 2 \cdots M$$

From (5.61), we get

$$u(t) = \frac{x(t) + \displaystyle\sum_{i=0}^{M} \sigma_i \alpha_i}{\displaystyle\sum_{i=0}^{M} \sigma_i \beta_i} = \frac{x(t) + \displaystyle\sum_{i=0}^{M-1} \sigma_i \alpha_i + \sigma_M \alpha_M}{\displaystyle\sum_{i=0}^{M-1} \sigma_i \beta_i + \sigma_M \beta_M} \quad (5.62)$$

Using simple but tedious calculations, we derive the following calculations:

$$u(t) = \frac{x(t) + \displaystyle\sum_{i=0}^{M-1} \sigma_i \alpha_i + \sigma_M \alpha_M}{\displaystyle\sum_{i=0}^{M-1} \sigma_i \beta_i} \times \frac{\displaystyle\sum_{i=0}^{M-1} \sigma_i \beta_i}{\displaystyle\sum_{i=0}^{M} \sigma_i \beta_i}$$

$$= \left[\frac{x(t) + \displaystyle\sum_{i=0}^{M-1} \sigma_i \alpha_i}{\displaystyle\sum_{i=0}^{M-1} \sigma_i \beta_i} + \frac{\sigma_M \alpha_M}{\displaystyle\sum_{i=0}^{M-1} \sigma_i \beta_i}\right] \times \left[1 - \frac{\sigma_M \beta_M}{\displaystyle\sum_{i=0}^{M} \sigma_i \beta_i}\right]$$

$$
\begin{aligned}
&= \frac{x(t) + \displaystyle\sum_{i=0}^{M-1} \sigma_i \alpha_i}{\displaystyle\sum_{i=0}^{M-1} \sigma_i \beta_i} - \frac{x(t) + \displaystyle\sum_{i=0}^{M-1} \sigma_i \alpha_i}{\displaystyle\sum_{i=0}^{M-1} \sigma_i \beta_i} \times \frac{\sigma_M \beta_M}{\displaystyle\sum_{i=0}^{M} \sigma_i \beta_i} + \frac{\sigma_M \alpha_M}{\displaystyle\sum_{i=0}^{M-1} \sigma_i \beta_i} \times \frac{\displaystyle\sum_{i=0}^{M-1} \sigma_i \beta_i}{\displaystyle\sum_{i=0}^{M} \sigma_i \beta_i} \\[2ex]
&= f_{M-1}^{-1}(x(t)) - \frac{\sigma_M \beta_M}{\displaystyle\sum_{i=0}^{M-1} \sigma_i \beta_i \sum_{i=0}^{M} \sigma_i \beta_i} \left[x(t) + \sum_{i=0}^{M-1} \sigma_i \alpha_i - \frac{1}{\beta_M} \sum_{i=0}^{M-1} \sigma_i \beta_i \alpha_M \right] \\[2ex]
&= f_{M-1}^{-1}(x(t)) - \frac{\sigma_M \beta_M}{\displaystyle\sum_{i=0}^{M-1} \sigma_i \beta_i \sum_{i=0}^{M} \sigma_i \beta_i} \left[x(t) - \frac{1}{\beta_M} \left[-\sum_{i=0}^{M-1} \sigma_i \alpha_i \beta_M + \sum_{i=0}^{M-1} \sigma_i \beta_i \alpha_M \right] \right] \\[2ex]
&= f_{M-1}^{-1}(x(t)) - \frac{\sigma_M \beta_M}{\displaystyle\sum_{i=0}^{M-1} \sigma_i \beta_i \sum_{i=0}^{M} \sigma_i \beta_i} \left[x(t) - \frac{1}{\beta_M} \sum_{i=0}^{M-1} (\sigma_i \beta_i \alpha_M - \sigma_i \alpha_i \beta_M) \right] \\[2ex]
&= f_{M-1}^{-1}(x(t)) - \sigma_M \left(\tilde{\beta}_M x(t) - \tilde{\alpha}_M \right)
\end{aligned}
\tag{5.63}
$$

where $f_{M-1}^{-1}(x(t))$ is the inverse of the original function without the *M*-th max function.

Using a max function and the induction assumption, we continue to write (5.63) as

$$
\begin{aligned}
u(t) &= f_{M-1}^{-1}(x(t)) - \sigma_M \max\{ \tilde{\beta}_M x(t) - \tilde{\alpha}_M, 0 \} \\[1ex]
&= x(t) + \alpha_0 - \sum_{k=1}^{M-1} \sigma_k \max\{ \tilde{\beta}_k x(t) - \tilde{\alpha}_k, 0 \} - \sigma_M \max\{ \tilde{\beta}_M x(t) - \tilde{\alpha}_M, 0 \} \\[1ex]
&= x(t) + \alpha_0 - \sum_{k=1}^{M} \sigma_k \max\{ \tilde{\beta}_k x(t) - \tilde{\alpha}_k, 0 \}
\end{aligned}
\tag{5.64}
$$

which is equation (5.57).

By comparing (5.56) and (5.57), we see the original piecewise affine nonlinear function $x(t) = f(u(t))$ and its inverse function $u(t) = f^{-1}(x(t))$ have similar forms. So in the next recursive identification process, we only use the piecewise affine nonlinear function $x(t) = f(u(t))$, and not its inverse function $u(t) = f^{-1}(x(t))$, due to this equivalent property.

5.3.3 Recursive identification of unknown parameters

The main contribution of this section is to identify all unknown parameters in piecewise affine Hammerstein models. Three identification methods are proposed in different situations, for example, the recursive least squares method, multi-innovation recursive least squares method, and the projection algorithm with dead zone.

5.3.3.1 Recursive least squares method

Substituting (5.56) into (5.55), we obtain

$$A(q)y(t) = B(q)x(t)$$

Expanding it as the following form

$$\left[1 + \sum_{i=1}^{n_a} a_i q^{-i}\right] y(t) = \sum_{i=1}^{n_b} b_i q^{-i} \left[u(t) - \alpha_0 + \sum_{i=1}^{M} \sigma_i \max\{\beta_i u(t) - \alpha_i, 0\}\right] \quad (5.65)$$

To eliminate that max function operation, introduce the discrete variable $\delta_i(t) \in \{0,1\}$

$$\delta_i(t) = \begin{cases} 0 & \beta_i u(t) - \alpha_i < 0 \\ 1 & \beta_i u(t) - \alpha_i \geq 0 \quad i \in [1, M] \end{cases}$$

Then equation (5.65) can be rewritten as

$$y(t) + \sum_{i=1}^{n_a} a_i q^{-i} y(t) = \sum_{i=1}^{n_b} b_i q^{-i} \left[u(t) - \alpha_0\right] + \sum_{i=1}^{n_b} b_i q^{-i} \sum_{i=1}^{M} \sigma_i \delta_i(t) \left(\beta_i u(t) - \alpha_i\right)$$

$$(5.66)$$

To reduce the computational complexity of the identification problem, we assume that $\alpha_0 = 0$, then term $\sum_{i=1}^{n_b} b_i q^{-i} \alpha_0$ vanishes.

Concentrating on the second term of the right side, we rewrite it as

$$\sum_{i=1}^{M} \sigma_i \delta_i(t)\left(\beta_i u(t) - \alpha_i\right) = \sigma_1 \delta_1(t) \beta_1 u(t) + \sigma_2 \delta_2(t) \beta_2 u(t)$$

$$+ \cdots + \sigma_M \delta_M(t) \beta_M u(t) - \sigma_1 \delta_1(t)\alpha_1 - \sigma_2 \delta_2(t)\alpha_2 - \cdots \sigma_M \delta_M(t)\alpha_M$$

$$= \beta_1 u_1(t) + \beta_2 u_2(t) + \cdots + \beta_M u_M(t) - \delta_1(t)\alpha_1 - \sigma_2 \delta_2(t)\alpha_2 - \cdots \sigma_M \delta_M(t)\alpha_M$$

$$(5.67)$$

Continuing to do sum operation on all terms, for example

$$\sum_{i=1}^{n_b} b_i q^{-i} \beta_1 u_1(t) = \begin{bmatrix} u_1(t-1) & u_1(t-2) & \cdots & u_1(t-n_b) \end{bmatrix} \begin{bmatrix} b_1 \\ b_2 \\ \vdots \\ b_{n_b} \end{bmatrix} \beta_1$$

$$\sum_{i=1}^{n_b} b_i q^{-i} \beta_2 u_2 (t) = \begin{bmatrix} u_2 (t-1) & u_2 (t-2) & \cdots & u_2 (t-n_b) \end{bmatrix} \begin{bmatrix} b_1 \\ b_2 \\ \vdots \\ b_{n_b} \end{bmatrix} \beta_2$$

...

$$\sum_{i=1}^{n_b} b_i q^{-i} \beta_M u_M (t) = \begin{bmatrix} u_M (t-1) & u_M (t-2) & \cdots & u_M (t-n_b) \end{bmatrix} \begin{bmatrix} b_1 \\ b_2 \\ \vdots \\ b_{n_b} \end{bmatrix} \beta_M \quad (5.68)$$

Similarly, we continue to compute

$$\sum_{i=1}^{n_b} b_i q^{-i} \delta_1 (t) \alpha_1 = \begin{bmatrix} \delta_1 (t-1) & \delta_1 (t-2) & \cdots & \delta_1 (t-n_b) \end{bmatrix} \begin{bmatrix} b_1 \\ b_2 \\ \vdots \\ b_{n_b} \end{bmatrix} \alpha_1$$

$$\sum_{i=1}^{n_b} b_i q^{-i} \delta_2 (t) \alpha_2 = \begin{bmatrix} \delta_2 (t-1) & \delta_2 (t-2) & \cdots & \delta_2 (t-n_b) \end{bmatrix} \begin{bmatrix} b_1 \\ b_2 \\ \vdots \\ b_{n_b} \end{bmatrix} \alpha_2$$

...

$$\sum_{i=1}^{n_b} b_i q^{-i} \delta_M (t) \alpha_M = \begin{bmatrix} \delta_M (t-1) & \delta_M (t-2) & \cdots & \delta_M (t-n_b) \end{bmatrix} \begin{bmatrix} b_1 \\ b_2 \\ \vdots \\ b_{n_b} \end{bmatrix} \alpha_M$$

$$(5.69)$$

Substituting all above equalities into (5.66), we obtain

$$y(t) = \begin{bmatrix} -y(t-1) & -y(t-2) & \cdots & -y(t-n_a) \end{bmatrix} \begin{vmatrix} a_1 \\ a_2 \\ \vdots \\ a_{n_a} \end{vmatrix}$$

$$+ \begin{bmatrix} u(t-1) & u(t-2) & \cdots & u(t-n_b) \end{bmatrix} \begin{bmatrix} b_1 \\ b_2 \\ \vdots \\ b_{n_b} \end{bmatrix}$$

$$+\begin{bmatrix} u_1(t-1) & u_1(t-2) & \cdots & u_1(t-n_b) \end{bmatrix} \begin{bmatrix} b_1 \\ b_2 \\ \vdots \\ b_{n_b} \end{bmatrix} \beta_1$$

$$+\begin{bmatrix} u_2(t-1) & u_2(t-2) & \cdots & u_2(t-n_b) \end{bmatrix} \begin{bmatrix} b_1 \\ b_2 \\ \vdots \\ b_{n_b} \end{bmatrix} \beta_2$$

$$+\cdots$$

$$+\begin{bmatrix} u_M(t-1) & u_M(t-2) & \cdots & u_M(t-n_b) \end{bmatrix} \begin{bmatrix} b_1 \\ b_2 \\ \vdots \\ b_{n_b} \end{bmatrix} \beta_M$$

$$-\begin{bmatrix} \delta_1(t-1) & \delta_1(t-2) & \cdots & \delta_1(t-n_b) \end{bmatrix} \begin{bmatrix} b_1 \\ b_2 \\ \vdots \\ b_{n_b} \end{bmatrix} \alpha_1$$

$$-\begin{bmatrix} \delta_2(t-1) & \delta_2(t-2) & \cdots & \delta_2(t-n_b) \end{bmatrix} \begin{bmatrix} b_1 \\ b_2 \\ \vdots \\ b_{n_b} \end{bmatrix} \alpha_2$$

$$-\cdots-\begin{bmatrix} \delta_M(t-1) & \delta_M(t-2) & \cdots & \delta_M(t-n_b) \end{bmatrix} \begin{bmatrix} b_1 \\ b_2 \\ \vdots \\ b_{n_b} \end{bmatrix} \alpha_M \tag{5.70}$$

Defining the following parameter vectors as

$$\theta_1 = \begin{bmatrix} a_1 \\ a_2 \\ \vdots \\ a_{n_a} \end{bmatrix}, \theta_2 = \begin{bmatrix} b_1 \\ b_2 \\ \vdots \\ b_{n_b} \end{bmatrix}, \theta_3 = \begin{bmatrix} b_1 \\ b_2 \\ \vdots \\ b_{n_b} \end{bmatrix} \beta_1 \cdots \theta_{M+2} = \begin{bmatrix} b_1 \\ b_2 \\ \vdots \\ b_{n_b} \end{bmatrix}$$

$$\beta_M, \theta_{M+3} = \begin{bmatrix} b_1 \\ b_2 \\ \vdots \\ b_{n_b} \end{bmatrix} \alpha_1 \cdots \theta_{2M+2} = \begin{bmatrix} b_1 \\ b_2 \\ \vdots \\ b_{n_b} \end{bmatrix} \alpha_M \tag{5.71}$$

From above defined vectors, vector θ_2 can be used to identify other unknown parameters, i.e.

$$\beta_1 = \frac{\left(b_1 + b_2 + \cdots b_{n_b}\right)\beta_1}{b_1 + b_2 + \cdots b_{n_b}} = \frac{\sum\limits_{i=1}^{n_b}\theta_{3i}}{\sum\limits_{i=1}^{n_b}\theta_{2i}} \ or \ \beta_1 = \frac{\dfrac{\theta_{31}}{\theta_{21}} + \dfrac{\theta_{32}}{\theta_{22}} + \cdots \dfrac{\theta_{3n_b}}{\theta_{2n_b}}}{n_b} = \frac{\sum\limits_{i=1}^{n_b}\dfrac{\theta_{3i}}{\theta_{2i}}}{n_b} \quad (5.72)$$

where θ_{3i} denotes the ith element of vector θ_3.

Similarly, we give the closed relations among these defined vectors.

$$\theta_1, \theta_2, \theta_3 = \theta_2\beta_1, \theta_4 = \theta_2\beta_2 \cdots \theta_{M+2} = \theta_2\beta_M$$
$$\theta_{M+3} = \theta_2\alpha_1, \theta_{M+4} = \theta_2\alpha_2 \cdots \theta_{2M+2} = \theta_2\alpha_M \quad (5.73)$$

So based on identified parameter vectors $\{\theta_1, \theta_2 \cdots \theta_{2M+2}\}$, $\{\alpha_i, \beta_i\}_{i=1}^M$ can be obtained by using (5.72) or (5.73), and $\{a_i\}_{i=1}^{n_a}, \{b_i\}_{i=1}^{n_b}$ are included in θ_1 and θ_2 respectively.

To identify unknown parameter vectors $\{\theta_i\}_{i=1}^{2M+2}$, we combine the following vectors in (5.70).

$$y(t) = \varphi_1^T(t)\theta_1 + \varphi_2^T(t)\theta_2 + \cdots + \varphi_{M+2}^T(t)\theta_{M+2}$$
$$+ \varphi_{M+3}^T(t)\theta_{M+3} + \cdots + \varphi_{2M+2}^T(t)\theta_{2M+2}$$
$$= \underbrace{\left[\varphi_1^T(t) \quad \varphi_2^T(t) \quad \cdots \quad \varphi_{M+2}^T(t) \quad \varphi_{M+3}^T(t) \quad \cdots \quad \varphi_{2M+2}^T(t)\right]}_{\varphi^T(t)}$$

$$\times \underbrace{\begin{bmatrix} \theta_1 \\ \theta_2 \\ \vdots \\ \theta_{M+2} \\ \theta_{M+3} \\ \vdots \\ \theta_{2M+2} \end{bmatrix}}_{\theta} = \varphi^T(t)\theta$$

In above theoretical analysis, we do not cover external noise. However, in practical engineering problems, it is inevitable to suffer from uncertain noise or disturbance, such as ambient noise, manufacturing tolerance. Here we add noise $e(t)$.

$$y(t) = \varphi^T(t)\theta + e(t) \quad (5.74)$$

where $\varphi(t)$ is a regression vector. In the framework of classical probabilistic description on noise $e(t)$, the basic least squares method is applied to identify the unknown parameter vector θ directly, the parameter estimator $\hat{\theta}$ is given as

$$\hat{\theta} = \left[\sum_{t=1}^{N} \varphi(t)\varphi^T(t) \right]^{-1} \left[\sum_{t=1}^{N} \varphi(t)y(t) \right] \tag{5.75}$$

where N denotes the number of observed data.

Its recursive least squares method is easily obtained to reduce the computational complexity.

$$\begin{cases} P^{-1}(t) = \sum_{t=1}^{N} \varphi(t)\varphi^T(t) \\ L(t) = \dfrac{P(t-1)\varphi(t)}{1+\varphi^T(t)P(t-1)\varphi(t)} \\ P(t) = P(t-1) - \dfrac{P(t-1)\varphi(t)\varphi^T(t)P(t-1)}{1+\varphi^T(t)P(t-1)\varphi(t)} \\ \hat{\theta}(t) = \hat{\theta}(t-1) + L(t)\left[y(t) - \varphi^T(t)\hat{\theta}(t-1) \right] \end{cases} \tag{5.76}$$

where $\hat{\theta}(t)$ denotes the ith iterative estimator. Based on classical probabilistic description on noise $e(t)$.

5.3.3.2 Multi-innovation recursive least squares method

To further increase the efficiency of the above basic recursive least squares method, we modify it by expanding the length of innovation to obtain the multi-innovation recursive least squares method. As each element in regression vector $\varphi(t)$ is known, we assume p as the length of innovation and define the following matrices or vectors:

$$Y(t) = \begin{bmatrix} y(t) \\ y(t-1) \\ \vdots \\ y(t-p+1) \end{bmatrix}, E(t) = \begin{bmatrix} e(t) \\ e(t-1) \\ \vdots \\ e(t-p+1) \end{bmatrix}, \Phi(t) = \begin{bmatrix} \varphi^T(t) \\ \varphi^T(t-1) \\ \vdots \\ \varphi^T(t-p+1) \end{bmatrix}$$

Then using above matrices and vectors, we have relation

$$Y(t) = \Phi(t)\theta + E(t)$$

Define one cost function as

$$V(\theta) = \left\| Y(t) - \Phi(t)\theta \right\|^2$$

where matrix norm $\| \; \|$ is defined as the trace operation.

$$\left\| Y(t) - \Phi(t)\theta \right\|^2 = tr\left(Y(t) - \Phi(t)\theta \right)\left(Y(t) - \Phi(t)\theta \right)^T$$

Then the multi-innovation recursive least squares method is used to identify the unknown parameter vector.

$$\begin{cases} \hat{\theta}(t) = \hat{\theta}(t-1) + P(t)\Phi^T(t)\left[Y(t) - \Phi(t)\hat{\theta}(t-1) \right] \\ P^{-1}(t) = P^{-1}(t-1) + \Phi^T(t)\Phi(t) \\ P(0) = p_0 I \end{cases} \tag{5.77}$$

when $p \geq 2$, $E(t) = Y(t) - \Phi(t)\hat{\theta}(t-1) \in R^p$ is an innovation vector. During the initial step, choose p_0 as a very large number, for instance $p_0 = 10^3$, the initial value $\hat{\theta}_0$ is a very small vector, such as

$$\hat{\theta}_0 = \frac{1}{p_0} I$$

where I denotes a column vector which all elements are 1.

5.3.3.3 Projection algorithm with dead zone

As noise $e(t)$ is white noise, the parameter estimator obtained by recursive least squares method is unbiased. But this result does not hold for color noise. Further, white noise is an ideal case, it will not exist in engineering. In addition, deriving statistical properties of noise is often very difficult in practice as it is usually not possible to measure noise directly. On the other hand, assumption on the knowledge of noise bound is less restrictive as noise is bounded and the bound can be roughly calculated from the specification of the used sensor. Here we investigate the identification of unknown parameters in the presence of bounded noise, as the robustness of the method can be enhanced by introducing a dead zone in the parameter update equation.

Consider again (5.74), now noise $e(t)$ denotes a bounded noise term such that $\sup |v(t)| \leq \Delta$, where Δ is a bound. Introduce the following projection algorithm with dead zone to identify parameter vector θ.

$$\hat{\theta}(t) = \hat{\theta}(t-1) + \frac{a(t)\varphi(t)}{c + \varphi^T(t)\varphi(t)}\left[y(t) - \varphi^T(t)\hat{\theta}(t-1) \right] \tag{5.78}$$

where $\hat{\theta}_0$ is given, $c > 0$ and

$$a(t) = \begin{cases} 1 & \left| y(t) - \varphi^T(t)\hat{\theta}(t-1) \right| > 2\Delta \\ 0 & \text{otherwise} \end{cases} \tag{5.79}$$

When the prediction error is smaller than the size of noise, the choice of $\{a(t)\}$ is to terminate the iterative algorithm. To give an explicit analysis on projection algorithm with dead zone, we subtract true parameter vector θ_0 from both sides of (5.78) and give

$$\begin{cases} \tilde{\theta}(t) = \hat{\theta}(t) - \theta_0 \\ \tilde{\theta}(t) = \tilde{\theta}(t-1) - \dfrac{a(t)\varphi(t)}{c + \varphi^T(t)\varphi(t)} \left[\varphi^T(t)\tilde{\theta}(t-1) + e(t) \right] \end{cases} \qquad (5.80)$$

Noting that $a(t) = 0$ or 1.

$$\left\| \tilde{\theta}(t) \right\|^2 = \left\| \tilde{\theta}(t-1) \right\|^2 - \frac{2a(t)\left[w(t) - e(t)\right]w(t)}{c + \varphi^T(t)\varphi(t)} + \frac{a^2(t)\varphi^T(t)\varphi(t)w^2(t)}{\left[c + \varphi^T(t)\varphi(t)\right]^2}$$

$$\leq \left\| \tilde{\theta}(t-1) \right\|^2 + \frac{a(t)}{c + \varphi^T(t)\varphi(t)} \left[2w(t)e(t)\right] - \frac{a(t)w^2(t)}{c + \varphi^T(t)\varphi(t)} \qquad (5.81)$$

where $w(t)$ is model error, i.e.

$$w(t) = y(t) - \varphi^T(t)\hat{\theta}(t-1) = -\varphi^T(t)\tilde{\theta}(t-1) + e(t) \qquad (5.82)$$

Due to $2ab \leq ka^2 + b^2/k$, for any k, then

$$\left\| \tilde{\theta}(t) \right\|^2 = \left\| \tilde{\theta}(t-1) \right\|^2 - \frac{a(t)w^2(t)}{c + \varphi^T(t)\varphi(t)} + \frac{a(t)\varphi^T(t)\varphi(t)w^2(t)}{c + \varphi^T(t)\varphi(t)} \left[\frac{w^2(t)}{2} + 2w^2(t) \right]$$

$$\leq \left\| \tilde{\theta}(t-1) \right\|^2 - \frac{1}{2}\frac{a(t)w^2(t)}{c + \varphi^T(t)\varphi(t)} + \frac{2a(t)\Delta^2}{c + \varphi^T(t)\varphi(t)} \qquad (5.83)$$

From (5.83), $\left\{ \left\| \tilde{\theta}(t) \right\|^2 \right\}$ is a non-increasing sequence bounded blow by zero, which results in property 5.1.

Property 5.1:

$$\left\| \hat{\theta}(t) - \theta_0 \right\| \leq \left\| \hat{\theta}(t-1) - \theta_0 \right\| \leq \left\| \hat{\theta}(0) - \theta_0 \right\|, t \geq 1 \qquad (5.84)$$

Combing (5.78) and (5.82), we note that

$$\hat{\theta}(t) - \hat{\theta}(t-1) = \frac{-a(t)\varphi(t)}{c + \varphi^T(t)\varphi(t)} w(t) \qquad (5.85)$$

Hence squaring on both sides of (5.85).

$$\left\| \hat{\theta}(t) - \hat{\theta}(t-1) \right\|^2 = \frac{a^2(t)\varphi^T(t)\varphi(t)}{\left[c + \varphi^T(t)\varphi(t)\right]^2} w^2(t) \qquad (5.86)$$

Using the following inequalities:

$$\limsup_{t\to\infty} \frac{a(t)w^2(t)}{c+\varphi^T(t)\varphi(t)} \le \frac{4\Delta^2}{c}$$

or

$$\limsup_{t\to\infty} \frac{a(t)\left[c+\varphi^T(t)\varphi(t)\right]w^2(t)}{\left[c+\varphi^T(t)\varphi(t)\right]^2} \le \frac{4\Delta^2}{c}$$

Then

$$\limsup_{t\to\infty} \frac{a(t)\varphi^T(t)\varphi(t)w^2(t)}{\left[c+\varphi^T(t)\varphi(t)\right]^2} \le \frac{4\Delta^2}{c} \tag{5.87}$$

Substituting (5.87) into (5.86) and squaring on both sides, then property 5.2 is obtained.

Property 5.2:

$$\limsup_{t\to\infty} \left\| \hat{\theta}(t) - \hat{\theta}(t-1) \right\| \le \frac{2\Delta}{\sqrt{c}} \tag{5.88}$$

From these two properties, the parameter estimator will converge to its true value with the number of iterations increasing, and the approximate error between two adjacent parameter estimators is dependent on that bounded value corresponding to an uncertain region.

5.3.4 Conclusion

As piecewise affine Hammerstein models are one special hybrid systems that can model a large number of physical processes, in this chapter we study the problem of parameter identification of piecewise affine Hammerstein models. Given a piecewise affine nonlinear function, we prove that its inverse function is also a piecewise affine nonlinear function. To be able to identify all unknown parameters within these piecewise affine Hammerstein models, we propose three different recursive identification strategies according to the different descriptions of noise. But the performance analysis of these three recursive identification algorithms has not been studied here, it is our next subject.

5.4 Summary

In this section, we study the identification of the polynomial nonlinear state space system, which is the transformation form of many nonlinear systems. The equivalence between the nonlinear feedback system, block structure nonlinear system, and the polynomial nonlinear state space system are established. The cost function of the model error is constructed to solve the unknown parameter vector. The derivation process of the parallel distribution identification algorithm is given under unconstrained and constrained conditions.

References

[1]   Ljung, L. 1999. System Identification: Theory for the User. Prentice Hall.

[2]   Pintelon, R. and Schoukens J. 2001. System Identification: A Frequency Domain Approach. New York: IEEE Press.

[3]   Boyd, S. and Vandenberghe, L. 2008. Convex Optimization. UK: Cambridge University Press.

[4]   Bai, E.W. 2014. Kernal based approaches to local nonlinear nonparametric variable selection, Automatica, 50(1): 100-113.

[5]   Ohlsson, H. 2013. Identification of switched linear regression models using sum of norms regularization, Automatica, 49(4): 1045-1050.

[6]   Van Mulders, A. 2013. Identification of systems with localised nonlinearity: From state space to block structured models, Automatica, 49(5): 1392-1396.

[7]   Pillonetto, G. 2014. Kernel methods in system identification, machine learning and function estimation: A survey, Automatica, 50(3): 657-682.

[8]   Vanden Hof, P.M.J. 2013. Identification of dynamic models in complex networks with prediction error methods, Automatica, 49(10): 2994-3006.

[9]   Paduart, J. 2010. Identification of nonlinear systems using polynomial nonlinear state space models, Automatica, 46(4): 647-656.

[10] Van Mulders, A. 2010. Two nonlinear optimization methods for black box identification compared, Automatica, 46(10): 1675-1681.

[11] Zeilinger, M. 2011. Real time suboptimal model predictive control using a combination of explicit MPC and online optimization, IEEE Transactions of Automatic Control, 56(7): 1524-1534.

[12] Feller, C. 2013. An improved algorithm for combinatorial multi-parametric quadratic programming, Automatica, 49(5): 1370-1376.

[13] Pillonetto, G. 2010 A new kernel based approach for linear system identification, Automatica, 46(1): 81-93.

[14] Soderstorn, T. 2011. Accuracy analysis of a covariance matching approach for identify errors in variables systems, Automatica, 47(1): 272-282.

[15] Schoukens, M. 2014. Identification of Wiener Hammerstein systems by a nonparametric separation of the best linear approximation, Automatica, 50(2): 628-634.

[16] Novara, C., Formentin, S. and Savaresi, M. 2016. Data driven design of two degrees of freedom nonlinear controllers, Automatica, 72(10): 19-27.

[17] Bravo, J.M., Alamo, T. and Vasallo, M. 2017. A general framework for predictions based on bounding techniques and local approximation, IEEE Transaction on Automatic Control, 62(7): 3430-3435.

[18] Tanaskovic, M., Fagiano, L. and Smith, R. 2014. Adaptive receding horizon control for constrained MIMO systems, Automatica, 50(12): 3019-3029.

[19] Tanaskovic, M., Fagiano, L. and Novara, C. 2017. Data driven control of nonlinear systems: An on line direct approach, Automatica, 75(1): 1-10.

[20] Casini, M., Garulli, A. and Vicino, A. 2017. A linear programming approach to online set membership parameter estimation for linear regression models, International Journal of Adaptive Control and Signal Processing, 31(3): 360-378.

[21] Wang, J. 2018. Dynamic programming technique in multi UAVs formation anomaly detection, International Journal of Innovative Computing, Information and Control, 14(6): 1977-1981.

[22] Zhang, X. and Kamgarpour, M. 2017. A Georghiou. Robust optimal control with adjustable uncertainty sets, Automatica, 75(1): 249-259.

[23] Bravo, J.M., Alamo, T. and Camacho, E.F. 2016. Bounded error identification of systems with time varying parameters, IEEE Transactions on Automatic Control, 51(7): 1144-1150.

[24] Care, A., Csorji, B. and Campi, M.C. 2018. Finite sample system identification: An overview and a new correlation method, IEEE Control Systems Letters, 2(1): 61-66.

[25] Weyer, E. and Campi, M.C. 2017. Asymptotic properties of SPS confidence regions. Automatica, 82(8): 287-294.

CHAPTER 6

Data Driven Iterative Tuning Control

6.1 Introduction

In this chapter, the idea of data driven is applied in advanced control strategy, which corresponds to the one data driven control method- iterative correlation tuning control. For the convenience of understanding, the description process is given from system identification to advanced control, i.e. the content of this chapter is formulated as follows.

(1) The identification problem for the piecewise affine system is considered to be a special nonlinear system. As the difficulty in identifying a piecewise affine system is to determine each separated region and each unknown parameter vector simultaneously, we propose a multi class classification process to determine each separated region. This multi class classification process is similar to the classical data clustering process, and the merit of our strategy is that the first order algorithm of convex optimization can be applied to achieve this classification process. Furthermore, to relax the strict probabilistic description of external noise in identifying each unknown parameter vector, the zonotope parameter identification algorithm is proposed to compute a set that contains the parameter vector, consistent with the measured output and the given bound of the noise. To guarantee our derived zonotope not growing unbounded with iterations, a sufficient condition for this requirement to hold may be formulated as one linear matrix inequality.

(2) To design one unknown parameterized controller in a closed loop linear time invariant system, a new data driven method (iterative correlation tuning control) is proposed by using an idea of model structure validation process from system identification theory. This new method transforms the problem of identifying the unknown parameters in the parameterized controller into one finding roots process with respect to a cross correlation function. The common stochastic approximation algorithm is used to achieve the finding roots process, where an objective function corresponding to a cross correlation function is established when relaxing some prior conditions. Then a gradient algorithm is applied to solve an unconstrained optimization problem to

identify the unknown parameters, which is from the parameterized controller. Furthermore, the total number of the iteration steps is derived to terminate the gradient algorithm.

(3) For some complex systems with several controlled variables and with the interaction between these controlled variables, one complex mathematical model is established to include lots of elements in the control matrix. To reduce the number of elements in the control matrix and increase real time property in designing the unknown controllers, an idea of non-interaction property is introduced to simplify our mentioned closed loop system with many variables. To achieve the non-interaction property, some conditions are derived to guarantee one controlled input only influences one output. Based on this simplified model, the prediction error method coming from the system identification field is applied to design the optimal controllers. The advantage of our prediction error method is that the optimal controller is a constant ratio of two polynomials.

6.2 Zonotope parameter identification for piecewise affine system

The piecewise affine system considered in this chapter is one of the hybrid dynamical systems, as a piecewise affine system represents switching dynamics among a collection of linear differential or difference equations with state space being partitioned by a finite number of linear hyper-planes. Hybrid dynamical systems are a class of complex systems that involve interacting discrete events and continuous variable dynamics. They are important in applications in embedded systems, cyber physical systems, robotics, manufacturing systems, traffic management, and biomolecular networks, and have recently been at the center of intense research activity in control theory, computer aided verification, and artificial intelligence communities. But in the control theory, to control a system, one needs to know at least something about how it behaves and reacts to different actions taken on it. Hence we need a model of the system. A system can informally be defined as an entity that interacts with the rest of the world through more or less well defined input and output data. A model is then an approximate description of the system, and an ideal model may be simple, accurate, and general. This approximate description of the system can be constructed by system identification strategy, as the goal of system identification is to build a mathematical model of a dynamic system based on some initial information about the system and the measurement data collected from the system. According to [1], the process of system identification consists of designing and conducting the identification experiment to collect the measurement data, selecting the structure of the model and specifying the parameters to be identified, and eventually fitting the model parameters to the obtained data. Finally, the quality of the obtained model is evaluated through a model validation process. Generally, system identification is an iterative process and if the quality of the obtained model is not satisfactory, some or all of the listed phases can be repeated to obtain one satisfied model for that considered system.

Because of our complex world, all phenomena are described as nonlinear systems. But nonlinear systems cannot be convenient for other applications such as controller design, filter, prediction, etc. So during the system identification procedure, the most common models are linear difference equation descriptions, such as ARX and ARMAX models, as well as linear state space models. When linear models are not sufficient for describing accurately the dynamics of a system, nonlinear identification can be employed. A large number of nonlinear model structures have been constructed to investigate their properties, see [2], where some real time fast convex algorithms are proposed to identify model parameters. Many tools for identification, as well as for control, and stability analysis, have emerged in recent years. To be able to use these proposed tools, a mathematical model of the system is needed.

Identification of hybrid systems (for example, piecewise affine systems) is an area that is related to many other research fields within nonlinear system identification, as such, hybrid systems are sufficiently expressive to model a large number of physical processes, and can approximate nonlinear dynamics with arbitrary accuracy. In addition, given the equivalence between piecewise affine systems and several classes of hybrid systems [3], piecewise affine system identification techniques can be used to obtain hybrid models. The identification of a piecewise affine system is a challenging problem, as it involves the estimation of both the parameters of the affine sub-models and the coefficients of the hyper-planes defining the partition of the state-input set. This issue clearly underlies a classification problem such that each data point corresponds to one sub-model. In particular, one can find several different approaches which are applicable, or related to the piecewise affine system identification problem. Some examples of approaches for piecewise affine systems are neural networks with piecewise affine perceptions. In [4], mixed integer programming is used to solve the piecewise affine system identification problem. As the number of integer variables increases with the number of training samples, mixed integer programming is limited to problems with a small number of observations. To be able to reconstruct a possible discontinuous piecewise affine map with a multi-dimensional domain, [5] proposes to exploit the combined use of clustering, linear identification, and pattern recognition techniques, and allows to identify of both the affine sub-models and the polyhedral partition of the domain. In [6], the sub-model parameters are described through probability density functions, which are iteratively updated through particle filtering algorithms. The sum of norms regularization strategy in [7] can be computationally heavy in case of the appropriate step size. Piecewise affine system identification problem amounts to learning from a set of training data and the parameters defining each affine sub-model [8]. This piecewise affine system identification problem is an NP hard problem in general, see [9], for a detailed explanation of the complexity of piecewise affine system identification. For the sake of simplicity, the sparse property is imposed in piecewise affine systems [10], then the sparse optimization can greatly improve computational efficiency.

The strengths of the piecewise affine system identification problem of [11] are the computational efficiency and the ability to be run both in a batch and in a recursive way, where the combined use of recursive multiple least squares and linear multi-category discrimination is chosen for computing a solution of unconstrained optimization problems. When the identification error that exists in a piecewise affine system is bounded by a quantity, a three stage procedure of a bounded error approach for parametric identification of piecewise affine autoregressive exogenous models is proposed in [12]. But the performance of that bounded error approach is greatly dependent on noise, overestimated model orders, and classification accuracy. The conversion of piecewise affine models from state space input-output form was addressed by deriving necessary and sufficient conditions for a given piecewise affine state space model to admit equivalent representation [13]. From modern control theory, this state space input-output form could not be guaranteed to be a minimal realization, so piecewise affine models are widely used in nonlinear system identification or control. A convex relaxation, based on L1 regulation is proposed in [14] to approximate the underlying combinatorial problem appearing from piecewise affine regression. The statistical clustering technique in [15] first computes the parameters of the affine local models, then partitions the regressor space. The greedy algorithm of [16] to partition in feasible sets of linear inequalities can be computationally heavy in the case of large training sets.

The main limitation of the above approach is that the polyhedral partition of the regressor space is given by the Voronoi diagram, which will limit flexible capability [17]. The problem of finding a lower complexity estimation of piecewise affine models from noise corrupted input-output data is dealt with in [18], where an identification criterion formed by the average of a standard prediction error cost is combined with an L1 regularization term to promote sparse solution [19]. The piecewise linear Hammerstein model can be identified in the presence of a special excitation signal in [20], where it is convenient for the description of the processed with highly nonlinear or discontinuous memory less static functions. [21] employs identification method using pseudo random binary sequences input for decoupling the identification of a nonlinear static block with that of a piecewise affine dynamic block. New iterative algorithms to identify the Hammerstein system with piecewise linear nonlinearities are proposed in [22], further, normalized iterative method produces a convergent result with a smooth nonlinear part and finite impulse linear part. A new form of the Hammerstein model with modified parameterization, which will eliminate the main practical limitations, is introduced in [23], and a parameter estimation algorithm and a novel pole placement controller are presented to tune the identified model. Using a sparse overparameterization, the identification of nonlinear systems using piecewise linear models is turned into a convex optimization problem in [24], where a recursive likelihood based on methodology is proposed to penalize model complexity.

Our considered piecewise affine system is one popular modeling framework for hybrid systems proposed in control theory, furthermore piecewise affine system can be seen as a special case of switched systems with linear dynamics in each mode and mutual exclusive partitions of the state space. Based on the above descriptions of the piecewise affine system, in this chapter, we continue to study the detailed identification strategy for the piecewise affine system. It is well known that in such a piecewise affine system, the space is partitioned into many separate regions and a local linear form is used for each separate region. So the first step in identifying a piecewise affine system is to determine these separate regions. After the separate regions are given, the second identification problem is reduced to identify the linear submodels for each region. To deal with the above mentioned steps, we reformulate the problem of determining the separate regions as a multi class classification problem, which can be solved by a classical first order algorithm of convex optimization theory, such as the mirror descent algorithm or Nesterov's optimal algorithm. As a multi class classification problem coincides with a data clustering process into separate regions. When identifying the unknown parameter in each separate region, many classical identification algorithms can be used directly here, for example, least squares algorithm, maximal likelihood algorithm, Bayesian algorithm, etc. But all the classical identification algorithms hold that the considered noise may be a zero mean random signal. This condition corresponds to the classical probabilistic description of noise. To relax this strict probabilistic description of noise, we investigate the zonotope parameter identification algorithm in the presence of bounded noise. This bounded noise is considered in the set membership identification field widely, and it is a new deterministic identification algorithm. The zonotope parameter identification algorithm computes a set that contains the parameters consistent with the measured output and the given bound of the disturbance. To guarantee our derived zonotope not growing unbounded with iterations, some contracting properties can be imposed. In this work, a sufficient condition for these contracting properties to hold may be formulated as one linear matrix inequality. By solving one optimization problem with linear matrix inequality constraint, one approximate feasible solution set is obtained to contain the parameters.

6.2.1 Piecewise affine system

Consider an affine model as follows:

$$y(t) = -\sum_{i=1}^{n_a} a_i y(t-i) + \sum_{j=1}^{n_b} b_j u(t-i) + e(t) \tag{6.1}$$

where $u(t)$ and $y(t)$ are input and output respectively, $\{a_i\}$ and $\{b_i\}$ are the unknown model parameters, $e(t)$ is an external noise. Two orders n_a and n_b are priori known. This affine model can be rewritten as a linear regression form, after a regression vector $\phi(t)$ is introduced.

$$\begin{cases} \phi(t) = \left[-y(t-1) \quad \cdots \quad -y(t-n_a) \quad u(t-1) \quad \cdots \quad u(t-n_b) \right]^T \\ y(t) = \phi^T(t)\theta + e(t) \end{cases} \quad (6.2)$$

where the unknown parameter vector is stacked as

$$\theta = \left[a_1 \quad \cdots \quad a_{n_a} \quad b_1 \quad \cdots \quad b_{n_b} \right]^T \quad (6.3)$$

For large enough orders n_a and n_b, that affine model can be approximated any linear system. Although the affine model (6.1) is beneficial for the local approximation of some nonlinear systems, it cannot capture any nonlinear properties, so to introduce an affine model into the nonlinear system, a piecewise affine system is obtained. It means that the parameter vector θ is dependent on the region in the regression space, where regression vector $\phi(t)$ lies. Then the regression space is divided into n separate regions $R_1 \cdots R_n$, our considered piecewise affine system can be defined as

$$y(t) = \phi^T(t)\theta_i + e(t) \quad \text{if} \quad \phi(t) \in R_i \quad (6.4)$$

where the parameter vector θ_i depends on its separate region R_i. The problem of identifying a piecewise affine system is reformulated as that, after output and input $\{u(t), y(t)\}$ are collected, how to identify those unknown parameter vectors $\{\theta_i\}_{i=1}^n$? Due to the fact that regression vector $\phi(t)$ is constituted by output and input $\{u(t), y(t)\}$, so the first step is to judge which region the regression vector belongs to.

6.2.2 Multi-class classification process

As there exist n separate regions $R_1 \cdots R_n$, the problem of determining which region the regression vector lies is in conjunction with a multi class classification process. We observe N data points $\{z(t)\}_{t=1}^N$.

$$z(t) = \left[y(t) \quad \phi(t) \right]^T$$
$$= \left[y(t) \quad -y(t-1) \quad \cdots \quad -y(t-n_a) \quad u(t-1) \quad \cdots \quad u(t-n_b) \right]^T \quad (6.5)$$

where N denotes the number of observed data points, and each data point $z(t)$ belongs to one of n non-overlapping classes, along with labels $\lambda_t \in R^n$ which are basic orths in R^n; the index of the only nonzero entry in λ_t is the number of class to which $z(t)$ belongs. We want to build a multi class analogy of the standard linear classifier as follows: a multi class classifier is specified by a matrix A and a vector $a \in R^n$. Given a data point $z(t)$, we compute the n dimensional vector $Az(t) + a$, identify its maximal component, and treat the index of this component as our guess for the serial number of the class to which $z(t)$ belongs.

Let $\bar{\lambda}_t = 1 - \lambda_t$ be the component of λ_t. Given a data point z and the corresponding label λ, let us set

$$h = h(A, a, z, \lambda) = [Az + a] - [\lambda(Az + a)] + \bar{\lambda} \qquad (6.6)$$

If i_* is the index of the only nonzero entry in λ, then the i_*th entry in h is zero. And h is nonpositive if and only if the classifier, given by A, a and evaluated at z, recovers the class i_* of z with margin 1, i.e. we have

$$[Az + a]_j \leq [Az + a]_{i_*} - 1 \quad \text{for} \quad j \neq i_* \qquad (6.7)$$

On the other hand if the classifier fails to classify z correctly, that is

$$[Az + a]_j \geq [Az + a]_{i_*} \quad \text{for some} \quad j \neq i_* \qquad (6.8)$$

Then the maximal entry in h is ≥ 1. So we set

$$\eta(A, a, z, \lambda) = \max_{1 \leq j \leq n} [h(A, a, z, \lambda)]_j \qquad (6.9)$$

We get a nonnegative function which vanishes for the pairs (z, λ), which are quite reliably-with margin $\geq$ 1-classified by (A, a), and is ≥ 1 for the pairs (z, λ) with z not classified correctly. Thus the function

$$H(A, a) = E\{\eta(A, a, z, \lambda)\} \qquad (6.10)$$

The expectation being taken over the distribution of the pairs (z, λ), is an upper bound on the probability for classifier (A, a) to misclassify a data point. What we would do is to minimize $H(A, a)$ over A and a. To do this, since $H(A, a)$ is not observable, we replace the expectation by its empirical counterpart.

$$H_N(A, a) = \frac{1}{N} \sum_{t=1}^{N} \eta(A, a, z(t), \lambda_t) \qquad (6.11)$$

For the sake of simplicity, imposing an upper bound on some norm $\|A\|$ of A, one optimization problem is obtained.

$$\min_{A, a} \frac{1}{N} \sum_{t=1}^{N} \max_{i \leq n} [Az(t) + a - \lambda_t (Az(t) + a) + \bar{\lambda}_t]_i$$
$$\text{subject to} \quad \|A\| \leq 1 \qquad (6.12)$$

A natural choice of the norm $\|A\|$ is the maximum of the $\|A\|_2$ norm. The classical first order algorithm of the convex optimization theory can be used to solve that optimization problem (6.12), such as the mirror descent algorithm or Nesterov's optimal algorithm. Once optimization variables A and a are obtained, then the linear classifier $Az(t) + a$ is got. From the above multi class classification process, we see that once one data point is collected, we can cluster it with a linear classifier. So based on the above linear classifier, all data points can be clustered together, then those data points clustering together as one class can be used in the second identification problem for unknown parameters.

6.2.3   Zonotope parameter identification algorithm

After all collected data points are clustered as *n* classes, then those data points belonging to one same class can be used to identify one unknown parameter. Here, we only rewrite the following piecewise affine system in the *i*th separate region.

$$y(t) = \phi^T(t)\theta_i + e(t) \quad \phi(t) \in R_i \qquad (6.13)$$

The mission of this section is to identify the unknown parameter vector θ_i in case of unknown but bounded noise. It is well known that in equation (6.13), there are two ways to represent uncertainties: the statistical approach and the deterministic approach. In the statistical approach, the uncertainty or disturbance is modeled by a random process with a known statistical property, when estimates of the probability distributing of the uncertainty or disturbance are available. But in many applications, there are situations when then the probability distributing of the uncertainty or disturbance is not known and only the bound of the uncertain domain can be fixed, then the probabilistic assumptions on the uncertainty are no longer valid. In the deterministic approach, the disturbance is assumed to belong to a set. Different families of classical sets are used depending on their properties. The main advantage of the deterministic approach is that disturbance is assumed to be unknown but bounded and this is often simpler to verify than the criterion on the probability disturbance. This is the main reason why we choose the deterministic approach to model the disturbance affecting the system behavior. Based on this remark, one of the deterministic approach-zonotope parameter identification algorithms has been chosen in this section to identify two unknown parameter vector θ_i. This identification algorithm computes a set that contains the parameters consistent with the measured output and the given bound of the disturbance. This set is represented by a zonotope, which is an affine map of a unitary hypercube.

Observing equation (6.13) again, as *e*(*t*) represents the considered noise, this noise belongs to a bounded set, i.e.

$$e(t) \in \{e \in R : |e| \leq \sigma\} \qquad (6.14)$$

where $\sigma \in R$ is an upper bound and *e*(*t*) is unknown, but has known bound

From the set membership identification theory [25], given a set of measured outputs, the feasible solution set is defined as the set of parameters that are consistent with measured outputs and the given bounds. More precisely, the following definitions are given in this section.

Definition 6.1 (Feasible Solution Set): Suppose that the pairs {*y*(*t*), *ϕ*(*t*)}, *t* = 1, 2 ⋯ *N* are given. The vector θ_i is said to belong to the feasible solution set if there exists θ_i such that

$$\left| y(t) - \phi^T(t)\theta_i \right| \leq \sigma, t = 1, 2 \cdots N \qquad (6.15)$$

Definition 6.2 (Information set): Given the pairs $\{y(t), \phi(t)\}$, $t = 1, 2 \cdots N$ at time instant t, the information set I_t is a set of all feasible parameters, that are consistent with the model (6.13), the measured output $y(t)$ and the known bound at time instant t, namely

$$I_t = \left\{\theta_i \in R^{n_a+n_b} : -\sigma \leq y(t) - \phi^T(t)\theta_i \leq \sigma\right\} \tag{6.16}$$

Geometrically I_t represents a strip that consists of $\{y(t), \phi(t)\}$, $t = 1, 2 \cdots N$. Feasible solution set at time instant $t + 1$, denoted as FSS_{t+1}, can be computed exactly from the one corresponding to time instant t by the following recursion:

$$FSS_{t+1} = FSS_t \cap I_t \tag{6.17}$$

It is difficult to compute the feasible solution set [26], so an outer bound of the feasible solution set can be used.

Definition 6.3 (Approximated Feasible Solution Set): An approximated feasible solution set denoted AFSS, is a set that satisfies that FSS. The intersection $FSS_t \cap I_t$ is approximated by means of the intersection between a zonotope and a strip at time instant t.

Definition 6.4 (Zonotope of order m): Given a vector $p \in R^{n_a+n_b}$ and a matrix $H \in R^{(n_a+n_b)\times m}$, a zonotope of order m is a set of $n_1 = n_a + n_b$ dimensional vectors defined by

$$Z = \left\{\theta_i \in R^{n_1} : \theta_i \in p \oplus HB^m\right\} \tag{6.18}$$

where HB^m is a linear projection of B^m into $n_1 = n_a + n_b$ dimensional parameter space, B^m is a unit hypercube of order m, and $\oplus$ denotes Miniowski sum.

Using the approximated feasible solution set on the intersection (6.17), then

$$FSS_{t+1} = FSS_t \cap I_t \subseteq AFSS_{t+1} \tag{6.19}$$

If in equation (6.19) feasible solution set FSS_t is denoted by a zonotope and the information set I_t is a strip, then a family of zonotopes that over bound the intersection between a zonotope and a strip, is derived as the following Theorem 6.1.

Theorem 6.1: Suppose FSS at time instant t, denoted as a zonotope

$$FSS_t = \hat{p}_t \oplus \hat{H}_t B^r \subset R^{n_1} \tag{6.20}$$

The information set or a strip

$$I_t = \left\{\theta_i \in R^{n_a+n_b} : -\sigma \leq y(t) - \phi^T(t)\theta_i \leq \sigma\right\} \tag{6.21}$$

and a scalar γ, define

$$\begin{cases} \hat{p}_t(\gamma) = \hat{p}_t + \gamma\left(y(t) - \phi^T(t)\hat{p}_t\right) \\ \hat{H}_t(\gamma) = \left[\left(I - \gamma\phi(t)\right)\hat{H}_t \quad \sigma\gamma\right] \end{cases} \tag{6.22}$$

Thus we have

$$FSS_{t+1} = FSS_t \cap I_t \subseteq AFSS_{t+1} = \hat{p}_t(\gamma) \oplus \hat{H}_t(\gamma) B^{r+1} \tag{6.23}$$

where I is an identity matrix.

The optimization-based method is used to choose a scalar $\gamma \in R$, by minimizing the volume of a zonotope. Now the minimization of the P-radius of a zonotope is applied, as the P-radius criterion allows to guarantee the non-increasing property of the guaranteed zonotope at each time instant. It tells us that to guarantee the approximated feasible solution set does not grow unbounded with iteration steps, the following inequality relation between two neighboring zonotope is imposed:

$$l_t \le \beta l_{t-1} + \sigma^2 \tag{6.24}$$

where $\beta \in (0,1]$ is a contraction rate, and l_t is the P-radius of zonotope parameter estimation set at time instant t, which is defined by

$$l_t = \max_{\theta_i \in FSS_t} \left(\|\theta_i - \hat{p}_t\|_P^2\right) \tag{6.25}$$

where P is an n_1-dimensional positive definite matrix.

Substituting the definition (6.25) into the inequality relation (6.24), we have

$$\max_{\hat{z} \in B^{r+1}} \left(\left\|\hat{H}_t(\gamma)\hat{z}\right\|_P^2\right) \le \max_{z \in B^r} \beta\left(\left\|\hat{H}_{t-1}z\right\|_P^2\right) + \max_{\eta \in B^1} \|\sigma\eta\|_2^2 \tag{6.26}$$

Expanding equation (6.26) to obtain

$$\hat{z}^T \hat{H}_t(\gamma) P \hat{H}_t(\gamma)\hat{z} - \beta z^T \hat{H}_{t-1} P \hat{H}_{t-1} z - \eta^2 \sigma^2 \le 0 \tag{6.27}$$

Due to the recursion of $\hat{H}_t(\gamma)$ in equation (6.22), we compute

$$\hat{H}_t(\gamma)\hat{z} = \left(I - \gamma\phi(t)\right)\hat{H}_{t-1}z + \sigma\gamma\eta = \left(I - \gamma\phi(t)\right)\bar{z} + \sigma\gamma\eta \tag{6.28}$$

where we set

$$\begin{cases} \bar{z} = \hat{H}_{t-1}z \\ \hat{z} = \begin{bmatrix} z & \eta \end{bmatrix}^T \end{cases} \tag{6.29}$$

Applying equation (6.28) in (6.27), we get

$$\left[\bar{z}^T\left(I - \gamma\phi(t)\right)^T + \sigma\gamma\eta\right] P\left[\left(I - \gamma\phi(t)\right)\bar{z} + \sigma\gamma\eta\right] - \beta\bar{z}^T P\bar{z} - \eta^2\sigma^2 \le 0 \tag{6.30}$$

Formulating above inequality as that

$$\bar{z}^T \left(I - \gamma \phi(t) \right)^T P \left(I - \gamma \phi(t) \right) \bar{z} + \bar{z}^T \left(I - \gamma \phi(t) \right)^T P \sigma \gamma \eta$$

$$+ \sigma \gamma \eta P \left(I - \gamma \phi(t) \right) \bar{z} + \sigma^2 \gamma^2 \eta^2 - \beta \bar{z}^T P \bar{z} - \eta^2 \sigma^2 \leq 0 \qquad (6.31)$$

A sufficient condition for equation (6.31) to hold can be rewritten as one linear matrix inequality

$$\begin{bmatrix} \bar{z} \\ \eta \end{bmatrix} \begin{bmatrix} \left(I - \gamma \phi(t) \right)^T P \left(I - \gamma \phi(t) \right) - \beta P & \left(I - \gamma \phi(t) \right)^T P \sigma \gamma \\ P \sigma \gamma \left(I - \gamma \phi(t) \right) & \sigma^2 \left(\gamma^2 - \right) \end{bmatrix} \begin{bmatrix} \bar{z} \\ \eta \end{bmatrix} \leq 0, \forall^-, \quad (6.32)$$

Using the definition of positive definite matrix allows us to rewrite as that

$$\begin{bmatrix} \left(I - \gamma \phi(t) \right)^T P \left(I - \gamma \phi(t) \right) - \beta P & \left(I - \gamma \phi(t) \right)^T P \sigma \gamma \\ P \sigma \gamma \left(I - \gamma \phi(t) \right) & \sigma^2 \left(\gamma^2 - 1 \right) \end{bmatrix} \leq 0 \quad \forall \begin{bmatrix} \bar{z} \\ \eta \end{bmatrix} \neq 0 \quad (6.33)$$

The linear matrix inequality in (6.33) defines the feasible solution for scalar γ, i.e. γ can be computed by solving the following Eigenvalue problem:

$$\max_{\tau, \gamma} \tau$$

$$\text{subject to } \frac{(1 - \beta) P}{\max\limits_{\eta \in B} \|\gamma \eta\|_2^2} \geq \tau I, \quad \tau > 0$$

$$\begin{bmatrix} \left(I - \gamma \phi(t) \right)^T P \left(I - \gamma \phi(t) \right) - \beta P & \left(I - \gamma \phi(t) \right)^T P \sigma \gamma \\ P \sigma \gamma \left(I - \gamma \phi(t) \right) & \sigma^2 \left(\gamma^2 - 1 \right) \end{bmatrix} \leq 0 \qquad (6.34)$$

The above Eigenvalue problem can be solved by using the convex optimization algorithm, then based on this optimal scalar $\gamma \in R$, a zonotopic outer approximation of the intersection between a zonotope and a strip is obtained by using the matrix inequality optimization strategy. Finally, our zonotope parameter identification algorithm is formulated as follows:

Algorithm 6.1: Zonotope Parameter Identification Algorithm

(1) Obtain measured input-output data and construct regressor vector $\phi(t)$;
(2) Build a strip that bounds the consistent parameters, i.e. information set;

$$I_t = \left\{ \theta_i \in R^{n_a + n_b} : -\sigma \leq y(t) - \phi^T(t) \theta_i \leq \sigma \right\}$$

(3) Construct a zonotope $FSS_t = \hat{p}_t \oplus \hat{H}_t B^r \subset R^{n_1}$ to denote the feasible solution set at time instant t;
(4) Compute the intersection between a zonotope and a strip at time instant t and obtain a new zonotope

$$FSS_{t+1} = FSS_t \cap I_t \subseteq AFSS_{t+1} = \hat{p}_t(\gamma) \oplus \hat{H}_t(\gamma) B^{r+1}$$

to denote the approximated feasible solution set at time instant $t + 1$;

(5) Choose one optimal scalar γ through solving one matrix inequality optimization strategy;
(6) Repeat the above steps and terminate the recursive algorithm when the P-radius l_t is zero, then denote $\hat{p}^*$ as the vector in the last zonotope, so the unknown parameter vector θ_i is given by

$$\hat{\theta}_i = \hat{p}^* \tag{6.35}$$

Similarly, apply the above six steps to identify another unknown parameter vector.

6.2.4 Simulation example

In this section, one simple piecewise affine system is used to prove our strategies, such as the two class classification process and the zonotope parameter identification algorithm. This simple piecewise affine system is given as follows:

$$y(t) = \begin{cases} \phi^T(t)\theta_1 + e(t) & \text{if } \phi(t) > 0 \\ \phi^T(t)\theta_2 + e(t) & \text{if } \phi(t) \leq 0 \end{cases} \tag{6.36}$$

where regression vector $\phi(t)$ and two unknown parameter vectors (θ_1, θ_2) are described as that

$$\begin{cases} \phi(t) = \begin{bmatrix} -y(t) & -u(t) \end{bmatrix}^T \\ \theta_1 = \begin{bmatrix} 7 & 2 \end{bmatrix}^T; \theta_2 = \begin{bmatrix} 2 & 0.5 \end{bmatrix}^T \end{cases} \tag{6.37}$$

We exert the input signal $u(t)$ in the piecewise affine system. The actual input signal is given in Fig. 6.1(a), but this actual input signal is not suited for simulation. So we use its approximated input signal to replace the actual input signal in our simulation, where the approximated input signal is similar to the sinusoidal signal (Fig. 6.1(b)). Then we measure the output signal $y(t)$ by using some measuring devices, the observed output signal is plotted in Fig. 6.2.

Firstly, our mentioned multi class classification process is reduced to a two class classification problem in this simulation. Given one data point $(y(t), \phi(t))$, we need to determine which region this data point belongs to. Here, the number of given data points is $N = 500$, i.e. these 500 data points belong to one of the two classes. The clustering process can be seen in Fig. 6.3, where data points are clustering around two ellipsoids. As the three points deviate away from these two ellipsoids, they are regarded as outliers and we delete them. From Fig. 6.3, we see that all data points are classified correctly, except three data points.

Secondly, in the presence of bounded noise, choose upper bound $|e(t)| \leq \sigma = 0.5$, and all initial parameter values $\hat{\theta}_0 = \dfrac{1}{p_0} I$. Zonotope parameter

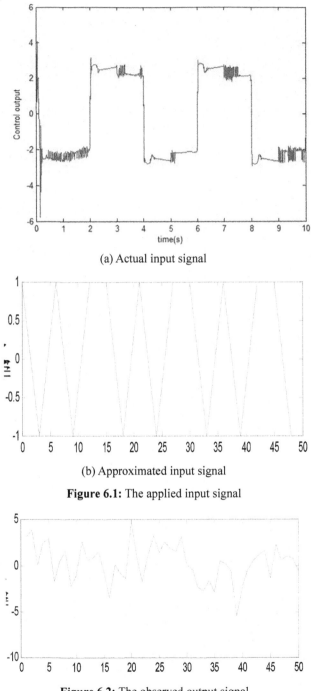

(a) Actual input signal

(b) Approximated input signal

Figure 6.1: The applied input signal

Figure 6.2: The observed output signal

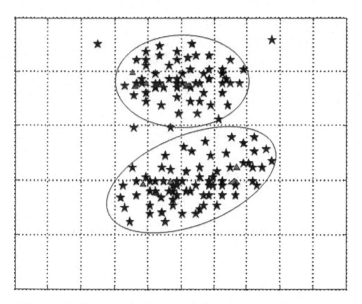

Figure 6.3: Whole clustering process for estimated collected data

identification algorithm is applied to identify those two unknown parameter vectors. Apply the above six steps to construct a sequence of candidate zonotopes, and after 20 iterations, these candidate zonotopes are given in Figs 6.4 and 6.5.

In Fig. 6.4, the black star denotes the optimal parameter vector as $\theta_1 = [7 \ 2]^T$, and a sequence of candidate zonotopes generated by zonotope parameter

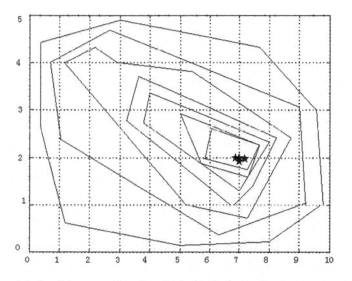

Figure 6.4: Candidate zonotopes iteratively for the first optimal parameter vector

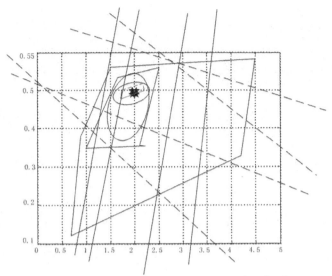

Figure 6.5: Candidate zonotopes iteratively for the second optimal parameter vector

identification algorithm include $\theta_1 = [7 \quad 2]^T$ as their interior point, as these candidate zonotopes have decreasing volumes with iterations, i.e. certain contracting properties hold. Generally, the unknown parameter vector corresponding to $\theta_1 = [7 \quad 2]^T$ can be chosen as the center of the smallest zonotope. Further, the black star is the optimal parameter vector as $\theta_2 = [2 \quad 0.5]^T$ in Fig. 6.5, and results are similar to them in Fig. 6.4

6.2.5 Conclusion

Above, we studied the problem of identifying a piecewise affine system, which combines the linear and nonlinear properties. As it is a nonlinear system that is piecewise affine in the regression space, so the parameter vector depends on the region in the regression space. The separate regions are determined as a multi class classification problem, which is solved by the classical first order algorithm of the convex optimization theory. In the presence of unknown but bounded noise, the zonotope parameter identification algorithm is proposed to identify the unknown parameter vector in each separated region.

6.3 Iterative correlation tuning control for closed loop linear time invariant system

From the previous section's description of the model validation process, we consider the reverse thinking idea to propose a new data driven method and apply the non-correlation test to design a feedback controller. Under the framework of a model reference adaptive control structure, we compute the expected closed loop output after exciting the given expected closed loop transfer function with the external input. After one subtraction operation between the expected closed loop output and true closed loop output, the closed loop matching error is obtained. Then by calculating the cross-correlation function between the above closed loop matching error and external input signal, this cross-correlation function would be equal to zero. It means that the above closed loop matching error is uncorrelated with the external input signal, then a closed loop controller is designed from this un-correlation property.

As the common controller structure is a parameterized form in engineering, such as a classical PID controller so the problem of designing this parameterized controller can be transformed to identify the unknown controller parameters. The whole process of identifying appropriate controller parameters is a process of finding roots, where the cross-correlation function is equal to zero. This process of finding roots can be solved by a stochastic approximation algorithm. When the step size in one stochastic approximation algorithm satisfices one constrained condition, the controller parameters identified by one stochastic approximation algorithm will converge to their true parameters. To improve the convergence of the stochastic approximation algorithm, we construct one criterion function corresponding to the cross-correlation function, and the controller parameters are identified by minimizing this criterion function. From the numerical optimization theory, as the gradient algorithm is very easily implemented in engineering, we use this gradient algorithm to solve that criterion function. Further, in our gradient algorithm, we not only give the important gradient iterative formula, but also derive an upper bound of the number of iteration steps that are required to achieve a specified accuracy. This upper bound quantifies the computational complexity and provides one reference value for us to terminate the iterative algorithm.

6.3.1 Model description

Consider the following closed loop linear time invariant system with a unit feedback in Fig. 6.6.

In Fig. 6.6, the plant is denoted by a rational transfer function form $P(z)$, and $P(z)$ is an unknown linear time invariant stable system, z is a time shift operator $(zu(t) = u(t-1))$, the forward controller $C(z, \theta)$ needs to be designed. Here, we use one unknown parameter vector $\theta \in R^n$ to parameterize this forward controller as follows:

$$\begin{cases} C(z,\theta) = \alpha^T(z)\theta \\ \alpha(z) = \begin{bmatrix} \alpha_1(z) & \alpha_2(z) & \cdots & \alpha_n(z) \end{bmatrix}^T \\ \theta = \begin{bmatrix} \theta_1 & \theta_2 & \cdots & \theta_n \end{bmatrix}^T \end{cases} \tag{6.38}$$

where $\alpha(z)$ denotes one known vector of linear discrete time transfer functions, $r(t)$ is external reference signal, $y(t)$ is true measured output, $M(z)$ is a given closed loop transfer function. The expected output $y_0(t)$ is obtained by exciting the given

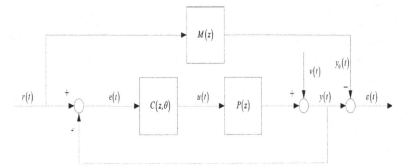

Figure 6.6: The closed loop linear time invariant system with unit feedback

closed loop transfer function $M(z)$ with an external reference signal $r(t)$. $v(t)$ is an additive noise, $u(t)$ is the input signal corresponding to plant $P(z)$, the closed loop matching error $\varepsilon(t)$ is obtained by one subtraction operation between closed loop expected output $y_0(t)$ and true output $y(t)$.

Given expected closed loop transfer function $M(z)$, the input-output measured data sequences $\{r(t), y(t)\}_{t=1}^{N}$ are collected by some sensors, N is the total number of data. The goal of our chapter is to use data sequences

$$\{r(t), \varepsilon(t) = y(t) - y_0(t)\}_{t=1}^{N}$$

to identify the unknown parameter vector $\theta \in R^n$ from controller $C(z, \theta)$. Then the problem of designing the controller can be transformed to identify the unknown parameter vector $\theta \in R^n$.

In Fig. 6.6, the input–output relations corresponding to the plant $P(z)$ are described as follows:

$$\begin{cases} y(t) = P(z)u(t) + v(t) \\ u(t) = C(z,\theta)e(t) = C(z,\theta)[r(t) - y(t)] \end{cases} \qquad (6.39)$$

Formulating (6.39) as that

$$y(t) = \frac{P(z)C(z,\theta)}{1 + P(z)C(z,\theta)}r(t) + \frac{1}{1 + P(z)C(z,\theta)}v(t) \qquad (6.40)$$

In (6.40), $y(t)$ is the true output in a whole closed loop linear time invariant system. Given the expected closed loop transfer function $M(z)$, the expected output $y_0(t)$ is obtained under the external reference signal $r(t)$.

$$y_0(t) = M(z)r(t) \qquad (6.41)$$

The classical model reference adaptive control means that the closed loop transfer function from $r(t)$ to $y(t)$ would approximate the expected transfer function

$M(z)$. The problem of designing feed forward controller $C(z, \theta)$ is formulated as an optimization problem.

$$\min_{\theta \in R^n} J_1(\theta) = \left\| \frac{P(z)C(z,\theta)}{1 + P(z)C(z,\theta)} - M(z) \right\|_2^2 \tag{6.42}$$

As plant $P(z)$ exists in (6.42), so if we use (6.42) to identify that unknown parameter vector θ, we need to construct the mathematical model of plant $P(z)$. To avoid this identification process and make plant $P(z)$ not appear in the optimization problem (6.42), here we introduce one data driven control-iterative correlation tuning control.

6.3.2   Iterative correlation tuning control

Iterative correlation tuning control comes from the idea of model validation in the system identification theory. The purpose of the model validation process is to quantify the accuracy of the identification model and provide one reference standard for accepting it. So firstly, we introduce the model validation process from the system identification theory, and then we extend it to our new iterative correlation tuning control. In the model validation process, assume that the controller $C(z, \theta)$ is known in the closed loop system, then one closed loop identification method is used to obtain the identified model $\hat{P}(z)$ with respect to plant $P(z)$. After choosing input signal $\{r(t)\}_{t=1}^{N}$, the expected output $y_0(t)$ and true output $y(t)$ are collected as $\{y_0(t), y(t)\}_{t=1}^{N}$, which correspond to plant $P(z)$ and identified model $\hat{P}(z)$ respectively under the same external input signal $\{r(t)\}_{t=1}^{N}$.

After subtraction operation between these two different outputs, the closed loop matching error is obtained as

$$\varepsilon(t) = y(t) - y_0(t) \tag{6.43}$$

Model validation process is to judge whether the closed loop matching error $\varepsilon(t)$ is uncorrelated with the external reference signal $r(t)$. So we do the following selection strategy

$$E\left[\varepsilon(t)r(t)\right] \begin{cases} = 0 & \text{accept } \hat{P}(z) \\ \neq 0 & \text{reject } \hat{P}(z) \end{cases} \tag{6.44}$$

Based on the above description of the model validation process and the feed forward controller $C(z, \theta)$ is known in prior, we accept that identified model $\hat{P}(z)$ by judging whether the cross correlation function in (6.44) is zero. But on the contrary, we consider that when the feed forward controller $C(z, \theta)$ is unknown in prior, can we tune the unknown parameter vector $\theta \in R^n$ in controller $C(z, \theta)$ to guarantee the cross correlation function be zero? This is the main essence of our new iterative correlation tuning control.

By observing the closed loop system in Fig. 6.6 again, as the external reference signal $r(t)$ can be chosen freely, i.e. $r(t)$ is known before, the closed loop transfer function $M(z)$ is also known, so according to (6.42) the expected closed loop output is obtained. The true output $y(t)$ can be collected by some sensors, then the closed loop matching error is computed directly as follows:

$$\varepsilon(t,\theta) = y(t) - y_0(t) = \left[\frac{P(z)C(z,\theta)}{1+P(z)C(z,\theta)} - M(z)\right]r(t) + \frac{1}{1+P(z)C(z,\theta)}v(t)$$

$$(6.45)$$

To avoid the appearance of unknown plant $P(z)$ in (6.45), assume there exists one ideal controller $C_0(z, \theta_0)$ such that

$$\frac{P(z)C_0(z,\theta_0)}{1+P(z)C_0(z,\theta_0)} = M(z) \qquad (6.46)$$

Reformulating (6.46) as

$$\frac{1}{1+P(z)C_0(z,\theta_0)} = 1 - M(z) \qquad (6.47)$$

Substituting (6.47) in (6.45), the closed loop matching error is rewritten as

$$\varepsilon(t,\theta) = \left[1-M(z)\right]v(t) + \left[1-M(z)\right]C(z,\theta)y(t) - M(z)r(t) \qquad (6.48)$$

where in (6.48), we use the law of large numbers from the probability theory to obtain the following equality.

$$\frac{1}{1+P(z)C(z,\theta)} \approx \frac{1}{1+P(z)C_0(z,\theta_0)}$$

and the unknown parameter vector θ is replaced by its true value θ_0, (6.48) holds whatever θ and θ_0. From the closed loop matching error (6.48), plant $P(z)$ vanishes and all variables are known except controller $C(z, \theta)$.

Combining (6.45) and (6.48), another form of $\varepsilon(t, \theta)$ is given as follows [11].

$$\varepsilon(t,\theta) = y(t) - y_0(t) = \left[1-M(z)\right]v(t) + \left[1-M(z)\right]C(z,\theta)y(t) - M(z)r(t)$$

$$(6.49)$$

From (6.49), the closed loop matching error $\varepsilon(t, \theta)$ can be computed by one subtraction operation between two outputs, and the difference can be denoted as an explicit relation among expected transfer function $M(z)$, controller $C(z, \theta)$, input-output measured data sequences $\{r(t), y(t)\}_{t=1}^{N}$ and disturb signal $v(t)$.

According to this explicit expression, the unknown parameterized controller $C(z, \theta)$ is included in this closed loop matching error $\varepsilon(t, \theta)$, so we construct the following cross-correlation function.

$$\bar{f}(\theta) = \frac{1}{N} \sum_{t=1}^{N} \xi(t) \varepsilon(t, \theta) \tag{6.50}$$

By using the law of large numbers from probability theory, we obtain that

$$\lim_{N \to \infty} \bar{f}(\theta) = E\{\bar{f}(\theta)\} = f(\theta) \tag{6.51}$$

In the cross-correlation function (6.50), $\xi(t)$ is an instrumental variable. This instrumental variable is usually composed of a part of the input-output measured data. But the choice of instrumental variable $\xi(t)$ must guarantee that it is fully related with the external reference signal, but is uncorrelated with the disturb signal. It means that

$$\begin{cases} E\big[\xi(t) r(t)\big] = \lim_{N \to \infty} \frac{1}{N} \sum_{t=1}^{N} \xi(t) r(t) \neq 0 \\[2mm] E\big[\xi(t) v(t)\big] = \lim_{N \to \infty} \frac{1}{N} \sum_{t=1}^{N} \xi(t) v(t) = 0 \end{cases} \tag{6.52}$$

Rewriting the cross correlation function in (6.14), we have that

$$\begin{aligned} f(\theta) = E\{\bar{f}(\theta)\} &= E\left\{ \frac{1}{N} \sum_{t=1}^{N} \xi(t) \varepsilon(t, \theta) \right\} \\[2mm] &= E\left\{ \frac{1}{N} \sum_{t=1}^{N} \xi(t) \begin{bmatrix} [1-M] v(t) \\ +(1-M) C(\theta) y(t) - Mr(t) \end{bmatrix} \right\} \\[2mm] &= E \frac{1}{N} \sum_{t=1}^{N} \xi(t) [1-M] v(t) + E\left\{ \frac{1}{N} \sum_{t=1}^{N} \xi(t) \big[+(1-M) C(\theta) y(t) - Mr(t) \big] \right\} \\[2mm] &= 0 + E\left\{ \frac{1}{N} \sum_{t=1}^{N} \xi(t) \big[+(1-M) C(\theta) y(t) - Mr(t) \big] \right\} \\[2mm] &= E\left\{ \frac{1}{N} \sum_{t=1}^{N} \xi(t) \big[+(1-M) C(\theta) y(t) - Mr(t) \big] \right\} \end{aligned} \tag{6.53}$$

where in (6.51), for the sake of simplification, variable z is neglected, and we use one condition that the instrumental variable $\xi(t)$ is uncorrelated with the disturb signal $v(t)$.

Assume that cross correlation function $f(\theta)$ in (6.53) be zero. As an unknown parameter vector θ is included in $f(\theta)$, so the estimation of parameter vector θ can be transformed to find roots process, i.e. we want to find roots with respect to the following equation

$$f(\theta) = 0$$

Considering this finding roots process, one stochastic approximation algorithm can be introduced here. It means that the roots of equation $f(\theta) = 0$ can be solved by one iterative formula. This iterative formula is given as follows:

$$\theta_i = \theta_i - \gamma_i \overline{f}(\theta_i) \tag{6.54}$$

where θ_{i+1} is the iterative value at time step $i + 1$, γ_i is a positive scale step size, $\overline{f}(\theta_i)$ is obtained after the iterative value θ_i at time step i is substituted into (6.40). From some knowledge about stochastic approximation algorithm, when that positive step size γ_i satisfies that

$$\sum_{i=1}^{\infty} \gamma_i = \infty, \quad \sum_{i=1}^{\infty} \gamma_i^2 < \infty$$

Then in the iterative formula (6.54), when $i \to \infty$, θ_∞ will converge to one solution which are the roots of the cross correlation function (6.53).

Generally, from the above descriptions of iterative correlation tuning control, this method uses input-output measured data and the expected transfer function to design a parameterized controller. The estimation of the unknown parameter vector is transformed into a finding roots operation concerning one cross correlation function. This iterative correlation process can not only be used to design a controller, but also guarantee the cross correlation function is zero. It satisfies the necessary condition of the model validation process.

6.3.3 Gradient algorithm for identifying unknown parameter vector

For applying the stochastic approximation algorithm in iterative correlation tuning control, some theoretical analyses are given in this section. The constrained condition about the step size is important to guarantee convergence. Here, we relax that requirement on equation $f(\theta) = 0$, and construct a new criterion function to obtain the unknown parameter vector θ. By comparing results, we see that $f(\theta) = 0$ is a special case of our new criterion function.

In (6.53), define a new variable as that

$$\varepsilon_{oe} = (1 - M)C(\theta)y(t) - Mr(t) \tag{6.55}$$

A new criterion function is constructed as the square of the cross correlation function.

$$J_2(\theta) = f^T(\theta)f(\theta) = E^2 \left\{ \frac{1}{N} \sum_{t=1}^{N} \xi(t)\varepsilon_{oe}(\theta) \right\} \tag{6.56}$$

The instrumental variable $\xi(t)$ is chosen as

$$\xi(t) = \left[r(t+n_z) \quad \cdots \quad r(t) \quad \cdots \quad r(t-n_z) \right]^T \tag{6.57}$$

where order n_z is greater than the model order of the closed loop system.
Formulating (6.56) as

$$J_2(\theta) = f^T(\theta) f(\theta) = E\{\overline{f}^T(\theta)\} E\{\overline{f}(\theta)\} = \sum_{\tau=-n_z}^{\tau=n_z} R^2(\tau) \tag{6.58}$$

where $R(\tau)$ is defined as

$$R(\tau) = E\{\varepsilon_{oe}(\theta,t) r(t-\tau)\} \tag{6.59}$$

when the number of data is finite, the new criterion function $J_2(\theta)$ cannot be expressed as a mathematical expectation of a random term. So the stochastic approximation algorithm cannot be used to minimize that new criterion function (6.58) directly.

As the following inequality holds

$$\left(\overline{f}(\theta) - f(\theta) \right)^T \left(\overline{f}(\theta) - f(\theta) \right) \geq 0$$

Expanding the left part of above inequality, we have

$$\overline{f}^T(\theta)\overline{f}(\theta) - \overline{f}(\theta)f(\theta) - f^T(\theta)\overline{f}(\theta) + f^T(\theta)f(\theta) \geq 0$$

Taking expectation operation on both sides, we see

$$E\{\overline{f}^T(\theta)\overline{f}(\theta)\} - f^T(\theta)f(\theta) \geq 0$$

It means that

$$J_3(\theta) = E\{\overline{f}^T(\theta)\overline{f}(\theta)\} \geq J_2(\theta) = f^T(\theta)f(\theta) \tag{6.60}$$

Constructing the following optimization problem

$$\min_{\theta} J_3(\theta) = \min_{\theta} E\{\overline{f}^T(\theta)\overline{f}(\theta)\}$$

$$\overline{f}(\theta) = \frac{1}{N} \sum_{t=1}^{N} \xi(t)\varepsilon_{oe}(\theta) \tag{6.60}$$

In the above unconstrained optimization problem (6.61), lots of algorithms can be introduced to solve it. As the gradient algorithm is very easy, and many software packages can be applied directly, so we use the gradient algorithm to solve optimization problem (6.61). But in the gradient algorithm, we need to compute the partial derivative of $J_3(\theta)$ with respect to unknown parameter vector θ, for example

$$\frac{\partial J_3(\theta)}{\partial \theta} = 2 \frac{\partial \overline{f}(\theta)}{\partial \theta} \overline{f}(\theta), \frac{\partial \overline{f}(\theta)}{\partial \theta} = \frac{1}{N} \sum_{t=1}^{N} \xi(t) \frac{\partial \varepsilon_{oe}(\theta)}{\partial \theta}; \frac{\partial \varepsilon_{oe}(\theta)}{\partial \theta}$$

$$= (1 - M) \frac{\partial C(\theta)}{\partial \theta} y(t) \tag{6.62}$$

Applying the gradient algorithm in optimization problem (6.61), the main steps are given as follows.

Gradient algorithm

Step 1. Choose initial value $\theta_0 \in R^n$

Step 2. At time step k, compute $J_3(\theta_k)$ and $J_3'(\theta_k)$

Step 3. Construct the following iterative formula

$$\theta_{k+1} = \theta_k - h J_3'(\theta_k) \tag{6.63}$$

Step 4. Give a small positive value $\zeta > 0$, check if the following inequality holds

$$\|\theta_{k+1} - \theta_k\| \le \zeta$$

If it holds, then terminate the above iterative formula, or turn to step 2.

The advantage of the iterative formula in a gradient algorithm is that the number of iteration corresponding to the given accuracy ζ can be derived, when the iterative formula is terminated.

Computing the smallest eigenvalue with respect to Hessian matrix $\frac{\partial J_3^2(\theta)}{\partial \theta^2}$, we obtain

$$L = \lambda_{\min} \left(\frac{\partial J_3^2(\theta)}{\partial \theta^2} \right) \tag{6.64}$$

Assume that

$$m_k = \|\theta_k - \theta^*\|$$

where θ^* denotes a global minimum of optimization problem (6.61). To compare the relationship between the iterative value and the global minimum, we continue to compute

$$m_{k+1}^2 = \|\theta_k - \theta^* - h J_3'(\theta_k)\|^2 = m_k^2 - 2h \langle J_3'(\theta_k), \theta_k - \theta^* \rangle$$

$$+ h^2 \|J_3'(\theta_k)\|^2 \le m_k^2 - h \left(\frac{2}{L} - h \right) \|J_3'(\theta_k)\|^2 \tag{6.65}$$

where in (6.65), we use the smallest eigenvalue and one optimality necessary condition

$$J_3'(\theta^*) = 0$$

From (6.65), we see that $m_k \le m_0$.

Continuing to compute the following process

$$J_3\left(\theta_{k+1}\right) \le J_3\left(\theta_k\right) + \left\langle J_3'\left(\theta_k\right), \theta_{k+1} - \theta_k\right\rangle + \frac{L}{2}\left\|\theta_{k+1} - \theta_k\right\|^2 = J_3\left(\theta_k\right) - \omega\left\|J_3'\left(\theta_k\right)\right\|^2$$

(6.66)

where

$$\omega = h\left(1 - \frac{L}{2}\right)h$$

Define the following notation as

$$\Delta_k = J_3\left(\theta_k\right) - J^*\left(\theta^*\right)$$

Then we see that

$$\Delta_k \le \left\langle J_3'\left(\theta_k\right), \theta_{k+1} - \theta^*\right\rangle \le m_0\left\|J_3'\left(\theta_k\right)\right\|$$

(6.67)

As $\Delta_{k+1} \le \Delta_k - \dfrac{\omega}{m_0^2}\Delta_k^2$ and

$$\frac{1}{\Delta_{k+1}} \ge \frac{1}{\Delta_k} + \frac{\omega}{m_0^2}\frac{\Delta_k}{\Delta_{k+1}} \ge \frac{1}{\Delta_k} + \frac{\omega}{m_0^2}$$

(6.68)

Taking summation operation based on (6.68), we obtain the following inequality

$$\frac{1}{\Delta_{k+1}} \ge \frac{1}{\Delta_0} + \frac{\omega}{m_0^2}(k+1)$$

(6.69)

After expanding (6.69) and assuming $h = \dfrac{1}{L}$, then we derive that

$$J_3\left(\theta_k\right) - J_3^*\left(\theta^*\right) \le \frac{2L\left(J_3\left(\theta_0\right) - J_3^*\left(\theta^*\right)\right)\left\|\theta_0 - \theta^*\right\|^2}{2L\left\|\theta_0 - \theta^*\right\|^2 + \left(J_3\left(\theta_0\right) - J_3^*\left(\theta^*\right)\right)h(2 - Lh)k}$$

(6.70)

In (6.70), when the iteration step is at time step k, one error bound between the original criterion function $J_3(\theta_k)$ and its optimal function $J_3^*(\theta^*)$ is derived here.

Based on (6.70), we derive the number of iteration steps used to satisfy the given accuracy ζ. Assume the right part of (6.70) is less than ζ, then

$$\frac{2L\left(J_3\left(\theta_0\right) - J_3^*\left(\theta^*\right)\right)\left\|\theta_0 - \theta^*\right\|^2}{2L\left\|\theta_0 - \theta^*\right\|^2 + \left(J_3\left(\theta_0\right) - J_3^*\left(\theta^*\right)\right)h(2 - Lh)k} \le \zeta$$

(6.71)

Through some basic algebraic operations, we derive one important result in our chapter.

$$k \geq \frac{\left[\dfrac{2L(J_3(\theta_0) - J_3^*(\theta^*)) \| \theta_0 - \theta^* \|^2}{\zeta} - 2L \| \theta_0 - \theta^* \|^2 \right]}{J_3(\theta_0) - J_3^*(\theta^*)} \tag{6.72}$$

The above result gives one appropriate value corresponding to the number of iteration steps. After the gradient algorithm is initiated, this appropriate number of iteration steps can guarantee that the approximate error between the original criterion function and its optimal function will be less than the given accuracy. Furthermore, this appropriate value can be also used in step 4 to terminate the gradient algorithm.

6.3.4   Simulation example

In this section, we apply our iterative correlation tuning control approach in one flight simulation and in one discrete-time linear closed loop system respectively.

(1) Firstly, our iterative correlation tuning control approach is used to design one PID controller in flight simulation.

Flight simulation is a speed servo system with high precision position. Flight simulation with six-degrees of freedom is seen in Fig. 6.7.

The driven element of flight simulation is an electric motor, and the essence of the control structure in flight simulation is a closed loop system corresponding to the position or speed of that electric motor. According to the analysis of the servo control system, one negative feedback part is added to reduce the sensitivity in

Figure 6.7: Flight simulation with six-degrees of freedom

the closed loop system, while the cascade regulator is introduced in each feedback control structure to reduce the dependence on the electric motor's parameter.

Here we give an example of the pitch position tracking loop from flight simulation to verify the feasibility of our iterative correlation tuning control approach in a precision servo control system. In the closed loop system of flight simulation, the photoelectric encoder is mounted on the outer pitch frame, and the angular position signal collected at the outer pitch frame is regarded as the position feedback part. After the difference between two angular positions goes through the position correlation part and power amplifier part, then this difference will make the electric motor start to rotate. The pitch position tracking loop from flight simulation is simplified in Fig. 6.8.

In Fig. 6.8, the input signal is the relative angular signal of the inner pitch loop, and this input signal is collected by one photoelectric encoder located in the inner pitch frame. It means that one photoelectric encoder collects the angular position signal to send one position feedback part. The transfer function model of that simplified pitch position tracking loop can be seen in Fig. 6.9.

In Fig. 6.9, we regard the encoder as a constant and merge it in the power amplifier, then the close loop system is a unit feedback. θ_{me} is the input signal with respect to the electric motor, the controller in this position tracking loop is the classical PID controller. The linear combination of each proportion (P), integral (I), and differential (D) of that difference is used to control the electric motor. The classical PID control structure is given in Fig. 6.10, $r(t)$ is the chosen input signal, and $y(t)$ is the true output, $e(t)$ is the difference or error.

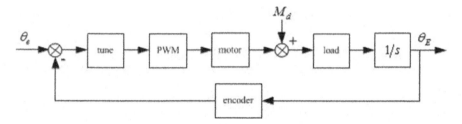

Figure 6.8: The simplified pitch position tracking loop

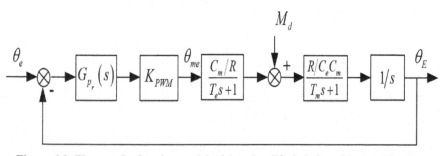

Figure 6.9: The transfer function model of that simplified pitch position tracking loop

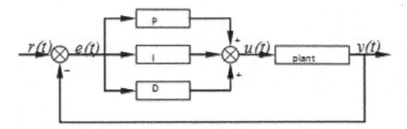

Figure 6.10: The classical PID control structure

PID controller is a kind of linear controller which includes proportion, integral, and differential of error. Its control law is that

$$u(t) = K_P \left[e(t) + \frac{1}{T_I} \int_0^t e(\tau) d\tau + \frac{T_D de(t)}{dt} \right]$$

Rewriting the above control law as the transfer function form

$$G(z) = \frac{U(z)}{E(z)} = K_P (1 + \frac{1}{T_I z} + T_D z) = K_P + \frac{K_i}{z} + K_d z$$

where K_P is the proportional coefficient, T_I is the integral coefficient, and T_D is the differential coefficient. After some computations, we obtain the expected transfer function of outer pitch loop.

$$M(z) = \frac{1.725}{(0.0005z + 1)(2.41z + 1)}, \theta_E = M(z)\theta_e$$

According to the actual parameters of the flight simulation, the electrical and mechanical part of the external pitch frame motor and load are given as follows:

$$\frac{C_m/R}{T_e z + 1} = \frac{0.499}{0.1111z + 1}, \quad \frac{R/C_e C_m}{T_m z + 1} = \frac{2.903}{0.0091z + 1}$$

By comparing this flight simulation example and our iterative correlation tuning control, we regard the motor and load as an integer part, and collect the input-output measured data $\{\theta_{me}, \theta_E\}$ relating to this integer part. θ_e is computed by measured data and the expected transfer function $M(s)$.

$$\theta_e = M^{-1}(z)\theta_E$$

The gradient algorithm is applied to identify those three unknown parameters. After 100 iterations, the three unknown parameters are obtained

$$K_p = 9.8100, K_d = 0.4039, K_i = 2.0325$$

The three PID parameters constitute the tuning part in Fig. 6.8. Applying these three PID parameters in the outer pitch position tracking loop from flight simulation, after ten seconds, the cut-off frequency before correction is very low.

$$\omega_c = 0.55 \text{ Hz}$$

The system bandwidth is narrow, and the dynamic response performance is poor. The open loop cut-off frequency after correlation is changed as

$$\omega_c = 19.8 Hz$$

The system bandwidth is improved significantly, and the overshoot is decreased as $\sigma = 0.81\%$, the settling time is $t_s = 0.442s$. All the indexes meet the real-time requirements. The step response of the system before and after correction is shown in Fig. 6.11.

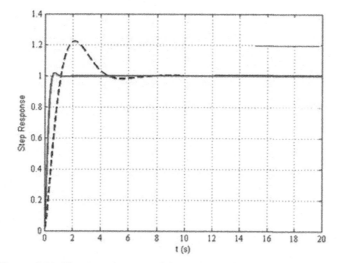

Figure 6.11: The step response of the system before and after correction

In our PID correction part, when the input signal is the sine wave, the real time position tracking response curves at different frequencies are drawn in Fig. 6.12. From Fig. 6.12, the PID correction part which is designed by the gradient algorithm can effectively extend the system bandwidth, accurately track the high frequency signal, and has a small overshoot. Moreover, the properties of no distortion waveform, fast tracking, and real-time are also obvious.

(2) Secondly, our iterative correlation tuning control approach is applied in the one closed loop linear discrete time invariant system.

Consider the linear discrete time invariant system, characterized by the following transfer function model:

$$P(z) = \frac{(z-1.2)(z-0.4)}{z(z-0.3)(z-0.8)}$$

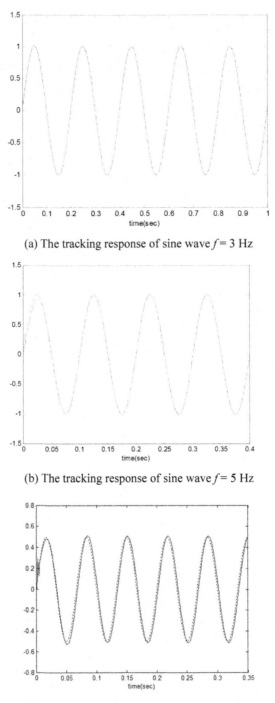

(a) The tracking response of sine wave $f = 3$ Hz

(b) The tracking response of sine wave $f = 5$ Hz

(c) The tracking response of sine wave $f = 15$ Hz

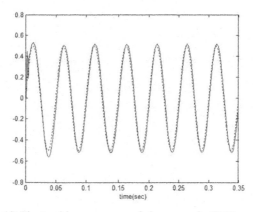

(d) The tracking response of sine wave $f = 20$ Hz

Figure 6.12: The real time position tracking response of the outer pitch loop under PID control strategy

The PID controller is parameterized as

$$C(\theta) = \alpha^T(z)\theta = \begin{bmatrix} \dfrac{z^2}{z^2 - z} & \dfrac{z}{z^2 - z} & \dfrac{1}{z^2 - z} \end{bmatrix} \begin{bmatrix} \theta_1 \\ \theta_2 \\ \theta_3 \end{bmatrix}$$

The true parameterized controller is that

$$C(\theta) = \begin{bmatrix} \dfrac{z^2}{z^2 - z} & \dfrac{z}{z^2 - z} & \dfrac{1}{z^2 - z} \end{bmatrix} \begin{bmatrix} 0.86 \\ 0.2 \\ 0.1 \end{bmatrix}$$

The desired closed-loop transfer function is

$$M(z) = \frac{z(z-1)(0.86z^4 - 1.1z^3 + 3.9z^2 + 0.8z + 0.48)}{z^7 - 3z^6 - 0.96z^5 - 0.72z^4 - 0.93z^3 + 3.9z^2 + 0.8z + 0.48}$$

The input-output data $\{u(t), y(t)\}_{t=1,2\cdots1000}$ in the closed loop system are collected, the number of data points is 1000. Then these data points are used to design the controller directly. The proposed gradient algorithm is used to solve that optimization problem (6.61), before the iterative algorithm starts, the initial values are chosen as

$$\theta = \begin{bmatrix} 0.75 & 0.25 & 0.15 \end{bmatrix}^T$$

The evolution process with the iteration steps when applied to the gradient algorithm is shown in Table 6.1. From Table 6.1, when the iteration steps tend to infinite, each parameter value will approximate to its true value respectively and the objective function in the optimization problem (6.61) will tend to zero.

Table 6.1: The estimation value of the iterative process

i	θ_1	θ_2	θ_3	$J_3\left(\hat{\theta}\right)$
Initial	0.75	0.25	0.15	26.7365
100	0.8492	0.1789	0.0688	12.9280
500	0.8577	0.1805	0.0698	1.82101
1000	0.8571	0.1821	0.0732	0.00861
1500	0.8593	0.1832	0.0776	0.00017
2000	0.8590	0.1954	0.0883	0.00012
2500	0.8622	0.1960	0.0928	0.00000
3000	0.8598	0.1998	0.0998	0.00000
True value	0.86	0.2	0.1	$\rightarrow 0$

To show the proximity of the closed loop transfer function $\dfrac{P(z)C_1(z,\theta)}{1+P(z)C_1(z,\theta)C_2(z,\eta)}$ to the true reference model $M(z)$. The Bode plots between the true closed loop transfer function and the proposed one by our iterative correlation tuning control approach are shown in Fig. 6.13. In Fig. 6.13, the lateral axis denotes the frequency point, and the vertical axis denotes the phase and magnitude of the transfer function. From Fig. 6.13, these Bode plots are coincided with each other. The simulation shows the controller designed by iterative correlation tuning control approach can replace the true controller accurately and the actual closed loop system can undertake the true one.

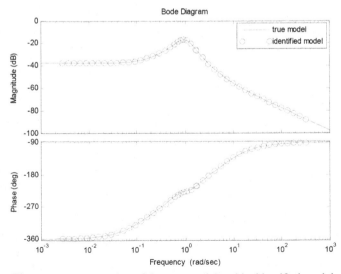

Figure 6.13: Comparing of the true model and its identified model

6.3.5   Conclusion

In this section, we propose a new data driven control-iterative correlation tuning control approach to design one unknown parameterized controller in a closed loop system by using only input-output data. This new approach transforms the problem of identifying the unknown parameter into one minimization problem about a cross correlation function. Further one error bound corresponding to the gradient algorithm is derived.

6.4   Controller design for many variables closed loop system under non-interaction condition

For complex systems with several controlled variables and with the interaction between these controlled variables, a new design criterion must be proposed. This is the criterion of non-interaction. The purpose of this short note is to give a general method for designing such non-interaction controllers with many variables in closed loop systems of arbitrary complexity. More specifically, consider one system with multi-input and multi-output for our considered system. Then, if the number of input variables is greater than the number of output variables, the controlled variables are all the corresponding output variables and partly input variables. Given the prescribed measurements for input variables, two kinds of error functions are obtained. Based on these two error functions, linear controllers are used to combine some linear relations within the transfer function forms. For the case of the non-interaction property during the input-output relation, we derive two conditions for satisfying the non-interaction property, which corresponds to two different elements of two control matrices. When these two conditions are satisfied for the controllers, then the problem of designing the controllers is solved. The classical prediction error method, coming from system identification research, is proposed to achieve the goal of designing the unknown controllers. Generally, in this short note, our mission is to combine the control theory and system identification theory into considering the non-interaction for one complex system with multi-inputs and multi-outputs.

6.4.1   Closed loop system with many variables

For the convenience of understanding our considered closed loop system with many variables, firstly we consider one simple closed loop system, where its controlled output and excitation input are $y(t)$ and $x(t)$ respectively, furthermore, their Laplace transformations are $Y(s)$ and $X(s)$. Here t is the time variable and s is the frequency variable. The simple closed loop system is plotted in Fig. 6.14, where $E(s)$ is the transfer function for the plant, $L(s)$ is the transfer function for measurement, i.e. measurement is used to constitute the named feedback effect. $S(s)$ is the transfer function for the servo motor, $C(s)$ is the transfer function for the designed controller. This controller is needed to be designed by the researcher. $V(s)$ is one random disturbance, coming from external influence.

Relation between input $W(s)$ and output $Y(s)$ is formulated as

$$Y(s) = E(s)W(s) = E(s)[S(s)U(s) + V(s)]$$

where $U(s)$ is the controller output and it holds that

$$U(s) = C(s)(X(s) - Z(z)) = C(s)(X(s) - L(z)Y(s))$$

Deleting $U(s)$ to get

$$Y(s) = \frac{E(s)U(s)C(s)L(s)}{E(s)U(s)C(s)L(s)+1}X(s) + \frac{E(s)}{E(s)U(s)C(s)L(s)+1}$$

The above equation is the Laplace transformation for output $y(t)$ and input $x(t)$ under one certain initial condition.

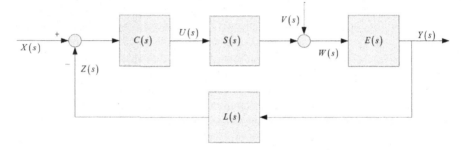

Figure 6.14: One simple closed loop system

In this simple closed loop system, the number of each physical variable is 1, so this system is also called the single input and single output closed loop system. Now similarly to Fig. 6.14, we start to consider one many-variable closed loop system with many controlled outputs $\{y_i(t)\}_{i=1}^m$ and many control settings, or input $\{w_i(t)\}_{i=1}^n$, where the number of controlled outputs be m and the number of inputs be n. Their Laplace transforms are $\{Y_i(s)\}_{i=1}^m$ and $\{W_i(s)\}_{i=1}^n$ respectively. Consider the control designed in Fig. 6.15, the relations between inputs $\{W_i(s)\}_{i=1}^n$ and the outputs $\{Y_i(s)\}_{i=1}^m$ are given by

$$\begin{cases} Y_1(s) = E_{11}(s)W_1(s) + E_{12}(s)W_2(s) + \cdots + E_{1n}(s)W_n(s) \\ Y_2(s) = E_{21}(s)W_1(s) + E_{22}(s)W_2(s) + \cdots + E_{2n}(s)W_n(s) \\ \vdots \\ Y_m(s) = E_{m1}(s)W_1(s) + E_{m2}(s)W_2(s) + \cdots + E_{mn}(s)W_n(s) \end{cases} \quad (6.73)$$

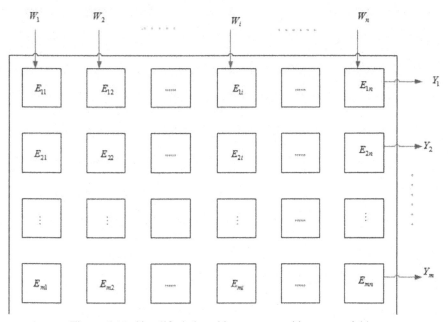

Figure 6.15: Simplified closed loop system with many variables

where in equation (6.73), $\{E_{ij}(s)\}_{i=1\cdots m,\, j=1\cdots n}$ are the closed loop transfer functions and each $E_{ij}(s)$ is the transfer function which, when operated on the input $W_i(s)$, gives a component of the output $Y_i(s)$. Commonly, each transfer function $E_{ij}(s)$ is a ratio of the two polynomials of the frequency variable s. For the sake of simplicity, equation (6.73) can be rewritten as

$$Y_i(s) = \sum_{j=1}^{n} E_{ij}(s)W_i(s), i = 1, 2 \cdots m \qquad (6.74)$$

In the process of designing controllers, our goals are to make all outputs $\{Y_i(s)\}_{i=1}^{m}$ vary with the prescribed values $\{X_i(s)\}_{i=1}^{m}$, and in addition set $E_\mu(s)$ for inputs $\{W_\mu(s)\}_{\mu=m+1}^{n}$, as the number of inputs is greater than the number of outputs. So the controlled quantities are the outputs $\{Y_i(s)\}_{i=1}^{m}$ and the inputs $\{W_\mu(s)\}_{\mu=m+1}^{n}$.

If the measurement for each input $\{W_i(s)\}_{i=1}^{n}$ is denoted by $\{\gamma_i(s)\}_{i=1}^{n}$, then the errors are $\{E_i(s) - \gamma_i(s)\}_{i=1}^{n}$.

Similarly the errors between the prescribed values $\{X_i(s)\}_{i=1}^{m}$ and their measurement values $\{Z_i(s)\}_{i=1}^{m}$ are defined as $\{X_i(s) - Z_i(s)\}_{i=1}^{m}$.

The essence of the control effect is to use these two kinds of errors to generate control signals $\{U_i(s)\}_{i=1}^{n}$, and always the control signals $\{U_i(s)\}_{i=1}^{n}$ depend on all errors linearly. As n error signals exist, then there are n control signals $\{U_i(s)\}_{i=1}^{n}$. Thus

$$
\begin{cases}
U_1(s) = C_{11}(X_1 - Z_1) + C_{12}(X_2 - Z_2) + \cdots + C_{1i}(X_i - Z_i) \\
\quad + C'_{1,i+1}(E_{i+1} - \gamma_{i+1}) + \cdots + C'_{1,n}(E_n - \gamma_n) \\
U_2(s) = C_{21}(X_1 - Z_1) + C_{22}(X_2 - Z_2) + \cdots + C_{2i}(X_i - Z_i) \\
\quad + C'_{2,i+1}(E_{i+1} - \gamma_{i+1}) + \cdots + C'_{2,n}(E_n - \gamma_n) \\
\vdots \\
U_n(s) = C_{n1}(X_1 - Z_1) + C_{n2}(X_2 - Z_2) + \cdots + C_{ni}(X_i - Z_i) \\
\quad + C'_{n,i+1}(E_{i+1} - \gamma_{i+1}) + \cdots + C'_{n,n}(E_n - \gamma_n)
\end{cases} \tag{6.75}
$$

Similarly, equation (6.75) is compressed as

$$
U_k(s) = \sum_{v=1}^{i} C_{kv}(X_v(s) - Z_v(s)) + \sum_{\mu=i+1}^{n} C'_{k\mu}(E_\mu(s) - \gamma_\mu(s)), k = 1, 2 \cdots n \tag{6.76}
$$

where in equation (6.76) the control matrices are separated into C and C', which correspond to two different kinds of error signals.

6.4.2 Conditions for non-interaction

From the engineering control theory, the measured value $Z_i(s)$ and $\gamma_\mu(s)$ are related to $Y_i(s)$ and $W_\mu(s)$ by using the transfer functions $L_{ii}(s)$ and $L_{\mu\mu}(s)$, i.e.

$$
\begin{cases}
Z_i(s) = L_{ii}(s)Y_i(s) \\
\gamma_\mu(s) = L_{\mu\mu}(s)W_\mu(s)
\end{cases} \tag{6.77}
$$

When consider external noise $V_k(s)$, then control signal $U_k(s)$ and external noise $V_k(s)$ are combined to generate input signal $W_k(s)$, it means that

$$
W_k(s) = S_{kk}(s)U_k(s) + V_k(s), k = 1, 2 \cdots n \tag{6.78}
$$

where $S_{kk}(s)$ is also a transfer function from control signal $U_k(s)$ to input signal $W_k(s)$. Equations (6.77) and (6.78) describe the control system with many variables.

Observing the above equations and substituting (6.78) into (6.74), it is that

$$
Y_i(s) = \sum_{j=1}^{n} E_{ij}(s)W_i(s) = \sum_{j-1}^{n} E_{ij}(s)\left[S_{kk}(s)U_k(s) + V_k(s)\right] \tag{6.79}
$$

Further substituting equation (6.76) into the above (6.79), it holds that

$$Y_i(s) = \sum_{j=1}^{n} E_{ij}(s) \left[S_{jj}(s) \left[\sum_{v=1}^{m} C_{jv}(X_v(s) - Z_v(s)) \right. \right.$$

$$\left. \left. + \sum_{\mu=m+1}^{n} C'_{j\mu}(E_\mu(s) - \gamma_\mu(s)) \right] + V_k(s) \right]$$

$$= \sum_{j=1}^{n} \left\{ \sum_{v=1}^{m} E_{ij}(s) S_{jj}(s) C_{jv}(X_v(s) - Z_v(s)) \right.$$

$$\left. + \sum_{\mu=m+1}^{n} E_{ij}(s) S_{jj}(s) C'_{j\mu}(E_\mu(s) - \gamma_\mu(s)) + E_{ij}(s) V_k(s) \right\} \quad (6.80)$$

Substituting equation (6.77) into the above (6.80) and changing the index parameter to give that

$$Y_i(s) = \sum_{k=1}^{n} \left\{ \sum_{v=1}^{m} E_{ik}(s) S_{kk}(s) C_{kv}(X_v(s) - L_{vv}(s) Y_v(s)) \right.$$

$$\left. + \sum_{\mu=m+1}^{n} E_{ik}(s) S_{kk}(s) C'_{k\mu}(E_\mu(s) - L_{\mu\mu}(s) W_\mu(s)) + E_{vk}(s) V_k(s) \right\} \quad (6.81)$$

Similarly, after simple but tedious calculation, substituting equation (6.76) into (6.78), then

$$W_k(s) = S_{kk}(s) \left[\sum_{v=1}^{m} C_{kv}(X_v(s) - Z_v(s)) + \sum_{\mu=m+1}^{n} C'_{k\mu}(E_\mu(s) - \gamma_\mu(s)) \right] + V_k(s)$$

$$= \sum_{v=1}^{m} S_{kk}(s) C_{kv}(X_v(s) - L_{vv}(s) Y_v(s))$$

$$+ \sum_{\mu=m+1}^{n} S_{kk}(s) C'_{k\mu}(E_\mu(s) - L_{\mu\mu}(s) W_\mu(s)) + V_k(s) \quad (6.82)$$

where we use equation (6.77) in deriving equation (6.82).

Combining equations (6.81) and (6.82), there are two kinds of controllers $\left\{ \{C_{kv}\}_{k=1,2\cdots n, v=1,2\cdots m}, \{C'_{k\mu}\}_{k=1,2\cdots n, \mu=m+1\cdots n} \right\}$ that are required to be designed in our later process. As the number of controllers $\left\{ \{C_{kv}\}_{k=1,2\cdots n, v=1,2\cdots m}, \{C'_{k\mu}\}_{k=1,2\cdots n, \mu=m+1\cdots n} \right\}$ is n^2, and in practical engineering the number n^2 will be large. So to reduce the computational complexity and increase the real time property, the non-interaction of control is now introduced here. The main goal of the non-interaction of control is to determine conditions on the elements of the control matrix C_{kv} and $C'_{k\mu}$, such that the setting $X_j(s)$ and

$E_\mu(s)$ will affect only their respective corresponding variables $Y_j(s)$ and $W_\mu(s)$, and nothing else.

To achieve the above non-interaction property, consider a special output $Y_g(s)$, $g \in \{1, 2 \cdots m\}$, then equations (6.81) and (6.82) can be rewritten as

$$Y_i(s) = \sum_{k=1}^{n} \left\{ \sum_{v=1, v \neq g}^{m} E_{ik}(s) S_{kk}(s) C_{kv}\left(X_v(s) - L_{vv}(s) Y_v(s)\right) \right.$$

$$\left. + \sum_{\mu=m+1}^{n} E_{ik}(s) S_{kk}(s) C'_{k\mu}\left(E_\mu(s) - L_{\mu\mu}(s) W_\mu(s)\right) + E_{vk}(s) V_k(s) \right\}$$

$$+ \sum_{k=1}^{n} E_{ik}(s) S_{kk}(s) C_{kg}\left(X_g(s) - L_{gg}(s) Y_g(s)\right) \tag{6.83}$$

and

$$W_k(s) = \sum_{v=1, v \neq g}^{m} S_{kk}(s) C_{kv}\left(X_v(s) - L_{vv}(s) Y_v(s)\right)$$

$$+ \sum_{\mu=m+1}^{n} S_{kk}(s) C'_{k\mu}\left(E_\mu(s) - L_{\mu\mu}(s) W_\mu(s)\right) + V_k(s)$$

$$+ S_{kk}(s) C_{kg}\left(X_g(s) - L_{gg}(s) Y_g(s)\right) \tag{6.84}$$

To guarantee that $X_g(s)$ will not influence $Y_i(s)$ or $W_\mu(s)$, except $Y_g(s)$, the last terms of the above equations must be zero for $i \neq g$ and for $k > i$. Then the condition for non-interaction is that

$$\begin{cases} \sum_{k=1}^{n} E_{ik}(s) S_{kk}(s) C_{kg} & \text{if } i \neq g \\ C_{kg} = 0 & \text{if } k > i \end{cases} \tag{6.85}$$

Equation (6.85) is the non-interaction condition on the elements of control matrix C_{kg}. To determine the non-interaction condition on the elements of another control matrix $C'_{k\mu}$, we need to consider the non-interaction condition for the controlled variables $W_\mu(s)$, $\mu = m + 1 \cdots n$, we rewrite two equations (6.81) and (6.82) as that

$$Y_j(s) = \sum_{k=1}^{n} \left\{ \sum_{v=1}^{m} E_{jk}(s) S_{kk}(s) C_{kv}\left(X_v(s) - L_{vv}(s) Y_v(s)\right) \right.$$

$$\left. + \sum_{\mu=m+1, \mu \neq r}^{n} E_{jk}(s) S_{kk}(s) C'_{k\mu}\left(E_\mu(s) - L_{\mu\mu}(s) W_\mu(s)\right) + E_{jk}(s) V_k(s) \right\}$$

$$+ \sum_{k=1}^{n} E_{jk}(s) S_{kk}(s) C'_{kr}\left(E_r(s) - L_{rr}(s) W_r(s)\right) \tag{6.86}$$

and

$$W_k(s) = \sum_{v=1}^{m} S_{kk}(s) C_{kv} \left(X_v(s) - L_{vv}(s) Y_v(s) \right)$$

$$+ \sum_{\mu=m+1, \mu \neq r}^{n} S_{kk}(s) C'_{k\mu} \left(E_\mu(s) - L_{\mu\mu}(s) W_\mu(s) \right) + V_k(s)$$

$$+ S_{kk}(s) C'_{kr} \left(E_r(s) - L_{rr}(s) W_r(s) \right) \qquad (6.87)$$

where

$$r \in \{m+1, m+2 \cdots n\}, j = 1, 2 \cdots m,$$

$$k \in \{m+1, m+2 \cdots n\}$$

From equations (6.86) and (6.87), we see that if control signal $E_r(s)$ only affects variables $W_r(s)$, it holds that the last terms must be zero, i.e.

$$\sum_{k=1}^{n} E_{jk}(s) S_{kk}(s) C'_{kr} = 0, j = 1, 2 \cdots m \qquad (6.88)$$

and

$$C'_{kr} = 0, k, r = m+1, m+2, \cdots n, k \neq r \qquad (6.89)$$

Equations (6.88) and (6.89) are the non-interaction condition on the elements of control matrix $C'_{k\mu}$.

Combining equations (6.85), (6.88) and (6.89), they are the conditions for non-interaction, corresponding to our considered closed loop system with many variables.

When above non-interaction conditions are satisfied, then equation (6.81) will simply be as

$$Y_i(s) = \sum_{v=1}^{m} \left(X_v(s) - L_{vv}(s) Y_v(s) \right) \sum_{k=1}^{n} E_{ik}(s) S_{kk}(s) C_{kv}$$

$$+ \sum_{\mu=m+1}^{n} \left(E_\mu(s) - L_{\mu\mu}(s) W_\mu(s) \right) \sum_{k=1}^{n} E_{ik}(s) S_{kk}(s) C'_{k\mu} + \sum_{k=1}^{n} E_{ik}(s) V_k(s)$$

$$(6.90)$$

According to equation (6.88), above equation (6.90) can be further simplified as

$$Y_i(s) = \left(X_i(s) - L_{ii}(s) Y_i(s) \right) \sum_{k=1}^{m} E_{ik}(s) S_{kk}(s) C_{kv} + \sum_{k=1}^{n} E_{ik}(s) V_k(s) \qquad (6.91)$$

The solution of equation (6.91) can be derived as follows:

$$Y_i(s) = \left(X_i(s) - L_{ii}(s)Y_i(s)\right)\sum_{k=1}^{m}E_{ik}(s)S_{kk}(s)C_{kv} + \sum_{k=1}^{n}E_{ik}(s)V_k(s)$$

$$\updownarrow$$

$$Y_i(s) = X_i(s)\sum_{k=1}^{m}E_{ik}(s)S_{kk}(s)C_{kv} - \sum_{k=1}^{m}E_{ik}(s)S_{kk}(s)C_{kv}L_{ii}(s)Y_i(s)$$

$$+ \sum_{k=1}^{n}E_{ik}(s)V_k(s)$$

$$\updownarrow$$

$$Y_i(s) = \left[1 + \sum_{k=1}^{m}E_{ik}(s)S_{kk}(s)C_{kv}L_{ii}(s)\right]^{-1}$$

$$\left[\sum_{k=1}^{m}E_{ik}(s)S_{kk}(s)C_{kv}X_i(s) + \sum_{k=1}^{n}E_{ik}(s)V_k(s)\right]$$

$$\updownarrow$$

$$Y_i(s) = R_{ii}(s)X_i(s) + \left[R_{ii}(s)L_{ii}(s) - 1\right]A_{ii}(s) \tag{6.92}$$

where

$$\begin{cases} R_{ii}(s) = \left[1 + \sum_{k=1}^{m}E_{ik}(s)S_{kk}(s)C_{kv}L_{ii}(s)\right]^{-1}\sum_{k=1}^{m}E_{ik}(s)S_{kk}(s)C_{kv}X \\ A_{ii}(s) = \sum_{k=1}^{n}E_{ik}(s)V_k(s) \end{cases} \tag{6.93}$$

Similarly, we have that

$$W_\mu(s) = S_{\mu\mu}(s)C'_{\mu\mu}\left(E_\mu(s) - L_{\mu\mu}(s)W_\mu(s)\right) + V_\mu(s)$$

$$\updownarrow$$

$$W_\mu(s) = \left[1 + S_{\mu\mu}(s)C'_{\mu\mu}L_{\mu\mu}(s)\right]^{-1}S_{\mu\mu}(s)C'_{\mu\mu}E_\mu(s)$$

$$+ \left[1 + S_{\mu\mu}(s)C'_{\mu\mu}L_{\mu\mu}(s)\right]^{-1}V_\mu(s)$$

$$\updownarrow$$

$$W_\mu(s) = R'_{\mu\mu}(s)E_\mu(s) + \left[R'_{\mu\mu}(s)L_{\mu\mu}(s) - 1\right]V_\mu(s) \tag{6.94}$$

where

$$R'_{\mu\mu}(s) = \left[1 + S_{\mu\mu}(s)C'_{\mu\mu}L_{\mu\mu}(s)\right]^{-1}S_{\mu\mu}(s)C'_{\mu\mu} = \frac{S_{\mu\mu}(s)C'_{\mu\mu}}{1 + S_{\mu\mu}(s)C'_{\mu\mu}L_{\mu\mu}(s)} \tag{6.95}$$

Consider the above mathematical derivations, the relations for calculating the controlled variables from the control signal and external noise are formulated as

$$\begin{cases} Y_i(s) = R_{ii}(s)X_i(s) + \left[R_{ii}(s)L_{ii}(s) - 1 \right] A_{ii}(s) \\ W_\mu(s) = R'_{\mu\mu}(s)E_\mu(s) + \left[R'_{\mu\mu}(s)L_{\mu\mu}(s) - 1 \right] V_\mu(s) \end{cases} \tag{6.96}$$

where in equation (6.96), $Y_i(s)$ is the controlled output, $W_\mu(s)$ is the input, $X_i(s)$ and $E_\mu(s)$ are reference values corresponding to $Y_i(s)$ and $W_\mu(s)$ respectively. $A_{ii}(s)$ is the filtered noise, whose filter is the transfer function $E_{ik}(s)$, $V_\mu(s)$ is the external noise.

But in the process of designing controllers $\{C_{ki}, C'_{\mu\mu}\}$, these two controllers do not exist in equation (6.96) explicitly. They are only included in transfer functions $R_{ii}(s)$ and $R'_{\mu\mu}(s)$ implicitly. It means that if we can derive the solutions of $R_{ii}(s)$ and $R'_{\mu\mu}(s)$, then these two controllers will be obtained easily. For example, we can use the following method to obtain controller $C'_{\mu\mu}$ from $R'_{\mu\mu}(s)$.

$$R'_{\mu\mu}(s) = \left[1 + S_{\mu\mu}(s)C'_{\mu\mu}L_{\mu\mu}(s) \right]^{-1} S_{\mu\mu}(s)C'_{\mu\mu} = \frac{S_{\mu\mu}(s)C'_{\mu\mu}}{1 + S_{\mu\mu}(s)C'_{\mu\mu}L_{\mu\mu}(s)}$$

$$\updownarrow$$

$$R'_{\mu\mu}(s) + R'_{\mu\mu}(s)S_{\mu\mu}(s)C'_{\mu\mu}L_{\mu\mu}(s) = S_{\mu\mu}(s)C'_{\mu\mu}$$

$$\updownarrow$$

$$C'_{\mu\mu}\left(S_{\mu\mu}(s) - R'_{\mu\mu}(s)S_{\mu\mu}(s)L_{\mu\mu}(s) \right) = R'_{\mu\mu}(s)$$

$$\updownarrow$$

$$C'_{\mu\mu} = \frac{R'_{\mu\mu}(s)}{S_{\mu\mu}(s) - R'_{\mu\mu}(s)S_{\mu\mu}(s)L_{\mu\mu}(s)} \tag{6.97}$$

where in the solution $C'_{\mu\mu}$, $S_{\mu\mu}(s)$ and $L_{\mu\mu}(s)$ are all known transfer functions, only $R'_{\mu\mu}(s)$ is unknown, so our next problem is how to solve $R'_{\mu\mu}(s)$ from the relations (6.96).

6.4.3 Designing controllers

Here we use the prediction error method to obtain the transfer function $R'_{\mu\mu}(s)$, then controller $C'_{\mu\mu}$ is given by equation (6.97). The derivation about the transfer function $R_{ii}(s)$ and controller C_{ki} is similar, so in this section we only give a detailed derivation on the transfer function $R'_{\mu\mu}(s)$.

Rewriting the relation as that

$$W_\mu(s) = R'_{\mu\mu}(s)E_\mu(s) + \left[R'_{\mu\mu}(s)L_{\mu\mu}(s) - 1 \right] V_\mu(s) \tag{6.98}$$

One step ahead prediction value on $W_\mu(s)$ is compressed as

$$\begin{aligned} \hat{W}_\mu(s) &= \frac{1}{1 - R'_{\mu\mu}(s)L_{\mu\mu}(s)} R'_{\mu\mu}(s)E_\mu(s) + \left[1 - \frac{1}{1 - R'_{\mu\mu}(s)L_{\mu\mu}(s)} \right] W_\mu(s) \\ &= \frac{R'_{\mu\mu}(s)}{1 - R'_{\mu\mu}(s)L_{\mu\mu}(s)} E_\mu(s) - \frac{R'_{\mu\mu}(s)L_{\mu\mu}(s)}{1 - R'_{\mu\mu}(s)L_{\mu\mu}(s)} W_\mu(s) \end{aligned} \tag{6.99}$$

Constructing one step ahead prediction error or residual as

$$\varepsilon(t) = W_\mu(s) - \hat{W}_\mu(s)$$

$$= W_\mu(s) - \frac{R'_{\mu\mu}(s)}{1 - R'_{\mu\mu}(s)L_{\mu\mu}(s)}E_\mu(s) + \frac{R'_{\mu\mu}(s)L_{\mu\mu}(s)}{1 - R'_{\mu\mu}(s)L_{\mu\mu}(s)}W_\mu(s)$$

$$= \frac{1 - R'_{\mu\mu}(s)L_{\mu\mu}(s) + R'_{\mu\mu}(s)L_{\mu\mu}(s)}{1 - R'_{\mu\mu}(s)L_{\mu\mu}(s)}W_\mu(s) - \frac{R'_{\mu\mu}(s)}{1 - R'_{\mu\mu}(s)L_{\mu\mu}(s)}E_\mu(s)$$

$$= \frac{1}{1 - R'_{\mu\mu}(s)L_{\mu\mu}(s)}W_\mu(s) - \frac{R'_{\mu\mu}(s)}{1 - R'_{\mu\mu}(s)L_{\mu\mu}(s)}E_\mu(s)$$

$$= \frac{1}{1 - R'_{\mu\mu}(s)L_{\mu\mu}(s)}\left[W_\mu(s) - R'_{\mu\mu}(s)E_\mu(s)\right] \tag{6.100}$$

In the standard prediction error method, using only measured data set $\{W_\mu(s), E_\mu(s)\}$, the unknown transfer function is given by solving one numerical optimization problem

$$\underset{R'_{\mu\mu}(s)}{\arg\min}\frac{1}{N}\sum_{s_1}^{s_N}\varepsilon^2(s) = \underset{R'_{\mu\mu}(s)}{\arg\min}\frac{1}{N}\sum_{s_1}^{s_N}\frac{1}{\left[1 - R'_{\mu\mu}(s)L_{\mu\mu}(s)\right]^2}$$

$$\left[W_\mu(s) - R'_{\mu\mu}(s)E_\mu(s)\right]^2 \tag{6.101}$$

where the number of frequency variables is N, i.e. $s \in \{s_1, s_2 \cdots s_N\}$.
    Set

$$\begin{cases} W_\mu = \begin{bmatrix} W_\mu(s_1) & W_\mu(s_2) & \cdots & W_\mu(s_N) \end{bmatrix}^T \\ E_\mu = \begin{bmatrix} E_\mu(s_1) & E_\mu(s_2) & \cdots & E_\mu(s_N) \end{bmatrix}^T \end{cases} \tag{6.102}$$

Then

$$\sum_{s_1}^{s_N}\left[W_\mu(s) - R'_{\mu\mu}(s)E_\mu(s)\right]^2 = \left(W_\mu - R'_{\mu\mu}E_\mu\right)^T\left(W_\mu - R'_{\mu\mu}E_\mu\right) \tag{6.103}$$

Then the original numerical optimization problem is compressed as

$$\underset{R'_{\mu\mu}(s)}{\arg\min}\frac{1}{N}\sum_{s_1}^{s_N}\varepsilon^2(s) = \underset{R'_{\mu\mu}(s)}{\arg\min}\frac{1}{N}\sum_{s_1}^{s_N}\frac{1}{\left[1 - R'_{\mu\mu}L_{\mu\mu}\right]^2}\left(W_\mu - R'_{\mu\mu}E_\mu\right)^T\left(W_\mu - R'_{\mu\mu}E_\mu\right)$$

$$\tag{6.104}$$

Applying the necessary and sufficient condition of optimality, by differentiating with respect to $R'_{\mu\mu}$ and by setting the derivative equal to zero.

$$\frac{E_\mu^T \left(W_\mu - R_{\mu\mu}' E_\mu \right) \left[1 - R_{\mu\mu}' L_{\mu\mu} \right]^2}{\left[1 - R_{\mu\mu}' L_{\mu\mu} \right]^4} = 0 \qquad (6.105)$$

Reformulating equation (6.105) to give that

$$E_\mu^T \left(W_\mu - R_{\mu\mu}' E_\mu \right) \left[1 - R_{\mu\mu}' L_{\mu\mu} \right]^2 + \left(W_\mu - R_{\mu\mu}' E_\mu \right)^T$$
$$\left(W_\mu - R_{\mu\mu}' E_\mu \right) L_{\mu\mu} \left[1 - R_{\mu\mu}' L_{\mu\mu} \right] = 0 \qquad (6.106)$$

Continuing to reformulate the above equation

$$E_\mu^T \left(W_\mu - R_{\mu\mu}' E_\mu \right) \left[1 - R_{\mu\mu}' L_{\mu\mu} \right] + \left(W_\mu - R_{\mu\mu}' E_\mu \right)^T \left(W_\mu - R_{\mu\mu}' E_\mu \right) L_{\mu\mu} = 0 \quad (6.107)$$

Applying the finding roots formula on above equation (6.105), the optimal transfer function $R_{\mu\mu}'$ is given by

$$\hat{R}_{\mu\mu}' = \frac{E_\mu^T E_\mu + 3 L_{\mu\mu} W_\mu^T E_\mu \pm \left(E_\mu^T E_\mu - L_{\mu\mu} W_\mu E_\mu \right)}{4 L_{\mu\mu} E_\mu^T E_\mu}$$

$$= \begin{cases} \dfrac{2 E_\mu^T E_\mu + 2 L_{\mu\mu} W_\mu^T E_\mu}{4 L_{\mu\mu} E_\mu^T E_\mu} = \dfrac{E_\mu^T E_\mu + L_{\mu\mu} W_\mu^T E_\mu}{2 L_{\mu\mu} E_\mu^T E_\mu} \\[4mm] \dfrac{W_\mu^T E_\mu}{E_\mu^T E_\mu} \end{cases} \qquad (6.108)$$

Comparing the two forms of the optimal transfer function $\hat{R}_{\mu\mu}'$, as the second form is a constant ratio of two polynomials, so in practical engineering we always choose the second form as the optimal transfer function, i.e.

$$\hat{R}_{\mu\mu}' = \frac{W_\mu^T E_\mu}{E_\mu^T E_\mu} \qquad (6.109)$$

Based on the optimal transfer function $\hat{R}_{\mu\mu}'$, then the optimal controller $\hat{C}_{\mu\mu}'$ is given as

$$\hat{C}_{\mu\mu}' = \frac{\hat{R}_{\mu\mu}'(s)}{S_{\mu\mu}(s) - \hat{R}_{\mu\mu}'(s) S_{\mu\mu}(s) L_{\mu\mu}(s)} \qquad (6.110)$$

The detailed derivation about the optimal transfer function $\hat{R}_{\mu\mu}'$ and optimal controller $\hat{C}_{\mu\mu}'$ are similar to above process, due to space limit, here we neglect them.

6.4.4 Conclusion

To reduce the computational complexity and the number of elements in designing the unknown controllers, the idea of non-interaction property is introduced to derive some conditions for achieving the non-interaction property. Based on this simplified model, satisfying the non-interaction property, the prediction error method is applied to design the optimal controllers. In deriving the conditions for non-interaction property, some prescribed values are given in priori. These necessities are similar to the popular model predictive control, so our next paper aims at connecting one closed bridge between non-interaction property and model predictive control.

6.5 One improvement on zonotope guaranteed parameter estimation

The automatic control field includes three aspects: prediction, filtering, and smoothing. One common property that exists in these three aspects is to estimate the state of the system. The difference is that different states of the system at different sample instants are estimated by using different input and output observed variables. State estimation is very important in closed loop feedback control and target tracking process, because some information about the linear system or nonlinear system is included in the state estimation. The problem of state estimation is similar to the system identification theory. Their specific process is formulated as follows: given a mathematical model concerning the real system and collecting some observed input-output measurements, the state of the real system has to be estimated from these measurements by applying some statistic analysis. Then one example is given to illustrate the importance of the state estimation in the target tracking process. The goal of target tracking is to estimate some parameters corresponding to the considered target with lots of observed data that are collected by sensors. These unknown parameters include the information on position and velocity. The target tracking can be defined as follows: using some prior probability knowledge and obtaining the state estimation of the target from observed sequences. The state estimation can be used as the state variable in the movement process. Then the state estimation is of great interest for the next design flight controller.

The methods used for estimating the state are divided into two parts: stochastic methods and deterministic methods. The difference between these two methods is whether the prior information about noise is known. Based on some probabilistic assumptions on noise, the stochastic methods (such as Kalman filter, maximum likelihood, and Bayes estimate) apply the minimum mean square state estimation error to obtain the state estimation. But the probabilistic assumptions on noise are not realistic and it means these probabilistic assumptions are not realized in reality. So to relax the probabilistic assumptions on noise, the deterministic methods are proposed to assume that the noises are unknown but bounded. This unknown but bounded assumption is weaker than the formal probabilistic assumption, as it needs not any prior distribution of noise.

6.5.1 Problem formulation

Consider one uncertain nonlinear discrete time system of the following form:

$$\begin{cases} x_{k+1} = f(x_k, w_k) \\ y_k = g(x_k, v_k) \end{cases} \tag{6.111}$$

where in equation (6.111) $x_k \in R^n$ is the state of the system and $y_k \in R^p$ is the measured output vector at sample time k. The vector $w_k \in R^{n_w}$ is the time varying process parameters and process perturbation vector. $v_k \in R^{p_v}$ is the measurement noise vector. Here, we assume that the uncertainties and the initial state are bounded by known compact sets.

$$w_k \in W, v_k \in V, x_0 \in X_0$$

To easily understand the concepts of zonotope, the definitions of zonotope and its other state sets are given as follows.

Definition 6.5 (Zonotope): Given a vector $p \in R^n$ and a matrix $H \in R^{n \times m}$, the set

$$p \oplus HB^m = \left\{ p + Hz : z \in B^m \right\} \tag{6.112}$$

is called a zonotope of order m. Note that $\oplus$ means the minkowski sum operation.

Definition 6.6 (Consistent state set): Given system (6.111) and a measured output y_k, the consistent state set at time k is defined as

$$X_{y_k} = \left\{ x \in R^n : y_k \in g(x, v) \right\}.$$

Definition 6.7 (Exact uncertain state set): Consider a system given by (6.111), the exact uncertain state set X_k is equal to the set of states that are consistent with the measured outputs $y_1, y_2 \cdots y_k$ and the initial state set X_0.

$$X_k = f(X_{k-1}, W) \cap X_{y_k}, k \geq 1 \tag{6.113}$$

6.5.2 Some improvements on interval analysis

From the interval arithmetic analysis, the natural interval extension is the fundamental theorem which replaces each occurrence of each variable by its corresponding interval variable. The natural interval extension is a particular and efficient way to compute an interval enclosure. By combining the natural interval extension and the Taylor series expansion, the improved mean value theorem can be obtained.

Theorem 6.2: Consider a function $f: R^n \to R$ with continuous derivatives about x and w, where $x \in R^n$ and $w \in R^{n_w}$. The two known compact sets X and W are given as the following zonotopes:

$$X = p \oplus HB^m, \quad W = c_w \oplus C_w B^{s_w} \tag{6.114}$$

Then

$$f(X,W) \subseteq f(p,c_w) \oplus MB^{m+s_w}$$

Proof: When applying the Taylor series expansion with respect to two variables, we obtain the following inclusion relation:

$$f(X,W) \subseteq f(p,c_w) \oplus \Box(\nabla f_x(X,W))(X-p) \oplus \Box(\nabla f_w(X,W))(W - c_w)$$

$$= f(p,c_w) \oplus \Box(\nabla f_x(X,W)) HB^m \oplus \Box(\nabla f_w(X,W)) C_w B^{s_w} \tag{6.115}$$

where $\Box$ means the natural interval extension, as $x_k \in R^n$, then $f(p, c_w)$ is a vector. It means that $f(p, c_w) \in R^n$. Three matrices are also defined as

$$M_1 = \Box(\nabla f_x(X,W))H, M_2 = \Box(\nabla f_w(X,W))C_w$$
$$M = [M_1, M_2]$$

Equation (6.115) can be continued to compute

$$f(X,W) \subseteq f(p,c_w) \oplus M_1 B^m \oplus M_2 B^{s_w}$$

$$= f(p,c_w) \oplus [M_1, M_2] \begin{bmatrix} B^m \\ B^{s_w} \end{bmatrix} = f(p,c_w) \oplus MB^{m+s_w} \tag{6.116}$$

From equation (6.116), we see that the state set is included in one zonotope. This improved mean value extension uses the continuous derivative in w, because there are two stochastic variables x and w in the state equation.

Consider the following interval extension

A zonotope $q \oplus SB^d$ such that $f(p,c_w) \subseteq q \oplus SB^d$

Then applying the above interval extension in equation (6), we obtain one zonotope which includes the state set. It means that

$$\begin{cases} f(X,W) \subseteq q \oplus SB^d \oplus MB^{m+s_w} = q \oplus [S \quad M] \begin{bmatrix} B^d \\ B^{m+s_w} \end{bmatrix} = q \oplus H_q B^{d+m+s_w} \\ H_q = [S \quad M] \end{cases} \tag{6.117}$$

The zonotopoe $q \oplus SB^d$ can be obtained by using a natural interval extension of $f(p, c_w)$. Now comparing equation (6.117), the difference is that the Taylor series expansion with two variables used here. As the stochastic variable w is considered in the derivation process, the obtained zonotope is more appropriate.

But there is no doubt that the computational complexity is increased. To reduce the computational complexity, the strategy of model reduction can be used to achieve this goal.

In the set membership estimation theory, the computation of the exact uncertain state set is very difficult. In practice, the state set is approximated by conservative outer bounds to reduce the computation complexity. This section presents one iterative method to compute an outer approximation using a zonotope. Consider an outer bound of the exact uncertain state set, denoted as $\hat{X}_{k-1}$, is available at time instant $k - 1$. Similarly, a measured output vector y_k is obtained at time instant k. Then the iterative guaranteed state estimation algorithm can be given as follows.

Iterative guaranteed state estimation algorithm

Step 1: Given the system (6.111), assume that the initial state x_0 is bounded by a known compact set:

$$x_0 \in X_0 \subseteq q_0 \oplus H_{q_0} B^{d+m+s_w}$$

Compute a state set at time instant k as

$$\overline{X}_k \subseteq q_{k-1} \oplus H_{q_{k-1}} B^{d+m+s_w} \tag{6.118}$$

where each variable is defined as

$$H_{q_{k-1}} = \begin{bmatrix} M_{1(k-1)} & M_{2(k-1)} \end{bmatrix}$$
$$M_{1(k-1)} = \Box\left(\nabla f_{x_{(k-1)}}\left(X_{(k-1)}, W_{(k-1)}\right)\right) H$$
$$M_2 = \Box\left(\nabla f_{w_{(k-1)}}\left(X_{(k-1)}, W_{(k-1)}\right)\right) C_w \tag{6.119}$$

Step 2: For $i = 1, 2 \cdots N$

Compute an outer bound of the consistent state set $X_{y_{k/i}}$ and $\hat{X}_{k/i-1}$ with $\hat{X}_{k/0} = X_0$

Step 3: Compute the intersection operation to obtain the state set $\overline{X}_k \cap X_{y_{k/i}}$.

End of algorithm

The iteration process is started from the initial time. The iteration means that the state set at the next time instant is dependent on the above state and the above measured output variable. This iteration process is similar to the classical Kalman filter theory. It also is divided into the prediction step, the measurement step, and the correction step.

6.6 Summary

Our proposed iterative correlation tuning control strategy is based on the system identification theory, whose model validation process is beneficial for this data driven control strategy. It means that the goal of identification for control is achieved in our continuous research about system identification, data driven control, model predictive control, etc. For the more general nonlinear system, zonotope identification is extended to identify the unknown parameter in each hyperplane, and then some interaction conditions are devised to simplify the complex system.

References

[1] Ljung, L. 1999. System Identification: Theory for the User. Prentice Hall Press, Upper Saddle River, New Jersey, USA.

[2] Boyd, S. and Vandenberghe, L. 2004. Convex Optimization. Cambridge University Press, Cambridge, UK.

[3] Pintelon R. and Schoulens, J. 2001. System Identification: A Frequency Domain Approach. IEEE Press, New York, NJ, USA.

[4] Roll, J., Bemporad, A. and Ljung, L. 2004. Identification of piecewise affine systems via mixed integer programming, Automatica, 40(1): 37-50.

[5] Trecate, G.F. and Muselli, M. 2003. A clustering technique for the identification of piecewise affine systems, Automatica, 39(2): 205-217.

[6] Jouloski, A.L. and Weiland, S. 2013. A Bayesian approach to identification of hybrid systems, IEEE Transactions on Automatic Control, 50(10): 1520-1532.

[7] Ohlsson, H. and Ljung, L. 2013. Identification of switched linear regression models using sum of norms regularization, Automatica, 49(4): 1045-1050.

[8] Lauer, F. and Bloch, G. 2011. A continuous optimization framework for hybrid system identification, Automatica, 47(3): 608-613.

[9] Lauer, F. 2016. On the complexity of switching linear regression, Automatica, 74(12): 80-83.

[10] Lauer, F. 2015 On the complexity of piecewise affine system identification, Automatica, 62(12): 148-153.

[11] Bako, L. 2011. Identification of switched linear systems via sparse optimization, Automatica, 47(4): 668-677.

[12] Breschi, V. and Piga, D. 2016. Piecewise affine regression via recursive multiple least squares and multi-category discrimination, Automatica, 73(11): 155-162.

[13] Bemporad, A. and Garulli, A. 2005, A bounded error approach to piecewise affine system identification, IEEE Transactions on Automatic Control, 50(10): 1567-1580.

[14] Paoletti, S. and Roll, J. 2010. On the input output representation of piecewise affine state space models, IEEE Transactions on Automatic Control, 55(1): 60-72.

[15] Bako, L. 2011. A recursive identification algorithm for switched linear/affine models, Nonlinear Analysis-Hybrid Systems, 5(2): 242-253.

[16] Nakada, H. and Takaka, K. 2005. Identification of piecewise affine systems based on statistical clustering technique, Automatica, 41(5): 905-913.

[17] Paoletti, S. and Jouloski A.L. 2008. Identification of hybrid systems: A tutorial, European Journal of Control, 13(2): 242-260.

[18] Bemporad, A. and Trecate, G.F. 2000. Observability and controllability of piecewise affine and hybrid systems, IEEE Transactions on Automatic Control, 45(10): 1864-1876.

[19] Calafiore, G.C. 2017. Leading impulse response identification via the elastic net criterion, Automatica, 80(4): 75-87.

[20] Xin, X. 2017. Distributed emitter parameter refinement based on maximum likelihood, International Journal of Intelligent Computing and Cybernetics, 10(1): 2-11.

[21] Dolanc, G. and Strmcnik, S. 2005. Identification of nonlinear systems using a piecewise-linear Hammerstein model, Systems & Control Letters, 54(2): 145-158.

[22] Bravo, J.M., Suarez, A. and Vasallo, M. 2016. Slide window bounded error time varying system identification, IEEE Transactions on Automatic Control, 61(8): 2282-2287.

[23] Tanaskovic, M., Fagiano, L. and Novara, C. 2017. Data driven control of nonlinear systems: An on line direct approach, Automatica, 75(1): 1-10.

[24] Dolanc, G. and Strmcnik, S. 2013. Identification and control of nonlinear systems using a piecewise-linear Hammerstein model. *In:* Case Studies in Control, Springer-Verlag London.

[25] Mattsson, P., Zacharian, D. and Stoica, P. 2016. Recursive identification method for piecewise ARX models: A sparse estimation approach, IEEE Transactions on Signal Processing, 64(19): 5082-5093.

[26] Bravo, J.M., Alamo, T. and Camacho, E.F. 2006. Bounded error identification of systems with time varying parameters, IEEE Transactions on Automatic Control, 51(7): 1144-1150.

Data Driven Applications

7.1   Introduction

There are lots of applications for our considered data driven identification and data driven control, however, in this chapter about the applications for data driven identification and control, we only give our research on this subject, for example, the state of charge estimation and model parameter identification.

(1) As the state of charge is one important variable to monitor the later battery management system, a traditional Kalman filter can be used to estimate the state of charge for a Lithium-ion battery based on a probability distribution of external noise. To relax this strict assumption on external noise, a set membership strategy is proposed to achieve our goal in case of unknown but bounded noise. External noise that is unknown but bounded is more realistic than white noise. After an equivalent circuit model is used to describe the Lithium-ion battery charging and discharging properties, one state space equation is constructed to regard state of charge as its state variable. Based on the state space model about the state of charge, two kinds of set membership strategies are put forth to achieve the state estimation, which corresponds to the state of charge estimation. Due to external noise being bounded, i.e. external noise is in a set, we construct an interval and ellipsoid estimation for the state estimation respectively in case the external noise is assumed in an interval or ellipsoid. Then the midpoint of an interval or center of the ellipsoid are chosen as the final value for the state of charge estimation.

(2) We summarize our previous work on aircraft flutter model parameters identification, and consider a new subject to complete previous identification strategies, i.e. optimal input signal design for flutter model parameters identification experiment, as input signal must be sufficiently informative to excite the considered stochastic model. More specifically, according to our constructed flutter stochastic model, whose input and output are corrupted by observed noises, separable least squares identification and set membership identification are proposed to identify those unknown model parameters in the sense of statistical noise and unknown but bounded noise respectively.

The common trace operation concerning the asymptotic variance matrix is minimized to solve the power spectral for the optimal input signal in the framework of statistical noise. Moreover, within the unknown bounded noise, the radius of information, corresponding to our established parameter uncertainty interval, is minimized to give that optimal input signal.

7.2 Applying set membership strategy in state of charge estimation for Lithium-ion battery

The Lithium-ion battery is the leading energy storage technology for many fields, such as electric vehicles, modern electric grids, transformation, etc. The main features of a Lithium-ion battery include energy density, a long time, and a lower self-discharge rate, so much research on these main features of Lithium-ion battery have been carried out in recent years from their different points of view. One interesting area of research is battery state estimation, especially named state of charge (SOC) estimation, as the State of charge can not only reflect the remaining capacity of a Lithium-ion battery, but also embody the performance and endurance mileage of an electric vehicle. Furthermore, the State of charge is the most important factor to be used in the battery management system, which is critical for the safety, efficiency and life expectancy of Lithium-ion batteries. Generally, the State of charge indicates the remaining battery capacity to show how long the battery will last. It helps the battery management system to protect the battery from overcharging and over-discharging and makes energy management system determine an effective dispatching strategy. But State of charge cannot be directly measured using physical sensors, then it must be estimated using some newly developed methods with the aid of measurable signals such as the voltage and current of the battery. Then here in this chapter, the State of charge estimation is our concern problem for Lithium-ion batteries.

State of charge estimation has widely been studied and lots of estimation algorithms have been proposed to acquire precise state of charge estimation. An improved extended Kalman filter method is presented to estimate the state of charge for the vanadium redox battery [1], using a gain factor. Some unknown parameters from the state space model are identified by the classical least squares method. The square root cubature Kalman filter algorithm has been developed to estimate the state of charge of a battery [1], where 2n points are calculated to give the same weight, according to cubature transform to approximate the mean of state variables. To improve the accuracy and reliability of state of charge estimation for battery, an improved adaptive cubature Kalman filter is proposed in [2], where the battery model parameters are online identified by the forgetting factor recursive least squares algorithm. An adaptive forgetting recursive least squares method is exploited to optimize the estimation alertness and numerical stability [3], as to achieve online adaptation of model parameters. To reduce the iterative computational complexity, a two stage recursive least squares approach is developed to identify the model parameters [4], then the measurement values of the open circuit voltage at varying relaxation periods and three temperatures are sampled to establish the relationships between the state of charge and open circuit voltage. In [5] multi-scale parameter adaptive method based on dual Kalman filters is applied to estimate multiple parameters. Based on the battery circuit model and battery model state equation, the real time recursive least squares method with the forgetting factor is used to identify unknown battery parameters [6]. After introducing the concept of the state of health, the average error of the obtained state of charge estimation is less than one given value. A novel state and parameter co-estimator is developed to concurrently estimate the state and model parameters of a Thevenin model for a Liquid mental battery [7], where the adaptive unscented Kalman filter (UKF) is employed for state estimation, including a battery state of charge. After performing Lithium-ion battery modelling and off-line parameter identification, a sensitivity analysis experiment is designed to verify which model parameter has the greatest influence on the state of charge estimation [8]. To improve the state of charge estimation accuracy under uncertain measurement noise statistics, a variational Bayesian approximation based adaptive dual extended Kalman filter is proposed in [9], and the measurement noise variances are simultaneously estimated in the state of charge estimation process. Actually to the best of our knowledge, these state of charge estimation methodologies can be roughly divided into data-driven methods and model-based methods [10]. In the model-based methods, the Kalman filter based state of charge estimation methods have the merits of self-correction, online computation, and the availability of dynamic state of charge estimation [11]. Kalman filter was firstly proposed to estimate the state of a linear system, then to apply it to a nonlinear system, the extended Kalman filter (EKF) and unscented Kalman filter were developed. Meanwhile, the date-driven methods typically include the look up table method, matching learning based method, artificial neural networks and support vector machine, etc. [12]. The data driven method means that in estimating the state whether in linear system or nonlinear system, no mathematical model is needed, i.e. the state is constructed only directly by observed data [13], so a large number of training data covering all of the operating conditions is collected to improve the estimation accuracy of the considered state.

From the above mentioned references, we see that only the Kalman filter is used to achieve the state estimation. Here we regard all kinds of Kalman filter's extended forms as the same category. To the best of our knowledge that no other new strategy is proposed to estimate the unknown state, except the Kalman filter or its extended forms [14]. Furthermore, through understanding the Kalman filter for state estimation carefully, roughly speaking, the Kalman filter holds for state estimation in case the considered external noise must be a zero mean random signal, i.e. white and normal noise. This condition corresponds to the classical probabilistic description of external noise. But this white noise is an ideal case, and it does not exist in reality.

To relax this strict probabilistic description of external noise, we propose to apply the set membership estimation strategy in estimating our considered state in the presence of unknown but bounded noise. Here, our considered unknown but bounded external noise is more realistic than white noise in engineering or other research fields. It means that our goal is to estimate the state in case of the unknown but bounded external noise in this chapter, where the estimated state corresponds to the state of charge in the constructed state space equation. As the classical Kalman filter or its extended forms are useless within the framework of unknown but bounded external noise, it is necessary to propose another estimation strategy to identify an unknown state based on the unknown but bounded external noise. The idea of set membership estimation is from the system identification theory or adaptive control, and to apply set membership estimation to deal with the problem of estimating the state of charge for a Lithium-ion battery, firstly we need to reformulate one state space equation for the state of charge estimation, by using one equivalent circuit model to replace the considered state of charge estimation for a Lithium-ion battery. Based on this constructed state space equation corresponding to the state of charge for the Lithium-ion battery, the idea of a set membership estimation strategy can be easily applied here. More specially, because the external noise is unknown but bounded, i.e. it is assumed to be in one set priori within the whole framework of the set membership estimation,

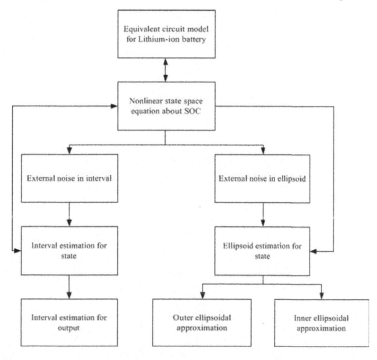

Figure 7.1: A flowchart of the first application

then two kinds of set membership estimation strategies are proposed here based on the used set, which includes the external noise. Without loss of generality, according to the commonly used interval and ellipsoid for the external noise, the interval estimation and ellipsoid estimation are derived for the considered state estimation respectively, which corresponds to the state of charge estimation for the Lithium-ion battery. This correspondence is from the equivalent state space equation. Based on our obtained interval estimation or ellipsoid estimation, the midpoint of the interval estimation can be chosen as the final state estimation, and similarly the center of the ellipsoid estimation can be also chosen.

A flowchart of our two proposed set membership strategies is given in Fig. 7.1, where the yellow parts are our two considered cases of external noise with interval or ellipsoid. The main contribution of this chapter is to derive the interval estimation and ellipsoid estimation for the state, which correspond to the above two considered unknown but bounded noises respectively.

7.2.1   Battery modelling

Our considered Lithium-ion battery has some merits in energy density and life, further, it is the leading development direction of power batteries for electric vehicles in the future. To give a brief introduction to Lithium-ion batteries, the internal states of a Lithium-ion battery are always divided into four parts, i.e. SOC, temperature, rate of current, and state of health. These four states reflect the internal relations of a Lithium-ion battery with a time variable. Here, our emphasis is on the internal structure of a Lithium-ion battery, which is shown in Fig. 7.2, whose cell generally comprises four parts: a polymer positive electrode, a diaphragm, a negative electrode, and an electrolyte. The positive electrode of a Lithium-ion battery is generally composed of a Lithium-ion polymer. Common cathode Lithium-ion polymer materials include lithium phthalate, Lithium-ion phosphate, barium acid strontium, Lithium-ion management, nickel diamond, and nickel-nickel aluminum ternary lithium. The diaphragm is in the process of the first charge and discharge of the liquid Lithium-ion battery. The electrode material reacts with the electrolyte at the solid-liquid phase interface to form a passivation layer covering the surface of the electrode material to isolate the electrode and the electrolyte, and the Lithium-ion can finish a chemical reaction with the diaphragm.

Actually, in all literature on the state of charge for a Lithium-ion battery, two commonly used battery models exist, i.e. the equivalent circuit model and the electrochemical model. As the electrochemical model is very complex, and it is very difficult to design the latter Kalman filter in the case of this electrochemical model, here, in modeling a Lithium-ion battery, the equivalent circuit model is recently used. The equivalent circuit model regards the battery's internal reactions as a circuit, containing some electronic components, so the equivalent circuit model consists of basic circuit components such as resistors, capacitors, and voltage sources. These four basic circuit components are widely explored due to their relatively simple mathematical structure and reduced computational complexity.

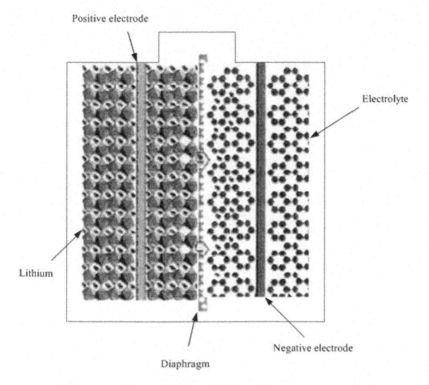

Figure 7.2: Battery internal structure

Equivalent circuit model is shown in Fig. 7.3, which is simple and clear in the physical meaning, and will be applied to describe the battery charging and discharging properties. By balancing the trade-off between model accuracy and computational complexity, one Thevenin equivalent circuit model is chosen for a Li-ion battery, which is regarded as our battery model.

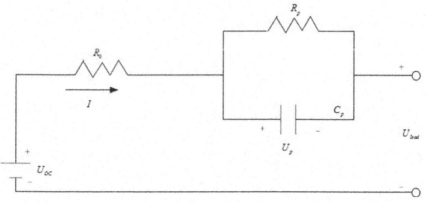

Figure 7.3: Equivalent circuit model

Applying Kirchhoff law, variable U_{load} is defined as that

$$U_{load} = U_{OC} - IR_0 - U_p \tag{7.1}$$

$$I = \frac{U_p}{R_p} + C_p \frac{dU_p}{dt} \tag{7.2}$$

where in equation (7.1) and (7.2), U_{load} is the terminal voltage, I is the load current, R_0 is the internal ohmic resistance, R_p and C_p are polarization resistence and polarization capacitance of the battery, U_p is the polarization voltage. U_{OC} is the open circuit voltage, which is monotonic with state of charge. Further U_{OC} is rewritten as the following polynomial form

$$U_{OC}(x) = d_5 + d_4 x + d_3 x^2 + d_2 x^3 + d_1 x^4 \tag{7.3}$$

where $\{d_i\}_{i=1}^5$ are the coefficients of the polynomial form, and x is the state of charge of the battery. The state of charge is defined as a ratio of the remaining capacity over the rated capacity. According to the ampere hour counting method, state of charge can be expressed as follows:

$$SOC(t) = SOC(t_0) - \eta \int_{t_0}^t \frac{Idt}{Q_N} \tag{7.4}$$

where t is the sample time, $SOC(t)$ is the state of charge of the battery at time instant t, $SOC(t_0)$ is the initial SOC, I is the load current, η is the coulombic efficiency, and Q_N is the nominal capacity of battery.

$$\begin{bmatrix} SOC_k \\ U_{p,k} \end{bmatrix} = \begin{bmatrix} 1 & 0 \\ 0 & \exp\left(-\dfrac{T_s}{R_p C_p}\right) \end{bmatrix} \begin{bmatrix} SOC_{k-1} \\ U_{p,k-1} \end{bmatrix} + \begin{bmatrix} -\eta \\ R_p \left(1 - \exp\left(-\dfrac{T_s}{R_p C_p}\right)\right) \end{bmatrix} I_{k-1} \tag{7.5}$$

$$U_{load,k} = U_{OC}(SOC_k) - U_{p,k} - I_k R_0 \tag{7.6}$$

where k is the sample time, SOC_k is the statue value at the kth sample time, T_s is the specified small sampling period. $U_{OC}(SOC_k)$ denotes a nonlinear function of SOC_k.

The parameters in above state space equation (7.5) and (7.6) can be identified by the classical least squares method, then our goal in this chapter is to estimate state of charge (SOC_k) at time instant k.

7.2.2   Interval estimation for SOC

In this section, we start to apply the set membership filter algorithm to estimate the SOC by combining equation (7.5) and (7.6), we see that SOC_k at time instant k is one state variable in that state space equation. Furthermore, we also want to testify which parameter will influence SOC estimation, then this parameter will be added as the new state variables in the extended state space equation.

As the main model parameter R_0 is classified as a new state variable with U_p and SOC, then an extended state space equation for set membership filter can be given as that

$$
\begin{bmatrix} SOC_k \\ U_{p,k} \\ R_{0,k} \end{bmatrix} = \begin{bmatrix} 1 & 0 & 0 \\ 0 & \exp\left(-\dfrac{T_s}{R_p C_p}\right) & 0 \\ 0 & 0 & 1 \end{bmatrix} \begin{bmatrix} SOC_{k-1} \\ U_{p,k-1} \\ R_{0,k-1} \end{bmatrix}
$$

$$
+ \begin{bmatrix} -\eta \\ R_p\left(1-\exp\left(-\dfrac{T_s}{R_p C_p}\right)\right) \\ 0 \end{bmatrix} I_{k-1} + \begin{bmatrix} w_{1,k-1} \\ w_{2,k-1} \\ w_{3,k-1} \end{bmatrix} \tag{7.7}
$$

$$
U_{load,k} = U_{OC}\left(SOC_k\right) - U_{p,k} - I_k R_0 + v_k \tag{7.8}
$$

Observing the above equation (7.7) and (7.8), the problem of state of charge for the Lithium-ion battery is to estimate the first state variable SOC_k at every time instant k. As the state of charge for a Lithium-ion battery is the first element of the state vector in equation (7.7), then this problem is similar to the state filtering in the modern control theory. So if the state noise or external noise is a white noise, then the classical Kalman filter can be well applied to deal with the filter problem. But if the probability distribution of the state noise or external noise is unknown, then Kalman filter strategy is useless here, because white noise is an ideal case in reality. To consider a more general case about the state noise or external noise, where the property of state noise or external noise is unknown but bounded.

Then in order to apply set membership algorithm into above state space equation to estimate the first state variable, we rewrite the above two equations (7.7) and (7.8) as follows:

$$
\begin{cases} x(k+1) = Ax(k) + Bu(k) + Dw(k) \\ y(k) = Cx(k) + v(k) \end{cases} \tag{7.9}
$$

where

$$
x(k+1) = \begin{bmatrix} SOC_k \\ U_{p,k} \\ R_{0,k} \end{bmatrix}, A = \begin{bmatrix} 1 & 0 & 0 \\ 0 & \exp\left(-\dfrac{T_s}{R_p C_p}\right) & 0 \\ 0 & 0 & 1 \end{bmatrix}, B = \begin{bmatrix} -\eta \\ R_p\left(1-\exp\left(-\dfrac{T_s}{R_p C_p}\right)\right) \\ 0 \end{bmatrix},
$$

$$
w(k) = \begin{bmatrix} w_{1,k-1} \\ w_{2,k-1} \\ w_{3,k-1} \end{bmatrix}, u(k) = I_{k-1}, D = I
$$

and

$$y(k) = U_{load,k}, \quad C = \frac{d\left(U_{OC}\left(SOC_k\right) - U_{p,k} - I_k R_0\right)}{d\left(SOC_k\right)},$$

where in equation (7.9) k is time instant, $x(k)$ is the state of this system at time instant k with its initial state $x(0)$, $y(k)$ is the observed output at time instant k. $u(k)$ is the control input, $w(k)$ and $v(k)$ are two unknown but bounded state noise and observed noise respectively. All matrices A, B, C, D are some matrices with compatible dimensions, i.e.

$$A \in R^{n \times n}, \; B \in R^{n \times m}, \; D \in R^{n \times n_w}, \; C \in R^{n_y \times n} \tag{7.10}$$

Our considered linear discrete time invariant system is one state space equation, whose structure can be seen in Fig. 7.4.

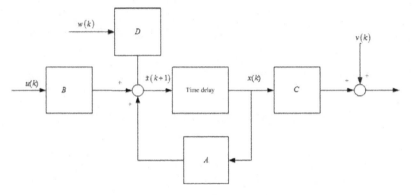

Figure 7.4: The structure of the state space equation

Let $\underline{x}$ and $\bar{x}$ be two vectors such that $\underline{x} \leq \bar{x}$ with the inequality holding component wise. An interval $[\underline{x}, \bar{x}]$ is defined by

$$[\underline{x}, \bar{x}] = \left\{ x \in R^n : \underline{x} \leq x \leq \bar{x} \right\} \tag{7.11}$$

Then first of all, we give the assumptions about initial state $x(0)$, state noise $w(k)$ and observed noise $v(k)$.

Assumption 7.1: There exist three kinds of unknown but bounded signals $[\underline{x}(0), \bar{x}(0)]$, $[\underline{w}(k), \bar{w}(k)]$, $[\underline{v}(k), \bar{v}(k)]$ respectively, such that three uncertainties in the state space equation (7.1) be

$$\begin{cases} \underline{x}(0) \leq x(0) \leq \bar{x}(0) \\ \underline{w}(k) \leq w(k) \leq \bar{w}(k) \quad \text{for all} \quad k \in R^+ \\ \underline{v}(k) \leq v(k) \leq \bar{v}(k) \end{cases} \tag{7.12}$$

where the inequalities are regarded as component wise.

As interval $[\underline{x}, \overline{x}]$ cannot be used in the latter computational process, its other equivalent form is defined.

Definition 7.1: The interval $[\underline{x}, \overline{x}]$ can be equivalently represented by the following equivalent form:

$$C(c_x, p_x) = \{c_x + P_x \alpha_x : \alpha_x \in R^n, \|\alpha_x\|_\infty \le 1\} \qquad (7.13)$$

where

$$c_x = \frac{\overline{x} + \underline{x}}{2}, P_x = diag(p_x), p_x = \frac{\overline{x} - \underline{x}}{2} \qquad (7.14)$$

Similarly

$$\overline{x} = c_x + p_x, \underline{x} = c_x + p_x \qquad (7.15)$$

Here the notation $\| \ \|_\infty$ is the infinite norm of one vector, and $diag()$ is the notation section. Also in Definition 7.1, c_x is the center of the interval $[\underline{x}, \overline{x}]$, and p_x its radius, i.e. $C(c_x, p_x) = [\underline{x}, \overline{x}]$.

Using the above Definition 7.1, the equivalent forms in Assumption 1 are given as the following Assumption 7.2.

Assumption 7.2: There exist three equivalent forms for three intervals $[\underline{x}(0), \overline{x}(0)], [\underline{w}(k), \overline{w}(k)], [\underline{v}(k), \overline{v}(k)]$ respectively.

$$\begin{cases} [\underline{x}(0), \overline{x}(0)] = C(c_x(0), p_x(0)) \\ [\underline{w}(k), \overline{w}(k)] = C(c_w(k), p_w(k)) \\ [\underline{v}(k), \overline{v}(k)] = C(c_v(k), p_v(k)) \end{cases} \qquad (7.16)$$

Here, the first contribution of our current chapter is to construct one interval $[\underline{x}(k)\ \overline{x}(k)]$ for state estimation $x(k)$ in equation (7.9), then after substituting $[\underline{x}(k), \overline{x}(k)]$ into the observed equation, the interval $[\underline{y}(k), \overline{y}(k)]$ corresponding to the prediction output can be obtained, while considering three uncertainties about initial state $x(0)$, state noise $w(k)$, and observed noise $v(k)$.

Firstly to obtain one interval $[\underline{x}(k), \overline{x}(k)]$ for state estimation $x(k)$ at time instant k, we take z transformation on both sides of the state equation, i.e.

$$zX(z) - zx(0) = AX(z) + BU(z) + DW(z) \qquad (7.17)$$

where z is the variable in frequency domain, and $X(z)$, $U(z)$, $W(z)$ are the transformation results in frequent domain, corresponding to their forms in time domain $x(k)$, $u(k)$, $w(k)$.

Formulating equation (7.17) to give that

$$X(z) = (zI - A)^{-1} zx(0) + (zI - A)^{-1} BU(z) + (zI - A)^{-1} DW(z) \qquad (7.18)$$

Taking inverse z transformation on both sides of equation (7.18), it holds that

$$x(k) = A^k x(0) + \sum_{i=0}^{k-1} A^{k-i-1} Bu(i) + \sum_{i=0}^{k-1} A^{k-i-1} Dw(i)$$

$$= A^k x(0) + \sum_{i=0}^{k-1} A^{k-i-1} \left(Bu(i) + Dw(i) \right) \tag{7.19}$$

Based on equation (7.19), we proceed to construct intervals for state estimation and prediction output respectively.

Observing equation (7.19), two uncertainties exist, i.e. initial state $x(0)$ and state noise $w(i)$. From Assumption 7.2, two intervals about initial state $x(0)$ and state noise $w(i)$ are given as

$$\begin{cases} x(0) \in C\left(c_x(0), p_x(0)\right) \\ w(i) \in C\left(c_w(i), p_w(i)\right) \end{cases} \tag{7.20}$$

Then we can describe the uncertain initial state $x(0)$ and state noise $w(i)$ by

$$\begin{cases} x(0) = c_x(0) + P_x(0) \quad {}_x \\ w(i) = c_w(i) + P_w(i) \quad {}_w \end{cases} \tag{7.21}$$

where $\alpha_x \in R^n$ and $\alpha_w \in R^{n_w}$ such that

$$\|\alpha_x\| \le 1 \text{ and } \|\alpha_w\| \le 1 \tag{7.22}$$

Substituting equation (7.20) into $x(k)$, then it holds that

$$x(k) = A^k \left[c_x(0) + P_x(0)\alpha_x \right] + \sum_{i=0}^{k-1} A^{k-i-1} Bu(i)$$

$$+ \sum_{i=0}^{k-1} A^{k-i-1} D \left[c_w(i) + P_w(i)\alpha_w \right]$$

$$= A^k c_x(0) + \sum_{i=0}^{k-1} A^{k-i-1} Bu(i) + \sum_{i=0}^{k-1} A^{k-i-1} Dc_w(i) + A^k P_x(0)\alpha_x$$

$$+ \sum_{i=0}^{k-1} A^{k-i-1} DP_w(i)\alpha_w \tag{7.23}$$

Define the following pair $C(c_x(k), p_x(k))$ as

$$\begin{cases} c_x(k) = A^k c_x(0) + \sum_{i=0}^{k-1} A^{k-i-1} Bu(i) + \sum_{i=0}^{k-1} A^{k-i-1} Dc_w(i) \\ p_x(k) = \left| A^k \right| p_x(0) + \sum_{i=0}^{k-1} \left| A^{k-i-1} D \right| p_w(i) \end{cases} \tag{7.24}$$

where notation $| \ |$ is the absolute value.

Then it holds that

$$x(k) \in C\big(c_x(k), p_x(k)\big), k = 1, 2 \cdots N \tag{7.25}$$

Generally, the above derivations can be formulated as the following Theorem 7.1.

Theorem 7.1: Set $C(c_x(0), p_x(0))$ and $C(c_w(i), p_w(i))$ be center-radius representations of two uncertainties $x(0)$ and $w(i)$, the interval $\left[\underline{x}(k), \overline{x}(k)\right]$ for state estimation $x(k)$ in state space equation (7.1) is constructed as

$$\begin{cases} \underline{x}(k) = c_x(k) - p_x(k) \\ \qquad = A^k c_x(0) + \displaystyle\sum_{i=0}^{k-1} A^{k-i-1} Bu(i) + \sum_{i=0}^{k-1} A^{k-i-1} Dc_w(i) - \left|A^k\right| p_x(0) \\ \qquad\quad - \displaystyle\sum_{i=0}^{k-1} \left|A^{k-i-1}D\right| p_w(i) \\ \overline{x}(k) = c_x(k) + p_x(k) \\ \qquad = A^k c_x(0) + \displaystyle\sum_{i=0}^{k-1} A^{k-i-1} Bu(i) + \sum_{i=0}^{k-1} A^{k-i-1} Dc_w(i) + \left|A^k\right| p_x(0) \\ \qquad\quad + \displaystyle\sum_{i=0}^{k-1} \left|A^{k-i-1}D\right| p_w(i) \end{cases} \tag{7.26}$$

where $c_x(k)$ and $p_x(k)$ are defined in equation (7.24), then interval for state estimation is given that

$$x(k) \in C\big(c_x(k), p_x(k)\big) = \left[\underline{x}(k), \overline{x}(k)\right], k = 1, 2 \cdots N \tag{7.27}$$

To analyze the recursive relation between the kth interval $\left[\underline{x}(k), \overline{x}(k)\right]$ and its latter $k + 1$th interval $\left[\underline{x}(k+1), \overline{x}(k+1)\right]$, we list their centers as follows.

$$\begin{cases} c_x(k) = A^k c_x(0) + \displaystyle\sum_{i=0}^{k-1} A^{k-i-1} Bu(i) + \sum_{i=0}^{k-1} A^{k-i-1} Dc_w(i) \\ c_x(k+1) = A^{k+1} c_x(0) + \displaystyle\sum_{i=0}^{k} A^{k-i-1} Bu(i) + \sum_{i=0}^{k} A^{k-i-1} Dc_w(i) \end{cases} \tag{7.28}$$

Taking subtract operation, we find that

$$c_x(k+1) - c_x(k) = A^k c_x(0)(A-I) + \sum_{i=0}^{k-1} A^{k-i-1} Bu(i)(A-I) +$$

$$\sum_{i=0}^{k-1} A^{k-i-1} Dc_w(i)(A-I) + Bu(k) + Dc_w(k)$$

$$= \left[A^k c_x(0) + \sum_{i=0}^{k-1} A^{k-i-1} Bu(i) + \sum_{i=0}^{k-1} A^{k-i-1} Dc_w(i) \right]$$

$$(A-I) + Bu(k) + Dc_w(k)$$

$$= c_x(k)(A-I) + Bu(k) + Dc_w(k) \tag{7.29}$$

Then it holds that

$$c_x(k+1) = Ac_x(k) + Bu(k) + Dc_w(k) \tag{7.30}$$

Equation (7.30) is the recursive expression of the centers. Similarly, the recursive expression of the radius is that

$$p_x(k+1) = |A| p_x(k) + |B| p_w(k) \tag{7.31}$$

From these two recursive relations between the adjacent interval for state estimation, we see that the $k+1$th interval can be obtained from the kth interval and the knowledge of control input and state noise. The recursive computation for the interval for state estimation is seen in Fig. 7.5.

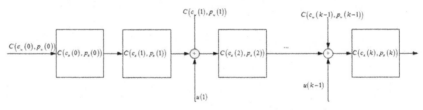

Figure 7.5: Recursive computation for interval

But the most important element in model predictive control is the prediction output, so the interval for the prediction output can be obtained by substituting interval $\left[\underline{x}(k), \bar{x}(k) \right]$ into the observed equation.

Due to $x(k) \in C\left(c_x(k), p_x(k) \right)$ and $v(k) \in C\left(c_v(k), p_v(k) \right)$, i.e.

$$\begin{cases} x(k) = c_x(k) + P_x(k)\alpha_x \\ v(k) = c_v(k) + P_v(k)\alpha_v \end{cases} \tag{7.32}$$

where $\alpha_x \in R^n$ and $\alpha_v \in R^{n_v}$ such that

$$\|\alpha_x\| \le 1 \text{ and } \|\alpha_v\| \le 1 \tag{7.33}$$

Substituting equation (7.32) into the observed equation (7.9), we have that

$$y(k) = C\left[c_x(k) + P_x(k)\alpha_x \right] + \left[c_v(k) + P_v(k)\alpha_v \right]$$

$$= \underbrace{Cc_x(k) + c_v(k)}_{c_y(k)} + CP_x(k)\alpha_x + P_v(k)\alpha_v \tag{7.34}$$

Similarly define the center and radius as

$$\begin{cases} c_y(k) = Cc_x(k) + c_v(k) \\ p_y(k) = Cp_x(k) + p_v(k) \end{cases} \tag{7.35}$$

Then we have that

$$y(k) \in C\big(c_y(k), p_y(k)\big) \tag{7.36}$$

Furthermore,

$$\begin{cases} \underline{y}(k) = c_y(k) - p_y(k) = Cc_x(k) + c_v(k) - Cp_x(k) - p_v(k) \\ \quad = C\big(c_x(k) - p_x(k)\big) + c_v(k) - p_v(k) \\ \overline{y}(k) = c_y(k) + p_y(k) = Cc_x(k) + c_v(k) + Cp_x(k) + p_v(k) \\ \quad = C\big(c_x(k) + p_x(k)\big) + c_v(k) + p_v(k) \end{cases} \tag{7.37}$$

Then it also means that

$$y(k) \in C\big(c_y(k), p_y(k)\big) = \big[\underline{y}(k), \overline{y}(k)\big] \tag{7.38}$$

The above equation (7.38) is our interval for the prediction output, which will be used for the next robust model predictive control.

To simplify the latter exposition in robust model predictive control, we need the explicit form of interval $y(k) \in C\big(c_y(k), p_y(k)\big) = \big[\underline{y}(k), \overline{y}(k)\big]$. To achieve this goal, some notations are introduced here.

$$c_x(k) = A^k c_x(0) + \sum_{i=0}^{k-1} A^{k-i-1} Bu(i) + \sum_{i=0}^{k-1} A^{k-i-1} Dc_w(i) = c_1(k) + \sum_{i=0}^{k-1} c_2(k-i-1)u(i)$$

$$p_x(k) = \left| A^k \right| p_x(0) + \sum_{i=0}^{k-1} \left| A^{k-i-1} D \right| p_w(i)$$

$$c_1(k) = A^k c_x(0) + \sum_{i=0}^{k-1} A^{k-i-1} Dc_w(i), \quad c_2(k-i-1) = A^{k-i-1} B \tag{7.39}$$

Substituting notation (7.39) into the expressions $\underline{y}(k)$ and $\overline{y}(k)$ respectively, we obtain

$$\begin{aligned} \overline{y}(k) &= c_y(k) + p_y(k) = Cc_x(k) + c_v(k) + Cp_x(k) + p_v(k) \\ &= C\big(c_x(k) + p_x(k)\big) + c_v(k) + p_v(k) \\ &= C\left[c_1(k) + \sum_{i=0}^{k-1} c_2(k-i-1)u(i) \right] + c_v(k) + p_v(k) + Cp_x(k) \\ &= a_1(k) + \sum_{i=0}^{k-1} a_2(k-i-1)u(i) \end{aligned} \tag{7.40}$$

where

$$\begin{cases} a_1(k) = Cc_1(k) + c_v(k) + p_v(k) + Cp_x(k) \\ a_2(k-i-1) = Cc_2(k-i-1) \end{cases} \tag{7.41}$$

Similarly,

$$\underline{y}(k) = C\big(c_x(k) - p_x(k)\big) + c_v(k) - p_v(k) = a_3(k) + \sum_{i=0}^{k-1} a_2(k-i-1)u(i) \tag{7.42}$$

where

$$a_3(k) = Cc_1(k) + c_v(k) - p_v(k) - Cp_x(k) \tag{7.43}$$

The advantage of reformulating $\underline{y}(k)$ and $\overline{y}(k)$ is that the explicit form can be divided as one linear affine function of the control input $u(i)$. Based on equation (7.40) and (7.43), we rewrite equation (7.38) as

$$y(k) \in C\big(c_y(k), p_y(k)\big) = \big[\underline{y}(k), \overline{y}(k)\big]$$

$$= \left[a_3(k) + \sum_{i=0}^{k-1} a_2(k-i-1)u(i), a_1(k) + \sum_{i=0}^{k-1} a_2(k-i-1)u(i) \right] \tag{7.44}$$

Then equation (7.44) will be used in the detailed computation of the other research field. From equation (7.32), the center or midpoint $c_x(k)$ can be chosen as the final state estimation, corresponding to our considered state of charge. Also combining equations (7.32) and (7.44), we derive not only the interval estimation for the state, but also the interval estimation for the prediction output. As the emphasis here is only on the interval estimation for the state, that interval estimation for the prediction output is not deeply studied here, and it is applied in learning model predictive control.

7.2.3   Ellipsoid estimation for SOC

In section 7.2.2, we assume the state noise and the initial state are in one interval (7.12). As in the research field, there are other sets to be used to denote the uncertainty, such as ellipsoid. It means that an interval and ellipsoid are two commonly used sets in a set membership estimation. So for completeness and comparison, in this section, we consider the ellipsoid estimation for the state, which corresponds to the state of charge for a Lithium-ion battery.

Observing only the first state equation in equation (7.9) again, matrices A, B, D can be identified by using the least squares method. Based on this state equation, our problem is to estimate the state $x(k)$ at different time instant $k = 1, 2 \cdots N - 1$. It is similar to section 7.2.2 that control input $u(k)$ is determined by researcher, and $u(k)$ is a deterministic value, not an uncertainty. So for convergence, we neglect this term $Bu(k)$ in the latter derivation. It means that if one ellipsoid estimation for the state is derived by our own derivation, then we can apply the translation transformation to give the true ellipsoid estimation for the state. Then we rewrite the considered state equation as follows:

$$\begin{cases} x(k+1) = Ax(k) + Dw(k) \\ x_0 = 0, k = 0, 1, \cdots N - 1 \end{cases} \tag{7.45}$$

This special state equation is driven by $w(k)$, satisfying the following norm bound:

$$\|w(k)\| \le 1, k = 0, 1, \quad N - 1 \tag{7.46}$$

Our goal here is to build an ellipsoid approximation of the state recursively. Let X_k be the set of all states where the system can be driven in time instant $k \in N$, and assume that we have built inner and outer ellipsoidal approximations E_{in}^k and E_{out}^k of the set X_k.

$$E_{in}^k \subset X_k \subset E_{out}^k \tag{7.47}$$

Let also

$$E = \left\{ x = Dw / w^T w \le 1 \right\} \tag{7.48}$$

Then the set

$$F_{in}^{k+1} = AE_{in}^k + E = \left\{ x = Aw_1 + w_2 / w_1 \in E_{in}^k, w_2 \in E \right\} \tag{7.49}$$

clearly cover X_{k+1}, and a natural recurrent way to define an outer ellipsoidal approximation of X_{k+1} is to take as E_{out}^{k+1} the smallest volume ellipsoid containing F_{out}^{k+1}. Note that the sets F_{in}^{k+1} and F_{out}^{k+1} are of the same structure: each of them is the arithmetic sum $\{x = w_1 + w_2 / w_1 \in W_1, w_2 \in W_2\}$ of two ellipsoids W_1 and W_2. Thus we come to the problem as follows: Given two ellipsoids W_1 and W_2, find the best inner and outer ellipsoidal approximations of their arithmetic sum $W_1 + W_2$. In fact, it makes sense to consider a problem.

Given two ellipsoids W_1 and W_2, find the best inner and outer ellipsoidal approximations of their arithmetic sum

$$W = \left\{ x = w_1 + w_2 / w_1 \in W_1, w_2 \in W_2 \right\} \tag{7.50}$$

of two ellipsoids W_1 and W_2.

(1) Outer ellipsoidal approximation

Let the ellipsoids W_1 and W_2 be represented as

$$W_i = \left\{ x / x^T D_i x \le 1 \right\} \tag{7.51}$$

Our strategy to approximate is that we want to build a parametric family of ellipsoids in such a way that, first, every ellipsoid from the family contains the arithmetic sum $W_1 + W_2$ of two given ellipsoids, and second, the problem of finding the smallest volume ellipsoid within the family is a simple problem.

Let us start with the observation that an ellipsoid $W[Z] = \{x/x^T Z x \leq 1\}$ contains $W_1 + W_2$ if and only if the following implication holds

$$\underbrace{\left\{ \{x^i\}_{i=1}^2 / \left[x^i\right]^T D_i \left[x^i\right] \leq 1, i = 1, 2 \right\}}$$

$$\Updownarrow$$

$$\left(x^1 + x^2\right)^T Z\left(x^1 + x^2\right) \leq 1 \tag{7.52}$$

Let D^i be one block diagonal matrix, such that all diagonal blocks, except the ith one, are zero, let $M[Z]$ be that

$$D^1 = \begin{bmatrix} D_1 & 0 \\ 0 & 0 \end{bmatrix}, D^2 = \begin{bmatrix} 0 & 0 \\ 0 & D_2 \end{bmatrix}, M[Z] = \begin{bmatrix} Z & Z \\ Z & Z \end{bmatrix} \tag{7.53}$$

Due to the fact that for every symmetric positive semidefinite matrix X such that $Tr(D^i X) \leq 1$, $i = 1, 2 \cdots m$, one has $Tr(M[Z]X) \leq 1$. Then we arrive at the following result.

Proposition 7.1: Let a positive definite matrix Z be such that the optimal value in the semidefinite program.

$$\max_X \left\{ Tr\left(M[Z]X\right) / Tr\left(D^i X\right) \leq 1, i = 1, 2, X \geq 0 \right\} \tag{7.54}$$

is ≤ 1. Then the ellipsoid $W[Z] = \{x/x^T Z x \leq 1\}$ contains $W_1 + W_2$ of two ellipsoids $W_i = \{x/x^T D_i x \leq 1\}$.

The above proposition is the first step to build a parametric family of ellipsoids, which contains the arithmetic sum $W_1 + W_2$. Then the second problem of finding the smallest volume ellipsoid within the parametric family can be reduced to one semidefinite program as that.

Proposition 7.2: Given two centered at the origin full dimensional ellipsoids.

$$W_i = \left\{ x / x^T D_i x \leq 1 \right\}, i = 1, 2$$

Let us associate with these two ellipsoids in the semidefinite program

$$\max_{t, Z, \lambda} \left\{ \begin{array}{l} t/t \leq Det^{\frac{1}{n_x}}(Z), \lambda_1 D^1 + \lambda_2 D^2 \geq M[Z] \\ \lambda_1 \geq 0, \lambda_2 \geq 0, \lambda_1 + \lambda_2 = 1 \\ Z \geq 0 \end{array} \right\} \tag{7.55}$$

Every feasible solution (t, Z, λ) to this semidefinite program with positive value of the objective produces ellipsoid $W[Z] = \{x/x^T Z x \leq 1\}$, which contains

$W_1 + W_2$, and the smallest volume ellipsoid is given by optimal solution of the semidefinite program (7.55).

(2) Inner ellipsoidal approximation

Let us represent the given centered at the origin ellipsoids W_i as

$$W_i = \left\{x / x = A_i w / w^T w \le 1\right\}, i = 1, 2$$

Due to the fact that an ellipsoid $E[Z] = \{x = Zw/w^Tw \le 1\}$ is contained in the sum $W_1 + W_2$ of the ellipsoids W_i if and only if one has

$$\forall x : \left\| Z^T x \right\|_2 \le \sum_{i=1}^{2} \left\| A_i^T x \right\|_2 \tag{7.56}$$

A natural way to generate ellipsoids satisfying equation (7.56) is to note that whenever matrix X_i satisfies the following the property about its special norms:

$$\left| X_i \right| = \sqrt{\lambda_{\max} \left(X_i^T X_i \right)} = \max_x \left\{ \left\| X_i x \right\|_2 / \left\| x \right\|_2 \right\} \le 1, i = 1, 2 \tag{7.57}$$

Then the matrix

$$Z = Z\{X_1, X_2\} = A_1 X_1 + A_2 X_2 \tag{7.58}$$

satisfies equation (7.56).

$$\left\| Z^T x \right\|_2 = \left\| [A_1 X_1 + A_2 X_2] x \right\|_2 \le \sum_{i=1}^{2} \left\| X_i^T A_i^T x \right\|_2 \le \sum_{i=1}^{2} \left| X_i^T \right| \left\| A_i^T x \right\|_2 \le \sum_{i=1}^{2} \left\| A_i^T x \right\|_2 \tag{7.59}$$

Thus every collection of square matrices X_i with spectral norms not exceeding 1 produces an ellipsoid satisfying equation (7.56) and thus is contained in W.

Similarly, the largest volume ellipsoid within the parametric family can also be reduced to the following semidefinite program.

Proposition 7.3: Let $W_i = \{x/x = A_i w/w^T w \le 1\}$, $A_i > 0$, $i = 1, 2$, consider the following semidefinite program:

$$\max \quad t$$

$$\text{subject to} \quad t \le \left(Det \left(\frac{1}{2} \sum_{i=1}^{2} \left[X_i^T A_i^T x \right] \right) \right)^{\frac{1}{n_x}};$$

$$\sum_{i=1}^{2} \left[X_i^T A_i^T x \right] \ge 0; \begin{bmatrix} I_n & -X_i^T \\ -X_i & I_n \end{bmatrix} \ge 0, i = 1, 2 \tag{7.60}$$

with design variables X_1, X_2. Every feasible solution (X_1, X_2, t) to this problem produces the ellipsoid

$$E(X_1, X_2) = \left\{ x = \left(\sum_{i=1}^{2} A_i X_i \right) w / w^T w \le 1 \right\}$$

contained in the arithmetic sum $W_1 + W_2$ of the original ellipsoids, and the largest volume ellipsoid which can be obtained in this way is associated with the optimal solution to the semidefinite program (7.60).

After solving these two semidefinite programs (7.55) and (7.60), we build inner and outer ellipsoidal approximations E_{in}^k and E_{out}^k of the set X_k, i.e. $E_{in}^k \subset X_k \subset E_{out}^k$.

Based on these two inner and outer ellipsoidal approximations E_{in}^k and E_{out}^k of the set X_k, then the final state estimation at time instant $x(k)$ can be chosen as the midpoint between the two centers of inner and outer ellipsoidal approximations E_{in}^k and E_{out}^k.

So generally, whether state noise or external noise is included in an interval or an ellipsoid, firstly, we apply our mentioned interval estimation or ellipsoid estimation to obtain the state estimation set. Secondly, the center or midpoint can be chosen as the final state estimation value, which corresponds to the state of charge for a Lithium-ion battery.

7.2.4 Simulation example

Here, we do not have the experimental platform yet, so this simulation example is based on references in the open literatures. To acquire experimental data such as current, voltage, and temperature from the battery, a battery test bench was established. The configuration of the battery test bench is shown in Fig. 7.6.

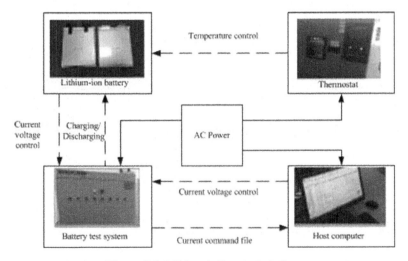

Figure 7.6: Lithium battery test platform

For convenience in the latter simulation example, the Lithium battery test needs to charge and discharge the Lithium ion battery at different temperatures and different rates. Therefore, the equipment required for the experimental bench includes a thermostat, a battery charging and discharging device, a ternary neon battery, and a host computer. Lithium battery test platform is plotted in Fig. 7.6, where the detailed processes are described as follows.

Step 1. The charging and discharging positive and negative terminals of the battery are respectively connected to the positive and negative electrodes of the battery through the wire harness, and the wire harness of the appropriate diameter is selected according to the allowable charging and discharging ratio of the battery to avoid the burning of the wire harness. One end of the voltage sampling line to the other end of the battery is connected to the voltage sampling and wiring port of the battery charging and discharging device. Finally, the temperature measuring line of the thermistor is attached to the surface of the battery, and the other side of the temperature detecting line is connected to the temperature detecting terminal of the battery charging and discharging device.

Step 2. Set the lithium battery in the incubator, and set the experimental ambient temperature.

Step 3. Start the battery charging and discharging equipment and incubator.

Step 4. In the online machine, we edit the charge and discharge test step or import the edited current test file into the host computer to automatically generate the test step, then set the sampling time and output file save address, and start the test.

Based on the experimental platform, the open-circuit voltage of the battery has a monotonic relationship with the state of charge. The relation between the open-circuit voltage and state of charge is established by running a test on the considered lithium-ion battery. Let all batteries be fully charged and rested for 3 hours, such that the internal chemical reactions attain a desired equilibrium state. Moreover, the discharge test includes a sequence of pulse current of 1 C with 6-min. discharge and 10-min. rest, then the discharge test can make the battery return back to its expected equilibrium state before running the next cycle, which is shown in Fig. 7.7.

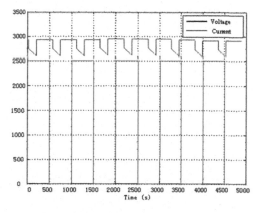

Figure 7.7: Voltage and current curves

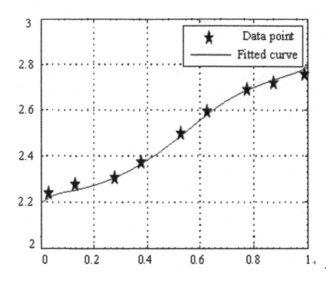

Figure 7.8: Polynomial form for U_{OC}

As U_{OC} is rewritten as the following polynomial form $U_{OC}(x) = d_5 + d_4 x + d_3 x^2 + d_2 x^3 + d_1 x^4$, to identify these unknown parameters in this polynomial form, the least squares method is used to achieve this goal. Then the identification result for this polynomial form is given in Fig. 7.8, which shows the relation between the true data point and its identified polynomial form.

In the whole simulation process, the true parameters can be identified by using some system identification strategy, for example the least squares method, instrumental variable method, maximum likelihood method, etc. Then identified parameter are obtained as follows:

$$R_0 = 0.0994\Omega, R_p = 0.030\Omega, C_p = 20773KF; I = 1.10A; T_s = 0.3s$$

Substituting the above values into the equation (7.7), (7.8) and (7.9), then each matrix is given as follows:

$$A = \begin{bmatrix} 1 & 0 & 0 \\ 0 & 0.68 & 0 \\ 0 & 0 & 1 \end{bmatrix}, B = \begin{bmatrix} -0.8 \\ 0.064 \\ 0.1 \end{bmatrix}, C = [2.5 \quad -1 \quad 1.2], D = 1$$

Consider the unknown but bounded noise in simulation, these two kinds of unknown but bounded signals in equation (7.9) are formulated as

$$\begin{cases} -0.5 \le x(0) \le 0.5 \\ 1 \le w(k) \le 1 \qquad \text{for all} \quad k \in R^+ \\ 1 \le v(k) \le 1 \end{cases}$$

In a simulation, we consider not only the state estimation, but also the output estimation, and the state estimation corresponds to the state of charge. The interval estimation on the observed output with bounded noise $v(k)$ can be used in other research fields, such as robust control, interval model predictive control, etc.

Firstly, we apply equation (7.27) to obtain the interval estimation for the state. The state trajectory can be easily obtained by using equation (7.19) in Matlab, where some priori information about the initial state and bounded noise is used. The simulation results are shown in Fig. 7.9, where the black curve is the true state trajectory and the two red curves denote the estimated curves. One curve consists of the upper bound, and the other curve is the lower bound. From Fig. 7. 9, we see that the true state trajectory lies in between the two red curves, so at each time instant, the midpoint of the upper bound and lower bound can be chosen as the final state estimation at the considered time instant. Similarly, the interval estimation for output is also given in Fig. 7.10, where the true output trajectory lies in between the two estimated curves. The interval estimation for output is obtained based on equation (7.44), and the final output estimation can also be selected as the midpoint at each time instant.

For the sake of completeness, the ellipsoid estimation for the state or state of charge is given in Fig. 7.11, where the true state trajectory is the same as that curve in Fig. 7.9. Twelve data points are sampled in the true state trajectory, and we need to construct twelve ellipsoids to include these twelve data points as their interior points. As an ellipsoid is used to denote the uncertainty, so at every time instant we will obtain one ellipsoid to include the true data point. From the simulation result in Fig. 7.11, twelve ellipsoids are constructed by using equation (7.47), and these twelve ellipsoids include the twelve data points as their own interior points

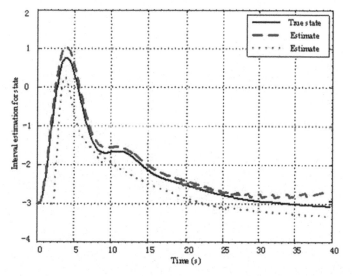

Figure 7.9: Interval state estimation

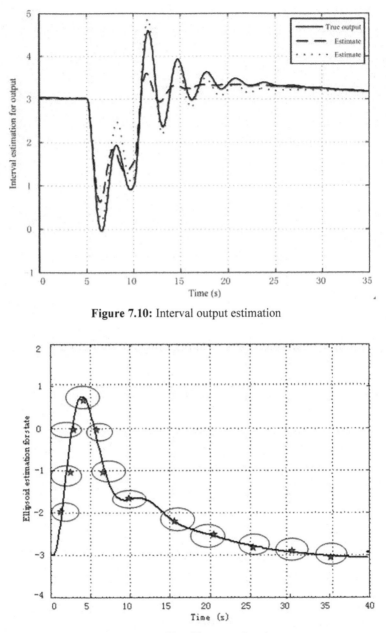

Figure 7.10: Interval output estimation

Figure 7.11: Ellipsoid state estimation

exactly. When the center of the corresponding ellipsoid is chosen as the final state estimation, we find that the error exists still, i.e. the center of the considered ellipsoid is not equal to the true state value at the considered time instant.

7.2.5 Conclusion

In this section, a set membership strategy is applied to estimate the state of charge for a Lithium-ion battery, so that the state estimation can be dealt with in case of unknown but bounded external noise. The goal of introducing a set membership strategy is to alleviate the shortcoming of the traditional Kalman filter algorithm. We formulate one state space equation for the state of charge estimation, by using one equivalent circuit model to replace the considered state of charge estimation for a Lithium-ion battery. According to the commonly used interval and ellipsoid for external noise, the interval estimation and ellipsoid estimation are derived for the considered state estimation respectively, which corresponds to the state of charge estimation for a Lithium-ion battery.

7.3 Optimal input signal design for aircraft flutter model parameters identification

Aircraft flutter is a complex aeroelastic instability phenomenon that poses a great threat to aircraft flight safety, as flutter is a self-excited vibration and is unstable. When the considered aircraft flies in the airflow and reaches a certain speed, the vibration will appear to cause the flutter under the condition of the interactions with unsteady aerodynamic force, inertial force, and elastic force. This flight speed, posing the aircraft flutter phenomenon, is called the flutter critical speed. Generally, the vibration with relatively large amplitude and non-attenuated aircraft components is the main cause of aircraft flutter. More specifically for an aircraft, its aerodynamics will increase as the flight speed increases, so there is one critical speed, at which the aircraft structure will become unstable. It means when the flight speed is lower than the critical speed, the vibration is attenuated, and if it is equal to the critical speed, then the vibration maintains the same amplitude. Moreover, if the vibration is divergent in case of high flight speed, then this divergent will damage the aircraft structure and flight security, so aircraft flutter is one typical phenomenon, that attracts the most attention for aircraft dynamic aeroelasticity.

As the aircraft flutter corresponds to the stability problem of the aeroelastic dynamics, it has a variety of phenomena. But in terms of its main aerodynamic causes, the aircraft flutter problem can be divided into two categories. The first type occurs in the potential flow, so flow separation and boundary layer effect do not influence the flutter process, i.e. this type of aircraft flutter mainly occurs in the streamlined profile lift system, such as the wing bending torsion coupled flutter, wing-aileron couples flutter, etc. This type of aircraft flutter is usually called the classical flutter. The second type of flutter problem depends on the flow separation and vortex formation, and it is referred to as stall flutter, being one kind of lifting surface flutter. If the flow is separated during all parts of the vibration time, then the aircraft is in a stall state. The flutter phenomenon at this time is different from the classical flutter, as its appearance neither depends on inertia, elasticity, and aerodynamic coupling, nor on the phase error between movement and motion. Since a large angle of attack is a necessary condition to separate the aircraft and the suction surface, then this stall flutter is related to the flight angle of attack. Therefore, stall flutter usually occurs in rotating mechanical components, for example, propellers, turbine blades, etc, as the above mentioned machines sometimes work near the steady stall angle of the blades. Consider the aircraft structures aircraft wing surfaces such as wings and tails rarely encounter stall flutter, i.e. the considered flutter belongs to the first type and is the most dangerous in practice. This kind of self-excited vibration can cause structural vibration damage in a few seconds or even less, then causing catastrophic consequences for the aircraft.

The flutter flight test is an essential part of verifying the flutter characteristics of the considered aircraft. Both the new model design and the modification of a certain model design must undergo the full scale flutter flight test at various speeds up to the designed dive speed envelope, then the designer can verify whether they are within the specified flight limit speed envelope. All critical configurations for each aircraft must not flutter on the altitude envelope, corresponding to the above dive speed envelope. The flutter is avoided based the equivalent airspeed at all points is increased by 15%. As a consequence, the flutter flight test is to verify whether the flutter safety margin exists for analyzing some estimations, for example, our considered flutter model parameters. Flight near the estimated flutter speed may cause damage to the aircraft structure, due to the occasional or potential flutter. Given this fact, the flight envelope is determined by gradually increasing the speed from a sufficiently low speed for the flight test, and analyzing the model frequency or damping coefficient. The identified model frequency is used to compare the results, coming from the ground resonance test, and then the model in response is later determined. Here, the above described model frequency and damping coefficient are combined to be our called aircraft flutter model parameters. The main purpose of the flutter flight test is to determine the aeroelastic stability boundary for the aircraft. The traditional flutter flight test uses the damping ratio of the flutter mode as an index to determine the flutter boundary. Due to the low signal-to-noise ratio for flutter flight test data, that damping ratio is one sensitive parameter, and also it is one nonlinear function with respect to the flight speed, so the traditional methods cannot give one credible flutter boundary.

Due to the important flutter model parameters (model frequency and damping coefficient) in predicting the flutter boundary, this predictive value can provide one reference value for later flutter suppression or flight control. In these recent years, many researchers all over the world start to study this identification or estimation problem of flutter model parameters from different views. For example, reference [15] uses the subspace method to identify the output predictions of one state space model, which is replaced by our considered frequency response function. Then the model parameters of a beam under free vibration are obtained by using singular value decomposition. In [16], a periodic sinusoidal wave is applied as the input signal during the whole model parameter identification test. From the experiment results, we see this periodic sinusoidal signal can improve the accuracy estimation for the frequency response function and reduce the leakage error. In [17], the frequency response function is applied in the detection of structural damage, further, a new weighted digital filter is constructed to perform the Fourier transform on the input signal-output data sequence [18]. This new weighted digital filter can smooth the effect of colored noise. Generally, the frequency response function estimation can be divided into parameter estimation and nonparameter estimation. Due to the additional unknown parameters in parameter estimation, then the solution of the unknown parameters will not be unique. On the contrary, as nonparameter estimation directly obtains the estimated value for the frequency response function, then it greatly simplifies the whole identification process. In [19], a transient impulse response modeling method is proposed to identify the frequency response function. This transient impulse response modeling method combines the approximation of the frequency response function and the leakage term as a function of the global parameters. Then these global parameters can be estimated by solving a global least squares problem. The use of the window function is based on the different forms of the input signal-output signal at some special frequency points. Although this window function method can reduce the influence of the leakage error, it also increases the interpolation error [20]. The identification of aircraft flutter model parameters is less studied in China. In China, the whole flutter test process was systematically studied in one published paper [21], where a wavelet method for flutter test data processing was proposed to improve the signal-to-noise ratio of flight test data under the condition of small rocket excitation. Then aiming at improving the effect of pulse excitation response, a method based on a support vector machine for flight test response data was proposed [8]. A dual time-frequency domain algorithm and a fractional Fourier domain method were proposed respectively for the rudder surface sweep excitation of the telex aircraft [23]. To make up for the shortcomings of the traditional least squares frequency domain fitting identification algorithm, the global least squares identification algorithm in the frequency domain is proposed in [24], which avoids the complex nonlinear optimization and the dependence on initial value in the iterative algorithm. According to the whole framework of the system identification theory [25], the above mentioned research contents correspond to two aspects of parameter estimation and experimental design in system identification theory. So far, there is a lack of works in the literature that seeks to solve the problem of parameter estimation of the flutter phenomenon.

The above paragraph about detailed descriptions of the identification or estimation problem for aircraft flutter model parameters is from our previously published paper [26], which formulates others' contributions on this interesting subject. In these recent years, the authors of this paper also complete some contributions to identifying the aircraft flutter model parameters. To give a clearer understanding of our existing contributions, we introduce our published results as follows next. When the observed input signal-output data are corrupted by the observed noises in the aircraft flutter stochastic model, the instrumental variable method and variance matching are combined to be one instrumental variable variance method, used to identify the unknown parameters in one constructed transfer function form [26]. This new proposed instrumental variable variance method achieves good estimation accuracy only on the condition of the independent and identically distributed random noises. The optimal filtering estimation problem for the actual input signal and output sequences in the aircraft flutter model parameter identification experiment is studied in our previous paper [27], where the estimation error and covariance expression for the estimations of the filtered input signal and output is derived by our derivations. The asymptotic covariance matrix expression is given to verify the effectiveness of our proposed instrumental variable covariance method [28], where an external excitation signal is devised. If the estimations of the noise variance are available priorly, the principle of the biased compensated method can readily be used to obtain a consistent parameter estimation [29]. A novel frequency domain subspace identification is discussed based on the forgetting factor [30]. The key idea of this novel frequency domain subspace identification is to guarantee the leakage free effect, calculated through the Discrete Fourier Transform.

Based on our above published papers [26–30] on aircraft flutter model parameter identification, in this paper here, we continue this research deeply based on our existing results. Without loss of generality, our previous work was only concerned with the identification problem for flutter model parameters, i.e. some new identification methods are proposed to identify the unknown model parameters in different cases of considered noise. It is well known that the topic of parameter identification is one process of the system identification theory field, i.e. the whole system identification theory includes four steps: experiment design or input signal design, model structure, parameter identification, and model validation. A good input signal must guarantee all the model responses are excited completely, in the sensor, and that the regressor matrix is full rank. The problem of input signal design responds to the condition of persistent estimation, as this property of persistent excitation is crucial in many adaptive schemes, where parameter convergence is guaranteed as parameter convergence is one of the primary objectives of the adaptive system. This condition of persistent excitation can be interpreted as a condition on the energy of input signal in all directions and then this persistent excitation can guarantee the parameter estimation to be exponentially fast in system identification and adaptive control. As the input signal is very important for the later parameter identification and model validation, in this chapter, we change our previous works on parameter identification to input signal design for aircraft flutter model parameter identification experiment. In [31], we study the input signal design for multi UAVs formation anomaly detection, which is proposed by the author, and now some researchers are still studying this new concept from different aspects, for example, formation control, formation communication, formation trajectory, etc. Reference [32] used subspace predictive control to suppress the effect, coming from the active noise. The difference between our previous work [33] and this new paper is that the considered systems are different. More specifically, the system, considered in [33], is one common form, whose output is corrupted by external noise, but the system is one, whose input signal and output are all corrupted by external noises. It means that system is in especial case of our considered system here. This case with two corrupted noises is more realistic in practice, as external noises exist whether in the input signal or output variables. The above description of the chosen system is similar to the model structure within the field of system identification. One complex model is always chosen in theory research, on the other hand, the simple model is used in engineering. Roughly speaking, the aircraft flutter model, proposed by the author, is one compromise model, whose input signal and output are all corrupted by external noises. Furthermore, this compromise model extends the classical linear form, and it is also suited for engineering analysis.

According to our previous works on the identification problem for the unknown flutter model parameters, our proposed identification algorithms are efficient on the condition that the considered external or internal noise is statistical noise, i.e. the statistical properties of the external noises are known, for example, the external noise may be independent and identically distributed random sequence with zero mean and unit variance. But it is well known that two methods are used to describe the considered external or internal noise, i.e. statistical description and deterministic description. According to the first common statistical description of external noise, this noise is always assumed to be white noise or colored noise with a priori known mean value and variance. On the contrary in deterministic description on noise, the priori knowledge on noise is nothing, except it is unknown but bounded. From the system identification research field, we see that statistical description of noise is one ideal case, as in practice the probabilistic distribution corresponding to external noise is not known priorly. It means that the commonly used white noise or colored noise is one strict assumption on noise, which is not considered in the practical industry. To relax this strict assumption on noise and consider more a practical situation in industry such as the aircraft flutter model, a deterministic description of noise is considered in this chapter, where the corrupted noise is unknown but bounded noise. For completeness, we consider the input signal design for statistical noise and unknown but bounded noise respectively. After constructing one aircraft flutter model, whose input signal and output variable are all corrupted by external noises, one separable lease squares identification algorithm is proposed to identify the unknown model parameters in case of statistical noise. Based on the obtained least squares parameter estimations, their asymptotic variance matrix expression is derived after simple but tedious calculation. In the case of identifying the aircraft flutter model parameters with unknown but bounded noise, a set membership strategy is proposed to achieve our goal. Furthermore, during our above mentioned references, only white noise is assumed to exist in the aircraft flutter model, so this section is one result and contribution, which extends our above research, i.e. this chapter is an extended version for unknown but bounded noise. The difference in the set membership identification strategy is that the identification result is not a constant value for each unknown parameter, but a parameter set or parameter uncertainty interval. This parameter set must be consistent with the corrupted input signal-output data, noise bounds, and our considered aircraft flutter model. Then the central estimate can be determined as the mid-point of each parameter set, and this central estimate is regarded as the terminal model parameter value, which is used for the latter analysis. When considering the input signal design, the optimal input signal can be easily yielded through minimizing the radius of information, which corresponds to the radius of the obtained parameter set or parameter uncertainty interval. Furthermore, the idea of the radius of information is also applied to construct one performance function, then the optimal input signal is obtained by minimizing the performance function. Generally, this optimality is satisfied for its corresponding performance function.

As a consequence, the main contributions of this second application are formulated as follows, and the structure of it is plotted in Fig. 7.12.

1. All of our previous works on aircraft flutter model parameters identification are reviewed.

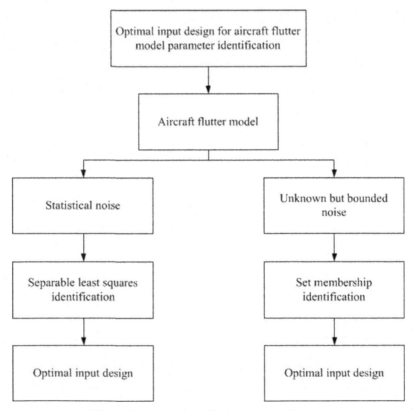

Figure 7.12: Structure of this second application

2. For an aircraft flutter statistical model with statistical noise, one separable least squares identification is proposed to identify the unknown model parameters. Then in the case of this statistical noise, the optimal input signal is designed to satisfy one given performance function.

3. For an aircraft flutter model with unknown but bounded noise, set membership identification is proposed to solve the parameter set for each unknown model parameter. Then in the case of this unknown but bounded noise, the optimal input signal is designed by applying the idea of the radius of information.

7.3.1 Aircraft flutter model

A necessary text for flutter model parameters identification is the flutter wind tunnel test, which includes pre-test preparation, excitation section, test operation, post-test model check, and data processing. In the flutter wind tunnel test, to correctly simulate the flight motion and support conditions of the flutter model, a support system, that meets the test requirements needs to be specially designed. When the aircraft is in the low speed test, the component model is usually supported on a rigid frame. Compared with the model, the rigid frame has a much larger stiffness and mass, and within the relevant flutter frequency range, involved in the test, the inherent rigid frame can not appear to be its natural frequency. When doing a full model test, generally support system can support the model and maintain the model with at least three rigid body motions, such as lifting, pitching, and rolling. On the other hand, in the high speed model test, the component model can be fixed on the side wall of the wind tunnel. To reduce the influence of the surface layer or the cave wall, then the width of the fuselage can be appropriately increased.

The experimenter realizes that to obtain some data with a high signal-to-noise ratio, it is important to achieve sufficient excitation, i.e. a sufficiently large energy excitation can excite the corresponding vibration mode, then the system stability can be seen from the response data. The requirements for the excitation system are not only to provide sufficient excitation force, but also within the relevant frequency range. The flutter excitation is seen in Fig. 7.13, where we can use the following excitation methods.

1. **Impulse excitation:** The pilot excites the control surface or uses a small rocket to generate an impulse excitation signal, applying to excite the aircraft in flight within a short action time. Impulse excitation is a relatively simple method, as its equipment is not complicated, and the time required for the flutter rest is relatively short. However, it is difficult to obtain a response signal with a high signal-to-noise ratio, so some complex data processing methods must be needed to obtain the model frequency and damping coefficient.

2. **Harmonic excitation:** Harmonic excitation can be realized by the electromagnetic exciter, for example, steady state excitation and attenuated oscillation. The main shortcomings of harmonic excitation are the excitation frequency is adjustable in the test, the excitation energy is large, and furthermore, the test time is longer, etc.

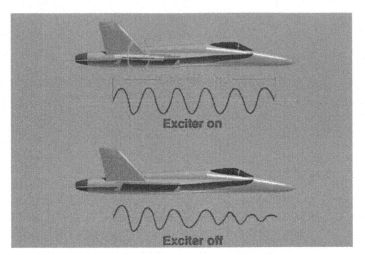

Figure 7.13: Flutter excitation

3. **Scan excitation:** A special signal generator is used to generate a specific signal, and within a selected frequency range, the frequency of the excitation signal is continuously changed, according to a designed law.

4. **Random excitation:** The function of random excitation is to use atmospheric turbulence or mechanical and aerodynamic methods to generate random signals. After exciting the aircraft structure, some response signals of the aircraft structure are recorded.

Before studying the identification problem for aircraft flutter model parameters, one statistical or stochastic model of the flutter test is needed. In modern control theory, state space form and transfer function form are two commonly used forms to describe the practical system. Due to the equivalence relation between transfer function and state space, so for convenience, we use that transfer function form to present our considered stochastic model of the aircraft flutter.

For clarity of presentation, here, we continue to use the transfer function model in analyzing the identification problem and input signal design in this paper, so consider the following stochastic model in Fig. 7.14, which corresponds to the flutter test experiment. Where in Fig. 7.14, variable $u(t)$ and $y(t)$ are input signal and output signal respectively, $u_0(t)$ represents the artificial excitation, applied to aircraft, $n_g(t)$ is atmospheric turbulence excitation. $G(q^{-1})$ is one transfer function of the considered aircraft, and it is unknown and needs to be identified, q^{-1} is one time shift operator, i.e. $qu(t) = u(t-1)$.

Since the flight test is inevitably affected by atmospheric turbulence excitation, then $n_g(t)$ is regarded as an unmeasured excitation, and the random response generated by $n_g(t)$ will be included as process noise in the measured response signal. $y_0(t)$ is the flutter acceleration signal, observed noises $\tilde{u}(t)$ and $\tilde{y}(t)$ are generated by the sensor. The processing method of the flutter test data

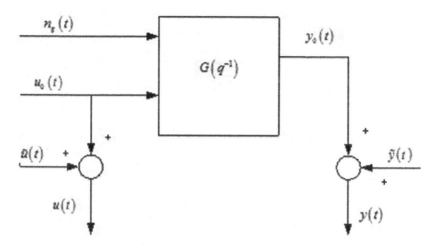

Figure 7.14: Stochastic model of the flutter test

and the choice of the excitation method are closely related to each other. At present, the commonly used excitation methods mainly include control surface pulse excitation, small rocket excitation, frequency sweep excitation, atmospheric turbulence excitation, etc. The excitation of the flight test uses rocket excitation. The principle of excitation point and sensor arrangement is to effectively stimulate the first-order symmetric bending mode of the wing of interest, the first-order anti-symmetric bending mode, and the first-order symmetric mode. And it is convenient to measure the response signal corresponding to these third-order modes. The flutter characteristics of an aircraft can be determined by estimating the frequency and damping of various flight states as a function of altitude and speed.

According to the relations shown in Fig. 7.14, we have that following relationships in time-domain.

$$\begin{cases} y(t) = y_0(t) + \tilde{y}(t) \\ u(t) = u_0(t) + \tilde{u}(t) \\ y_0(t) = G(q^{-1})\left[u_0(t) + n_g(t) \right] \end{cases} \tag{7.61}$$

After substituting the relation $y_0(t)$ into $y(t)$, it holds that

$$y(t) = y_0(t) + \tilde{y}(t) = G(q^{-1})\left[u_0(t) + n_g(t) \right] + \tilde{y}(t)$$
$$= G(q^{-1})u_0(t) + \underbrace{G(q^{-1})n_g(t) + \tilde{y}(t)}_{\tilde{y}_1(t)}$$

Introducing one new observed noise $\tilde{y}_1(t)$, then the above relation corresponding to output variable $y(t)$ is rewritten as

$$\begin{cases} y(t) = y_0(t) + \tilde{y}_1(t); u(t) = u_0(t) + \tilde{u}(t) \\ y_0(t) = G(q^{-1}) u_0(t) \end{cases}$$

Observing the above equations, we see that the effect, coming from unmeasured excitation $n_g(t)$, is included as process noise in the measured response signal, so for notational simplicity we neglect $n_g(t)$ in the above equation.

Then we have the following rational transfer function model, which is related to our considered aircraft structure.

$$G(q^{-1}) = \frac{B(q^{-1})}{A(q^{-1})},$$

$$A(q^{-1}) y_0(t) = B(q^{-1}) u_0(t)$$

$$A(q^{-1}) = 1 + a_1 q^{-1} + \cdots a_{n_a} q^{-n_a}; B(q^{-1}) = b_1 q^{-1} + \cdots b_{n_b} q^{-n_b} \qquad (7.62)$$

where a_i and b_i are coefficients of polynomial, n_a and n_b are orders of polynomial.

Define p_r: ($r = 1, 2 \cdots n^a$) as the poles of the transfer function $A(q^{-1})$, n^a is the number of poles, then it holds that

$$(p_r)^{n_a} + a_1 (p_r)^{n_a - 1} + \cdots a_{n_a - 1}(p_r) + a_{n_a} = 0, r = 1, 2 \cdots n^a$$

Applying the correspondence relation between the continuous time model and the discrete time model, the poles s_r: ($r = 1, 2 \cdots n^a$) are obtained in continuous time domain, i.e. $p_r = e^{s_r T}$, where T is the sampled period, so the poles s_r: ($r = 1,$ $2 \cdots n^a$) for continuous time model are given as $s_r = \ln\left(\dfrac{p_r}{T}\right)$.

Then the model frequency and damping coefficient can be derived as follows:

$$f_r = \mathrm{Im}(s_r)/2\pi, \quad \xi_r = -\mathrm{Re}(s_r)/|s_r| \qquad (7.63)$$

The main goal of the process of aircraft flutter model parameters identification is to identify the above model frequency and damping coefficient f_r, ξ_r ($r = 1, 2 \cdots n^a$). Obviously, accurate transfer function estimation is the premise of model parameter identification. Moreover, the goals of this chapter are not only to propose some identification methods to identify these unknown model parameters, but also to design the optimal input signal, such that the flutter statistical model is sufficiently excited.

7.3.2 Optimal input signal design for statistical noise

Based on our previous identification algorithms for aircraft flutter model parameters, in this section, we still study the problem of how to identify the model parameters in the sense of statistical noise. Then on basis of this identification algorithm for statistical noise, the latter optimal input signal is designed by minimizing the trace operation of the derived asymptotic variance matrix.

(1) Preliminary

Before introducing our proposed separable least squares algorithm, we need to assume these two observed noises $\tilde{u}(t)$ and $\tilde{y}(t)$ that are all independent and identically distributed white noises with zero mean and variance σ_u^2 and σ_y^2. These two observed noises $\tilde{u}(t)$ and $\tilde{y}(t)$ are independent of each other. Furthermore, the noises are independent of those two variables $u_0(t)$ and $y_0(t)$ as well.

Set

$$\phi_1(t) = \begin{bmatrix} Y_t & U_t \end{bmatrix}^T$$

$$Y_t = \begin{bmatrix} -y(t-1) & \cdots & -y(t-n_a) \end{bmatrix}; U_t = \begin{bmatrix} -u(t-1) & \cdots & -u(t-n_b) \end{bmatrix} \quad (7.64)$$

$$\theta = \begin{bmatrix} a_1 & \cdots & a_{n_a} & b_1 & \cdots & b_{n_b} \end{bmatrix}^T$$

where $\phi_1(t)$ is one regressor vector, θ denotes the unknown parameter vector.

Using the above defined vectors, equation (1) can be rewritten as

$$y(t) = \frac{B(q^{-1})}{A(q^{-1})} \left[u(t) - \tilde{u}(t) \right] + \tilde{y}(t) \quad (7.65)$$

It means that

$$A(q^{-1})y(t) = B(q^{-1})u(t) + A(q^{-1})\tilde{y}(t) - B(q^{-1})\tilde{u}(t) = B(q^{-1})u(t) + w(t)$$

$$w(t) = A(q^{-1})\tilde{y}(t) - B(q^{-1})\tilde{u}(t)$$

Rewriting equation (7.61) as the following linear regressor form:

$$y(t) = \phi_1(t)\theta_0 + w(t) \quad (7.66)$$

where θ_0 is the true parameter vector, corresponding to the unknown parameter vector θ, $w(t)$ is the stochastic disturb term.

(2) Separable least squares identification

To identify that unknown parameter vector θ, set on new instrumental variable $z(t) \in R^{n_z}, n_z = n_a + n_b + 2$, and consider the following over-parameterized system or equation.

$$\frac{1}{N}\sum_{t=1}^{N} z(t)w(t) = \frac{1}{N}\sum_{t=1}^{N} z(t)\left[y(t) - \phi_1(t)\theta \right] \quad (7.67)$$

where N denotes the number of observed data.

As that unknown parameter vector θ must satisfy the above over-parameterized system or equation, the elements of that instrumental variable $z(t)$ maybe independent of the stochastic disturb term $w(t)$, i.e. the classical instrumental variable estimation. But here due to those two unknown variances σ_u^2 and σ_y^2, we need to choose the elements of $z(t)$ to be related with the disturb term $w(t)$. From the existing results in the system identification field, the parameter estimation $\hat{\theta}$ with respect to that equation (7.66) is given as

$$\hat{\theta} = \left(R_{z\phi_1} - S(\sigma) \right)^+ \left(R_{zy} - \xi\sigma \right) \tag{7.68}$$

where $(R_{z\phi_1} - S(\sigma))^+$ denotes the moore penrose pseudo inverse matrix with full column rank, $S(\sigma)$ is one matrix function with respect to the noise variances $\sigma = \begin{bmatrix} \sigma_y^2 & \sigma_u^2 \end{bmatrix}$, ξ denotes one matrix with dimension $(n_z \times 2)$.

The cross-relation matrices $R_{z\phi_1}$ and R_{zy} are computed as

$$R_{z\phi_1} = \frac{1}{N}\sum_{t=1}^{N} z(t)\phi_1^T(t), \quad R_{zy} = \frac{1}{N}\sum_{t=1}^{N} z(t)y^T(t)$$

The different choice for the elements in instrumental variable $z(t)$ will determine the structure for matrix $S(\sigma)$ and ξ. Without loss of generality, we choose

$$z(t) = \begin{bmatrix} y(t) & Y_t & u(t) & U_t \end{bmatrix}^T \tag{7.69}$$

Then the dimension for this instrumental variable $z(t)$ is $(n_z = n_a + n_b + 2)$, then it holds that

$$R_{z\phi_1} = S(\sigma) = \frac{1}{N}\sum_{t=1}^{N} z(t)\phi_1^T(t) = \begin{bmatrix} 0 & 0 \\ \sigma_y^2 I_{n_a} & 0 \\ 0 & 0 \\ 0 & \sigma_u^2 I_{n_b} \end{bmatrix}$$

$$R_{zy} = \xi\sigma = \frac{1}{N}\sum_{t=1}^{N} z(t)y^T(t) = \sigma\begin{bmatrix} 1 & 0 \\ 0 & 0 \end{bmatrix} \tag{7.70}$$

where I_{n_a} and I_{n_b} are two unit matrix, with dimension $(n_a \times n_a)$ and $(n_b \times n_b)$ respectively.

Observing that parameter estimation $\hat{\theta}$, as noise variances σ exist in equation (7.68), it means the parameter estimation $\hat{\theta}$ depends on the variance estimation $\hat{\sigma}$. As here two unknown parameters $\left(\hat{\theta}, \hat{\sigma} \right)$ are needed to be identified simultaneously, so we use the separable least squares algorithm to achieve this goal.

Construct the following performance function with Euclidean norm $\| \ \|$.

$$f(\theta,\sigma) = \left\| R_{zy} - \xi\sigma - \left[R_{z\phi_1} - S(\sigma) \right]\theta \right\|^2 \tag{7.71}$$

The noise variance estimation $\hat{\sigma}$ is obtained by minimizing the above performance function $f(\theta, \sigma)$, i.e.

$$\hat{\sigma} = \arg\min_{\sigma}\left[\min_{\theta} f(\theta,\sigma)\right] \tag{7.72}$$

The main essence of separable least squares algorithm is to estimate θ and σ separately. After substituting equation (7.68) into (7.71), then the performance function is rewritten as

$$f_1(\sigma) = \left\| R_{zy} - \xi\sigma - \left(R_{z\phi_1} - S(\sigma)\right)\cdot\left(R_{z\phi_1} - S(\sigma)\right)^{+}\left(R_{zy} - \xi\sigma\right)\right\|^2 \tag{7.73}$$

Then that optimization problem (7.72) is requested to

$$\hat{\sigma} = \arg\min_{\sigma} f_1(\sigma) \tag{7.74}$$

After the noise variance estimation $\hat{\sigma}$ is solved from optimization problem (14), the parameter estimation $\hat{\theta}$ is easily obtained through substituting $\hat{\sigma}$ into equation (7.68).

Furthermore, we find that performance function (7.71) can be simplified as that

$$f(\theta,\sigma) = \left\| R_{zy} - R_{z\phi_1}\theta + \left((S_1\theta v_1 + S_2\theta v_2 - \xi)\sigma\right)\right\|^2 \tag{7.75}$$

where S_1 and S_2 are defined as

$$S_1 = \begin{bmatrix} 0 & 0 \\ I_{n_a} & 0 \\ 0 & 0 \\ 0 & 0 \end{bmatrix}, \quad S_2 = \begin{bmatrix} 0 & 0 \\ 0 & 0 \\ 0 & 0 \\ 0 & I_{n_b} \end{bmatrix}; v_1 = \begin{bmatrix} 1 & 0 \end{bmatrix}, v_2 = \begin{bmatrix} 0 & 1 \end{bmatrix}$$

According to equation (7.75), and given σ, then the minimum value for noise variance estimation $\hat{\sigma}$ is that

$$\hat{\sigma} = \left((S_1\theta v_1 + S_2\theta v_2) - \xi\right)^{+}\left(R_{z\phi_1}\theta - R_{zy}\right) \tag{7.76}$$

Substituting equation (7.76) into (7.75), we have

$$f_2(\theta) = \left\| R_{zy} - R_{z\phi_1}\theta + (S_1\theta v_1 + S_2\theta v_2 - \xi)\cdot\left((S_1\theta v_1 + S_2\theta v_2) - \xi\right)^{+}\left(R_{z\phi_1}\theta - R_{zy}\right)\right\|^2 \tag{7.77}$$

Its parameter estimation $\hat{\theta}$ is

$$\hat{\theta} = \arg\min_{\theta} f_2(\theta) \tag{7.78}$$

As there are two parameter functions $f_1(\sigma)$ and $f_2(\theta)$, and the minimization process for them are all nonlinear program, so some classical optimization methods can be directly used to solve $\hat{\theta}$ and $\hat{\sigma}$ separately, such as the Newton method, the first order gradient method, etc.

(3) Optimal input signal design

In parameter estimations (7.68) and (7.76), matrix inverse operations exist in the case they are full rank, but this condition may not hold in theory research or practice, so the way to satisfy this full rank condition depends on our studied optimal input design. To be precise, the goal of optimal input design is to guarantee the matrix inverse operation exist whether in equation (7.68) and (7.76). This section studies this process of optimal input signal design deeply, not superficially in the framework of statistical noises, i.e. those two observed noises $\tilde{u}(t)$ and $\tilde{y}(t)$ are all independent and identically distributed white noises with zero mean and variance σ_u^2 and σ_y^2.

More specifically, the main equation, used to design the optimal input signal is reviewed as

$$\frac{1}{N}\sum_{t=1}^{N} z(t)w(t) = \frac{1}{N}\sum_{t=1}^{N} z(t)\left[y(t) - \phi_1(t)\theta\right]$$

As the unknown parameter vector θ must satisfy the above over-parameterized equation, and after expanding and formulating it, we have the parameter estimation as follows:

$$\hat{\theta} = \left[\sum_{t=1}^{N}\phi_1^T(t)z(t)\right]^{-1}\left[\sum_{t=1}^{N} z(t)y(t)\right] \tag{7.79}$$

Substituting equation (7.66) and (7.79), then

$$\hat{\theta} = \left[\sum_{t=1}^{N}\phi_1^T(t)z(t)\right]^{-1}\left[\sum_{t=1}^{N} z(t)\left[\phi_1^T(t)\theta_0 + w(t)\right]\right]$$

$$= \left[\sum_{t=1}^{N}\phi_1^T(t)z(t)\right]^{-1}\left[\sum_{t=1}^{N} z(t)\phi_1^T(t)\theta_0 + \sum_{t=1}^{N} z(t)w(t)\right]$$

$$= \theta_0 + \left[\sum_{t=1}^{N}\phi_1^T(t)z(t)\right]^{-1}\left[\sum_{t=1}^{N} z(t)w(t)\right] \tag{7.80}$$

Comment: The true parameter vector θ_0 is used for later asymptotic analysis, in practice θ_0 is only regarded as one reference value. From equation (7.80), the parameter estimation $\hat{\theta}$ is one biased estimation, and its bias error is the second term $\left[\sum_{t=1}^{N}\phi_1^T(t)z(t)\right]^{-1}\left[\sum_{t=1}^{N} z(t)w(t)\right]$. When the chosen instrumental variable

$z(t)$ is not related with disturb $w(t)$, then parameter estimation $\hat{\theta}$ is reduced to be unbiased.

To study the problem of designing optimal input signal for statistical noises, the biased error is expanded on the basis of the instrumental variable $z(t)$ equal to the regressor variable $\phi_1(t)$, i.e.

$$z(t) = \begin{bmatrix} Y_t \\ U_t \end{bmatrix}$$

$$\hat{\theta} - \theta_0 = \left[\sum_{t=1}^{N} \phi_1^T(t) z(t) \right]^{-1} \left[\sum_{t=1}^{N} z(t) w(t) \right] \tag{7.81}$$

Continuing to compute the asymptotic variance, it holds that

$$cov(\hat{\theta}) = E\left[\hat{\theta} - \theta_0 \right]^2 = \left[\sum_{t=1}^{N} \phi_1^T(t) z(t) \right]^{-1} \left[\sum_{t=1}^{N} z(t) w(t) \right]$$

$$\left[\sum_{t=1}^{N} z(t) w(t) \right]^T \left[\sum_{t=1}^{N} \phi_1^T(t) z(t) \right]^{-1} = \left[\sum_{t=1}^{N} \phi_1(t) \phi_1^T(t) \right]^{-1} \tag{7.82}$$

where we use the instrumental variable $z(t)$ as its corresponding regressor variable, and the following statistical property of the assumed white noise:

$$E\left[w(t) w(s) \right] = \begin{cases} 1 & t = s \\ 0 & t \neq s \end{cases}$$

From equation (7.82), we know the statistical property of the white noise is beneficial for simplifying the derivation about the asymptotic variance matrix, i.e.

$$cov(\hat{\theta}) = E\left[\hat{\theta} - \theta_0 \right]^2 = \left[\sum_{t=1}^{N} \phi_1(t) \phi_1^T(t) \right]^{-1}$$

$$= \left[\sum_{t=1}^{N} \begin{bmatrix} Y_t \\ U_t \end{bmatrix} [Y_t \quad U_t] \right]^{-1} = \begin{bmatrix} \sum_{t=1}^{N} Y_t Y_t & \sum_{t=1}^{N} Y_t U_t \\ \sum_{t=1}^{N} U_t Y_t & \sum_{t=1}^{N} U_t U_t \end{bmatrix}^{-1} \tag{7.83}$$

Using the random process theory for equation (7.61), we have some known relations

$$\sum_{t=1}^{N} Y_t Y_t = \sum_{t=1}^{N} Y_t^2 = \phi_y(w) = |G(q)|^2 \phi_u(w) + 2|G(q)|^2 \sigma_u^2$$

$$\sum_{t=1}^{N} Y_t U_t = G(q) \phi_u(w); \sum_{t=1}^{N} U_t U_t = \phi_u(w) \tag{7.84}$$

where $\phi_u(w)$ and $\phi_y(w)$ are the power spectral of the observed input and output signal respectively, then equation (7.83) is reduced to

$$
\mathrm{cov}\left(\hat{\theta}\right)=\begin{bmatrix}\phi_y\left(w\right)I & G\phi_u\left(w\right)I \\ G\phi_u\left(w\right)I & \phi_u\left(w\right)\end{bmatrix}^{-1}=\dfrac{\begin{bmatrix}\phi_u\left(w\right)I & -G\phi_u\left(w\right)I \\ -G\phi_u\left(w\right)I & \phi_y\left(w\right)\end{bmatrix}}{\phi_y\left(w\right)\phi_u\left(w\right)-\left|G\right|^2\phi_u^2\left(w\right)}
\tag{7.85}
$$

where in equation (7.85), some basic matrix operations are used.

As the optimal input signal must be measured in the sense of one performance function, so here the performance function used to measure the optimality is chosen as the trace operation, which will be simplified as

$$
\mathrm{trace}\left[\mathrm{cov}\left(\hat{\theta}\right)\right]=\dfrac{n_a\phi_y\left(w\right)+n_b\phi_u\left(w\right)}{\phi_y\left(w\right)\phi_u\left(w\right)-\left|G\right|^2\phi_u^2\left(w\right)}
\tag{7.86}
$$

Observing equation (7.86), firstly we need to derive the following relations, such as

$$
\phi_y\left(w\right)=\left|G\right|^2\phi_u\left(w\right)+2\left|G\right|^2\sigma_u^2
$$

$$
\phi_y\left(w\right)\phi_u\left(w\right)-\left|G\right|^2\phi_u^2\left(w\right)=\phi_u\left(w\right)\left[\left|G\right|^2\phi_u\left(w\right)-\left|G\right|^2\phi_u\left(w\right)+2\left|G\right|^2\sigma_u^2\right]
$$

$$
=2\left|G\right|^2\sigma_u^2\phi_u\left(w\right)n_a\phi_y\left(w\right)+n_b\phi_u\left(w\right)
$$

$$
=n_a\left[\left|G\right|^2\phi_u\left(w\right)+2\left|G\right|^2\sigma_u^2\right]+n_b\phi_u\left(w\right)
$$

$$
=\left(n_a\left|G\right|^2+n_b\right)\phi_u\left(w\right)+2n_a\left|G\right|^2\sigma_u^2+n_b\phi_u\left(w\right)
$$

$$
=\left(n_a\left|G\right|^2+2n_b\right)\phi_u\left(w\right)+2n_a\left|G\right|^2\sigma_u^2
\tag{7.87}
$$

Substituting equation (7.87) into the performance function-trace $\left[\mathrm{cov}\left(\hat{\theta}\right)\right]$, we get

$$
\mathrm{trace}\left[\mathrm{cov}\left(\hat{\theta}\right)\right]=\dfrac{n_a\phi_y\left(w\right)+n_b\phi_u\left(w\right)}{\phi_y\left(w\right)\phi_u\left(w\right)-\left|G\right|^2\phi_u^2\left(w\right)}
$$

$$
=\dfrac{\left(n_a\left|G\right|^2+2n_b\right)\phi_u\left(w\right)+2n_a\left|G\right|^2\sigma_u^2}{2\left|G\right|^2\sigma_u^2\phi_u\left(w\right)}
$$

$$
=\dfrac{n_a\left|G\right|^2+2n_b}{2\left|G\right|^2\sigma_u^2}+\dfrac{2n_a\left|G\right|^2\sigma_u^2}{2\left|G\right|^2\sigma_u^2\phi_u\left(w\right)}
\tag{7.88}
$$

where to simplify notation, variable q is neglected.

In the performance function (7.88), the power spectral $\phi_u(w)$ of the input signal is regarded as one decision variable. Then the problem of optimal input signal design is formulated as the following Theorem 7.2.

Theorem 7.2: For the problem of optimal input signal design in the aircraft flutter stochastic model, the optimal input signal satisfies the following minimization problem with finite input signal energy constrained condition.

$$
\begin{cases}
\min\limits_{\phi_u(w)} \int_0^{2\pi} \dfrac{2n_a |G|^2 \sigma_u^2}{2|G|^2 \sigma_u^2 \phi_u(w)} \, dw = \int_0^{2\pi} \dfrac{n_a}{\phi_u(w)} \, dw \\[2ex]
\text{subject to} \quad \int_0^{2\pi} \phi_u(w) \, dw \le L_1
\end{cases}
\tag{7.89}
$$

where L_1 is one positive constant.

Then first term in equation (7.88) is neglected, as it does not include the decision variable $\phi_u(w)$. For the constrained optimization problem (7.89) with decision variable $\phi_u(w)$, the power spectral of the optimal input signal is obtained as that

$$
\phi_u^{opt}(w) = \frac{L_1}{\int_0^{2\pi} n_a \, dw} = \frac{L_1}{2n_a\pi}
\tag{7.90}
$$

The power spectral of the optimal input signal is one constant in case of statistical noises, this result is also similar to our previous work [31].

Comment: Theorem 7.2 gives an optimal input signal with constant power spectral $\dfrac{L_1}{2n_a\pi}$, but for the aircraft flutter stochastic model (Fig. 7.14), this input signal $u(t)$ is not the true excitation input, as it is corrupted by observed noise $\tilde{u}(t)$. Actually, the true external input signal is $u_0(t)$, so the power spectral of the optimal excitation input signal is given as

$$
\phi_{u_0}^{opt}(w) = \phi_u^{opt}(w) - \sigma_u^2 = \frac{L_1}{2n_a\pi} - \sigma_u^2
\tag{7.91}
$$

which is also one constant.

These detailed parameter estimation (7.68) and optimal excitation input signal (7.91) hold in case of independent and identically distributed white noise.

7.3.3 Optimal input signal design for bounded noise

All the above results of the identification algorithm and optimal input signal design hold on the condition of the two observed noises are white noise, but this condition is very strict in practice, as we can not obtain any statistical information about these two observed noises in engineering. So to relax this strict assumption on these two observed noises, in this section we apply the other way to describe these two observed noises, i.e. unknown but bounded noises. The main contents of this section 4 are to study set membership strategy to identify the unknown parameters under the unknown but bounded noises, and furthermore, devise the optimal input signal by using the radius of information, as the identification result is not a constant value, but a parameter set or parameter uncertainty interval.

Two observed noise $\tilde{u}(t)$ and $\tilde{y}(t)$ are assumed to be unknown but bounded with their known bounds $\Delta\tilde{u}(t)$ and $\Delta\tilde{y}(t)$ respectively, i.e.

$$\begin{cases} |\tilde{u}(t)| \leq \Delta\tilde{u}(t) \\ |\tilde{y}(t)| \leq \Delta\tilde{y}(t) \end{cases} \tag{7.92}$$

(1) Preliminary

The purpose of the identification problem in this section, is to determine the unknown parameter vector in the rational transfer function model, when given input signal-output observed data $\{u(1), y(1) \cdots u(N), y(N)\}$ with observed noises, N is the number of observed data, then the considered aircraft flutter model parameters can be identified from equation (7.62).

Rewrite the unknown parameter vector as that

$$\theta = \begin{bmatrix} a_1 & \cdots & a_{n_a} & b_1 & \cdots & b_{n_b} \end{bmatrix}^T$$

where the dimension of the unknown parameter vector θ is $(n_a + n_b)$.

The goal of this section is to identify this unknown parameter vector θ in case of unknown but bounded noise $(\tilde{u}(t), \tilde{y}(t))$ in equation (7.92).

Define one feasible parameter set D_θ for that unknown parameter vector θ as follows:

$$D_\theta = \begin{cases} \theta \in R^{n_a+n_b} : A(q^{-1})(y(t) - \tilde{y}(t)) = B(q^{-1})(u(t) - \tilde{u}(t)), \\ |\tilde{u}(t)| \leq \Delta\tilde{u}(t), |\tilde{y}(t)| \leq \Delta\tilde{y}(t), t = 1, 2 \cdots N \end{cases} \tag{7.93}$$

where N is also the number of observed data. Feasible parameter set D_θ is a set of all possible parameter vectors, which is consistent with the corrupted data, noise bounds and the aircraft flutter model. Further, define the parameter set or parameter uncertainty interval as

$$I_j = \begin{bmatrix} \underline{\theta}_j, \overline{\theta}_j \end{bmatrix} \tag{7.94}$$

where

$$\begin{cases} \underline{\theta}_j = \min_{\theta \in D_\theta} \theta_j \\ \overline{\theta}_j = \max_{\theta \in D_\theta} \theta_j \end{cases} \tag{7.95}$$

Then the computation of parameter set I_j is achieved by solving one constrained optimization problem. After obtaining this parameter set, the central estimate $\theta_c^j = \dfrac{\underline{\theta}_j + \overline{\theta}_j}{2}$ is chosen as the final parameter estimate.

(2) Set membership identification

Based on the constructed unknown parameter vector θ, define its corresponding regressor vector $\phi_2(t)$ as

$$\phi_2(t) = \begin{bmatrix} y_0(t-1) & y_0(t-2) & \cdots & y_0(t-n_a) \\ u_0(t-1) & u_0(t-2) & \cdots & u_0(t-n_b) \end{bmatrix}^T \tag{7.96}$$

where regressor vector $\phi_2(t)$ is full of input signal-output with free noise. The difference between $\phi_1(t)$ and $\phi_2(t)$ is that the elements of $\phi_2(t)$ are the input and output signal without observed noises.

Then the from set membership identification [33], the condition that parameter vector θ belongs to the parameter set, D_θ is formulated as follows.

$$\begin{cases} \left(\phi_2(t) - \Delta\phi_2(t)\right)\theta \le y(t) + \Delta\tilde{y}(t) \\ \left(\phi_2(t) + \Delta\phi_2(t)\right)\theta \ge y(t) - \Delta\tilde{y}(t), t = 1,2\cdots N \end{cases} \tag{7.97}$$

where

$$\Delta\phi_2(t) = \begin{bmatrix} \Delta\tilde{y}(t-1)\mathrm{sgn}(a_1) & \cdots & \Delta\tilde{y}(t-n_a)\mathrm{sgn}(a_{n_a}) \\ \Delta\tilde{u}(t-1)\mathrm{sgn}(b_1) & \cdots & \Delta\tilde{u}(t-n_a)\mathrm{sgn}(b_{n_b}) \end{bmatrix}^T \tag{7.98}$$

so the parameter set D_θ can be changed to the following other form.

$$D_\theta = \begin{cases} \theta \in R^{n_a + n_b} : \left(\phi_2(t) - \Delta\phi_2(t)\right)\theta \le y(t) + \Delta\tilde{y}(t), \\ \left(\phi_2(t) + \Delta\phi_2(t)\right)\theta \ge y(t) - \Delta\tilde{y}(t), t = 1,2\cdots N \end{cases} \tag{7.99}$$

Problem (7.95) can be reformulated as one constrained ploynomial optimization problem, i.e. computation of $\underline{\theta}_j$ and $\overline{\theta}_j$ computed by solving the following two constrained optimization problems.

$$\underline{\theta}_j = \min_{\theta,\tilde{u},\tilde{y}} \theta_j$$

subject to $y(t) = -\sum_{i=1}^{n_a}\left[y(t-i) - \tilde{y}(t-i)\right]a_i + \sum_{j=1}^{n_b}\left[u(t-j) - \tilde{u}(t-j)\right]b_j + \tilde{y}(t)$

$$\tilde{u}(t) \le \Delta\tilde{u}(t), -\tilde{u}(t) \le \Delta\tilde{u}(t)$$
$$\tilde{y}(t) \le \Delta\tilde{y}(t), -\tilde{y}(t) \le \Delta\tilde{y}(t), t = 1,2\cdots N \tag{7.100}$$

and

$$\overline{\theta}_j = \max_{\theta,\tilde{u},\tilde{y}} \theta_j$$

subject to $\quad y(t) = -\sum_{i=1}^{n_a}\left[y(t-i)-\tilde{y}(t-i)\right]a_i + \sum_{j=1}^{n_b}\left[u(t-j)-\tilde{u}(t-j)\right]b_j + \tilde{y}(t)$

$\tilde{u}(t) \le \Delta\tilde{u}(t), -\tilde{u}(t) \le \Delta\tilde{u}(t)$

$\tilde{y}(t) \le \Delta\tilde{y}(t), -\tilde{y}(t) \le \Delta\tilde{y}(t), t = 1,2\cdots N$ (7.101)

where

$$\begin{cases} \tilde{u} = \begin{bmatrix} \tilde{u}(1) & \tilde{u}(2) & \cdots & \tilde{u}(N) \end{bmatrix} \\ \tilde{y} = \begin{bmatrix} \tilde{y}(1) & \tilde{y}(2) & \cdots & \tilde{y}(N) \end{bmatrix} \end{cases}$$

Two constrained optimization problems (7.100) and (7.101) are used to compute $\underline{\theta}_j$ and $\overline{\theta}_j$ over $n_a + n_b + 2N$ optimization variables.

Now we are only concerned with the minimization problem (7.100), and give a detailed derivation. All the following derivations are applied to maximization problem (7.101).

Expanding the first constraint in equation (7.100) to get

$$y(t) = -\sum_{i=1}^{n_a} y(t-i)a_i + \sum_{j=1}^{n_b} u(t-j)b_j + \sum_{i=1}^{n_a}\tilde{y}(t-i)a_i - \sum_{j=1}^{n_b}\tilde{u}(t-j)b_j + \tilde{y}(t)$$

$$(7.102)$$

Reformulating the third term as

$$\sum_{i=1}^{n_a}\tilde{y}(t-i)a_i = \begin{bmatrix} \tilde{y}(t-1) & \tilde{y}(t-2) & \cdots & \tilde{y}(t-n_a) \end{bmatrix}\begin{bmatrix} a_1 \\ a_2 \\ \vdots \\ a_{n_a} \end{bmatrix}$$ (7.103)

Stacking the outputs with $t = 1, 2 \cdots N$ to obtain

$$Y = \begin{bmatrix} y(1) & y(2) & \cdots & y(N) \end{bmatrix}^T$$

Stacking N terms for observed noise $\tilde{y}(t)$ to get

$$\begin{bmatrix} \sum_{i=1}^{n_a}\tilde{y}(1-i)a_i + \tilde{y}(1) \\ \sum_{i=1}^{n_a}\tilde{y}(2-i)a_i + \tilde{y}(2) \\ \vdots \\ \sum_{i=1}^{n_a}\tilde{y}(N-i)a_i + \tilde{y}(N) \end{bmatrix} = \begin{bmatrix} \tilde{y}(0) & \tilde{y}(-1) & \cdots & \tilde{y}(1-n_a) \\ \tilde{y}(1) & \tilde{y}(0) & \cdots & \tilde{y}(2-n_a) \\ \vdots & & \vdots & \vdots \\ \tilde{y}(N-1) & \tilde{y}(N-2) & \cdots & \tilde{y}(N-n_a) \end{bmatrix}$$

$$\begin{bmatrix} a_1 \\ a_2 \\ \vdots \\ a_{n_a} \end{bmatrix} + \begin{bmatrix} \tilde{y}(1) \\ \tilde{y}(2) \\ \vdots \\ \tilde{y}(N) \end{bmatrix} \tag{7.104}$$

Applying the causality property on equation (7.104), then it reduced to

$$\begin{bmatrix} \sum\limits_{i=1}^{n_a} \tilde{y}(1-i)a_i + \tilde{y}(1) \\ \sum\limits_{i=1}^{n_a} \tilde{y}(2-i)a_i + \tilde{y}(2) \\ \vdots \\ \sum\limits_{i=1}^{n_a} \tilde{y}(N-i)a_i + \tilde{y}(N) \end{bmatrix} = \begin{bmatrix} \tilde{y}(1) & 0 & 0 & \cdots & 0 \\ \tilde{y}(2) & \tilde{y}(1) & 0 & \cdots & 0 \\ \tilde{y}(3) & \tilde{y}(2) & \tilde{y}(1) & \cdots & 0 \\ \vdots & & \vdots & & \vdots \\ \tilde{y}(N) & \tilde{y}(N-1) & \tilde{y}(N-2) & \cdots & \tilde{y}(N-n_a) \end{bmatrix} \begin{bmatrix} 1 \\ a_1 \\ a_2 \\ \vdots \\ a_{n_a} \end{bmatrix}$$

$$\tag{7.105}$$

Similarly stacking N terms for observed noise $\tilde{u}(t)$ to get

$$\begin{bmatrix} \sum\limits_{j=1}^{n_b} \tilde{u}(1-j)b_j \\ \sum\limits_{j=1}^{n_b} \tilde{u}(2-j)b_j \\ \vdots \\ \sum\limits_{j=1}^{n_b} \tilde{u}(N-j)b_j \end{bmatrix} = \begin{bmatrix} \tilde{u}(1) & \tilde{u}(2) & 0 & \cdots & 0 \\ \tilde{u}(2) & \tilde{u}(1) & \tilde{u}(0) & \cdots & 0 \\ \vdots & & & & \vdots \\ \tilde{u}(N-1) & \tilde{u}(N-2) & & \cdots & \tilde{u}(N-b_b) \end{bmatrix} \begin{bmatrix} b_1 \\ b_2 \\ \vdots \\ b_{n_b} \end{bmatrix}$$

$$\tag{7.106}$$

Consider the following N terms as

$$\sum_{i=1}^{n_a} y(t-i)a_i + \sum_{j=1}^{n_b} u(t-j)b_j, \quad t = 1,2\cdots N$$

Stacking them as

$$\begin{bmatrix} -y(0) & -y(-1) & \cdots & -y(1-n_a) & u(1) & u(0) & \cdots & u(1-n_b) \\ -y(1) & -y(0) & \cdots & -y(2-n_a) & u(2) & u(1) & \cdots & u(2-n_b) \\ \vdots & & & \cdots & & & & \vdots \\ -y(N-1) & -y(N-2) & & -y(N-n_a) & u(N) & u(N-1) & \cdots & u(N-n_b) \end{bmatrix}$$

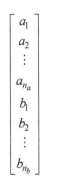

$$\begin{bmatrix} a_1 \\ a_2 \\ \vdots \\ a_{n_a} \\ b_1 \\ b_2 \\ \vdots \\ b_{n_b} \end{bmatrix} \tag{7.107}$$

Combining equation (7.104), (7.105), (7.106) and (7.107), we obtain

$$\underbrace{\begin{bmatrix} y(1) \\ y(2) \\ \vdots \\ y(N) \end{bmatrix}}_{Y}$$

$$= \underbrace{\begin{bmatrix} -y(0) & -y(-1) & \cdots & -y(1-n_a) & u(1) & u(0) & \cdots & u(1-n_b) \\ -y(1) & -y(0) & \cdots & -y(2-n_a) & u(2) & u(1) & \cdots & u(2-n_b) \\ \vdots & & & \cdots & & & & \vdots \\ -y(N-1) & -y(N-2) & & -y(N-n_a) & u(N) & u(N-1) & \cdots & u(N-n_b) \end{bmatrix}}_{A_1}$$

$$\underbrace{\begin{bmatrix} a_1 \\ a_2 \\ \vdots \\ a_{n_a} \\ b_1 \\ b_2 \\ \vdots \\ b_{n_b} \end{bmatrix}}_{\theta} + \underbrace{\begin{bmatrix} \tilde{y}(1) & 0 & 0 & \cdots & 0 \\ \tilde{y}(2) & \tilde{y}(1) & 0 & \cdots & 0 \\ \tilde{y}(3) & \tilde{y}(2) & \tilde{y}(1) & \cdots & 0 \\ \vdots & & \vdots & & \vdots \\ \tilde{y}(N) & \tilde{y}(N-1) & \tilde{y}(N-2) & \cdots & \tilde{y}(N-n_a) \end{bmatrix}}_{Hankel(\tilde{y})} \begin{bmatrix} 1 \\ a_1 \\ a_2 \\ \vdots \\ a_{n_a} \end{bmatrix}$$

$$- \underbrace{\begin{bmatrix} \tilde{u}(1) & \tilde{u}(2) & 0 & \cdots & 0 \\ \tilde{u}(2) & \tilde{u}(1) & \tilde{u}(0) & \cdots & 0 \\ \vdots & & \vdots & & \vdots \\ \tilde{u}(N-1) & \tilde{u}(N-2) & \cdots & & \tilde{u}(N-b_b) \end{bmatrix}}_{Hankel(\tilde{u})} \begin{bmatrix} b_1 \\ b_2 \\ \vdots \\ b_{n_b} \end{bmatrix} \tag{7.108}$$

where in equation (7.108) two simplified forms are used as $\text{Hankel}(\tilde{y})$ and $\text{Hankel}(\tilde{u})$.

Observing that

$$
\begin{bmatrix} a_1 \\ a_2 \\ \vdots \\ a_{n_a} \end{bmatrix} = \underbrace{\begin{bmatrix} 1 & 0 & \cdots & 0 & 0 & 0 & \cdots & 0 \\ 0 & 1 & \cdots & 0 & 0 & 0 & \cdots & 0 \\ \vdots & & \vdots & & & & \vdots \\ 0 & 0 & \cdots & 1 & 0 & 0 & \cdots & 0 \end{bmatrix}}_{B_1} \underbrace{\begin{bmatrix} a_1 \\ a_2 \\ \vdots \\ a_{n_a} \\ b_1 \\ b_2 \\ \vdots \\ b_{n_b} \end{bmatrix}}_{\theta}
\tag{7.109}
$$

and

$$
\begin{bmatrix} b_1 \\ b_2 \\ \vdots \\ b_{n_b} \end{bmatrix} = \underbrace{\begin{bmatrix} 0 & 0 & \cdots & 0 & 1 & 0 & \cdots & 0 \\ 0 & 0 & \cdots & 0 & 0 & 1 & \cdots & 0 \\ \vdots & & \vdots & & & & \vdots \\ 0 & 0 & \cdots & 0 & 0 & 0 & \cdots & 1 \end{bmatrix}}_{B_2} \underbrace{\begin{bmatrix} a_1 \\ a_2 \\ \vdots \\ a_{n_a} \\ b_1 \\ b_2 \\ \vdots \\ b_{n_b} \end{bmatrix}}_{\theta}
\tag{7.110}
$$

Then equation (7.109) can be rewritten as

$$
\begin{aligned}
Y &= A_1\theta + \text{Hankel}(\tilde{y})\,B_1\theta + \text{Hankel}(\tilde{u})\,B_2\theta \\
&= \left[A_1 \mid \text{Hankel}(\tilde{y})\,B_1 + \text{Hankel}(\tilde{u})\,B_2 \right]\theta
\end{aligned}
\tag{7.111}
$$

Similarly, other inequality constraints are reformulated as

$$
\begin{cases} \tilde{u}(t) \le \Delta\tilde{u}(t) \\ -\tilde{u}(t) \le \Delta\tilde{u}(t), & t = 1, 2 \cdots N \end{cases} \Rightarrow \begin{bmatrix} \tilde{u}(1) \\ \tilde{u}(2) \\ \vdots \\ \tilde{u}(N) \end{bmatrix} \le \begin{bmatrix} \Delta\tilde{u}(1) \\ \Delta\tilde{u}(2) \\ \vdots \\ \Delta\tilde{u}(N) \end{bmatrix},
$$

$$\text{and} \quad -\begin{bmatrix} \tilde{u}(1) \\ \tilde{u}(2) \\ \vdots \\ \tilde{u}(N) \end{bmatrix} \leq \begin{bmatrix} \Delta\tilde{u}(1) \\ \Delta\tilde{u}(2) \\ \vdots \\ \Delta\tilde{u}(N) \end{bmatrix} \tag{7.112}$$

i.e.

$$\begin{bmatrix} \tilde{u} \\ -\tilde{u} \end{bmatrix} \leq \begin{bmatrix} \Delta\tilde{u} \\ \Delta\tilde{u} \end{bmatrix} \tag{7.113}$$

Similarly, it holds that

$$\begin{bmatrix} \tilde{y} \\ -\tilde{y} \end{bmatrix} \leq \begin{bmatrix} \Delta\tilde{y} \\ \Delta\tilde{y} \end{bmatrix} \tag{7.114}$$

where

$$\begin{cases} \Delta\tilde{u} = \begin{bmatrix} \Delta\tilde{u}(1) & \Delta\tilde{u}(2) & \cdots & \Delta\tilde{u}(N) \end{bmatrix}^T \\ \Delta\tilde{y} = \begin{bmatrix} \Delta\tilde{y}(1) & \Delta\tilde{y}(2) & \cdots & \Delta\tilde{y}(N) \end{bmatrix}^T \end{cases}$$

Generally based on our mathematical derivations given above, the minimization problem (40) is reduced to

$$\underline{\theta}_j = \min_{\theta,\tilde{u},\tilde{y}} \theta_j$$

$$\text{subject to} \quad Y = \begin{bmatrix} A_1 + Hankel(\tilde{y})B_1 + Hankel(\tilde{u})B_2 \end{bmatrix}\theta$$

$$\begin{bmatrix} \tilde{u} \\ -\tilde{u} \end{bmatrix} \leq \begin{bmatrix} \Delta\tilde{u} \\ \Delta\tilde{u} \end{bmatrix}, \begin{bmatrix} \tilde{y} \\ -\tilde{y} \end{bmatrix} \leq \begin{bmatrix} \Delta\tilde{y} \\ \Delta\tilde{y} \end{bmatrix} \tag{7.115}$$

Then equation (7.115) is one simplified form of the minimization optimization problem. Many classical optimization methods can be directly applied to solve it here, such as the Newton method, the fast gradient method, the ellipsoidal method, etc.

Similarly, the maximization problem (7.101) is reduced to

$$\underline{\theta}_j = \max_{\theta,\tilde{u},\tilde{y}} \theta_j$$

$$\textit{subject to} \quad Y = \begin{bmatrix} A_1 + Hankel(\tilde{y})B_1 + Hankel(\tilde{u})B_2 \end{bmatrix}\theta$$

$$\begin{bmatrix} \tilde{u} \\ -\tilde{u} \end{bmatrix} \leq \begin{bmatrix} \Delta\tilde{u} \\ \Delta\tilde{u} \end{bmatrix}, \begin{bmatrix} \tilde{y} \\ -\tilde{y} \end{bmatrix} \leq \begin{bmatrix} \Delta\tilde{y} \\ \Delta\tilde{y} \end{bmatrix} \tag{7.116}$$

In the above mathematical derivations of the set membership identification, the subscript j in notation θ_j means the jth iterative sequence. When the iterative algorithm terminates, then that final interval is deemed as the terminal parameter set or the parameter uncertainty interval.

(3) Optimal input signal design

Without loss of generality, rewrite the above parameter uncertainty interval as

$$\theta \in I = \left[\underline{\theta}, \overline{\theta}\right] \tag{7.117}$$

Based on this parameter uncertainty interval I, lots of definitions are given in reference [34], such as the optimal estimate, tightest bound, identification error, etc. More importantly, one quantity R_I, named the radius of information, helps us to measure the identification accuracy, furthermore, this radius of information is applied to quantify the informative for one dataset, so the radius of information for that parameter uncertainty interval I is defined as

$$R_I = \left\|\frac{1}{2}\left[\overline{\theta} - \underline{\theta}\right]\right\|_p \tag{7.118}$$

where $\left\|\frac{1}{2}\left[\overline{\theta} - \underline{\theta}\right]\right\|_p$ means one L_p norm.

As the mission of optimal input signal design is to help us to maximize the information extracted from the observed data, and within the set membership identification, the parameter uncertainty is measured by the above radius of information, so the radius of information is minimized to design the optimal input signal for set membership identification.

Let $u_t^T = \{u\}_{t=1}^{T-1}$ be an input sequence from time t to time $T - 1$, and design an input sequence $u_t^T = \{u\}_{t=1}^{T-1}$ that applied to the flutter stochastic model yields a minimal radius of information $R_I = \left\|\frac{1}{2}\left[\overline{\theta} - \underline{\theta}\right]\right\|_p$.

Ideally, an easy solution to this optimal input signal design problem is formulated as

$$u_t^{T,opt} = \arg\min_u R_I = \left\|\frac{1}{2}\left[\overline{\theta} - \underline{\theta}\right]\right\|_p$$

$$\text{subject to} \quad Y = \left[A_1 + Hankel\left(\tilde{y}\right)B_1 + Hankel\left(\tilde{u}\right)B_2\right]\theta \tag{7.119}$$

Consider the flutter stochastic model again, that solving the optimal input signal design problem is easier with respect to the general dynamic case, since the output variable, corrupted by observed noise, depends only on the current input and not on the past input and output values. The numerical algorithm for solving the optimization problem (7.119) is iterative, at each iteration, a point in the regressor domain where the uncertainty is maximum is considered.

7.3.4   Numerical examples

In this section, we give three numerical examples to verify our identification strategies and optimal input signal in the case of statistical noise and bounded noise respectively.

(1) Firstly the proposed set membership identification strategy is applied to identify the aircraft flutter model parameters.

The flutter test data of a certain aircraft is used to verify the effectiveness of the proposed method. The aircraft flutter model is very complex, as it is full of flexible structure and aerodynamics. So in this simulation part, we only use the wind tunnel test to construct the flutter mathematical model of the two-dimensional wing. In the whole wind tunnel test for a two-dimensional wing, the input signal is chosen as an artificially applied excitation signal, and the output is an accelerometer measurement, collected from many sampling data points. The simulation is based on 100 independent experiments and 500 data points.

True system is that

$$y_0(t) = \frac{B(q^{-1})}{A(q^{-1})} u_0(t)$$

$$A(q^{-1}) = 1 - 0.8q^{-1} + 0.9q^{-2}; B(q^{-1}) = 0.5q^{-1} + 0.4q^{-2} \qquad (7.120)$$

Define the input signal without noise as that

$$u_0(t) = \frac{1}{1 - 0.2q^{-1} + 0.5q^{-2}} e(t)$$

where $e(t)$ is a white noise source, and the colored noise model is defined as follows.

$$C(q^{-1}) = 1 - 0.2q^{-1}$$

Define the variances of white noises as $\sigma_u^2 = 0.14$ and $\sigma_y^2 = 1.45$, which correspond to *SNR* of input signal-output signal be 10dB.

$$SNR = 10 \log_{10} \left(\frac{P_{x_0}}{P_{\tilde{x}}} \right)$$

where P_x is average power of the signal.

True parameter vector θ_0 is

$$\theta_0 = \begin{bmatrix} a_1 & a_2 & b_1 & b_2 \end{bmatrix}^T = \begin{bmatrix} -0.8 & 0.9 & 0.5 & 0.4 \end{bmatrix}^T$$

Two external noises $\tilde{u}(t)$ and $\tilde{y}(t)$ are assumed to be unknown but bounded, i.e.

$$\left|\tilde{u}(t)\right| \le 0.5, \quad \left|\tilde{y}(t)\right| \le 0.5$$

In applying the set membership identification strategy to construct a parameter set for unknown parameter vector, two constrained optimization problems are used to obtain the upper and lower bound for each parameter set.

To show the identification results for each pair $(a_1 \quad b_1) = (-0.8 \quad 0.5)$ and $(a_2 \quad b_2) = (0.9 \quad 0.4)$, two parameter sets are plotted in Fig 7.4, where two pairs of true values are the centers of two parameter sets. Since the model frequency and damping coefficient can be obtained from the pole of the transfer function. From Fig. 7.15, two parameter sets embrace the true parameter values. To improve the efficiency of our proposed method, we only calculate the pole of the transfer function during online flutter analysis i.e. the poles of the transfer function are used to be a comparison criterion to verify the validity of our proposed method. From Fig. 7.16, the true poles corresponding to the transfer function are very consistent with the identified poles from our proposed method.

(2) Secondly, the proposed separable least squares identification strategy is applied to identify the aircraft flutter model parameters.

To measure the identification accuracy for the flutter model parameters, the flutter stochastic model in (7.120) is also chosen as the system similarly, but in this second case, those two observed noises $\tilde{u}(t)$ and $\tilde{y}(t)$ are all assumed to be two white noises with zero mean and unit variance.

The input signal is chosen as one sine wave sequence, i.e. one sine wave signal is used to excite the above system (7.120), which is corrupted by white noise, then the output data are collected through some equivalent devices, such as

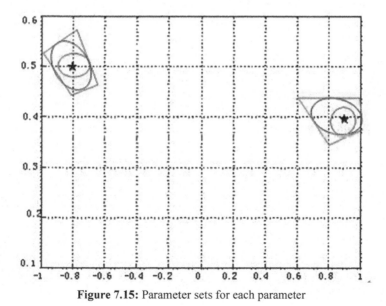

Figure 7.15: Parameter sets for each parameter

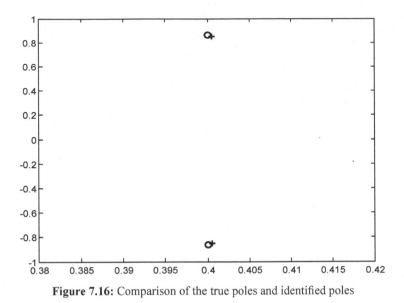

Figure 7.16: Comparison of the true poles and identified poles

sensor, oscilloscope, etc. these chosen input signal and collected output data are plotted in Fig. 7.17, where the identification output and identification error are also shown. More specifically, the sine wave input and collected output are applied to identify the parameter vector θ, then substituting this parameter estimation $\hat{\theta}$ into the original system (7.120) and plot its identification output. From Fig. 7.17, the identification output coincides with the collected output data, which means the identification error converges to zero with sample points increase.

(3) Thirdly the proposed optimal input signal design is applied for the aircraft flutter model parameters identification experiment.

A real aircraft, produced by our lab in Fig. 7.18, is used for the flutter experiment, and two excitation motors are installed on the left wing. An excitation motor is installed on the left front beam, and the other one is on the rear beam, so all flutter models are excited. During the whole flutter experiment, an important closed loop structure is the current loop, shown in Fig. 7.19. The function of the current loop is to control the current of the motor and not exceed the maximum locked-rotor current of the motor. At the same time, it is also necessary to make the armature current strictly follow the change of the control voltage command, so that we can accurately control the torque output by the motor to eliminate the effect of back-EMF on the output torque. To make the closed loop structure for the aircraft flutter experiment to be suited for our studied structure, we plot Fig. 7.19 as a similar open loop structure in Fig. 7.19, where the equivalent property means the outputs for Fig. 7.19 and Fig. 7.20 are the same. The detailed description of this aircraft flutter experiment and all blocks in Fig. 7.19 and Fig. 7.20 can be referred to the author's PhD thesis, published in 2011.

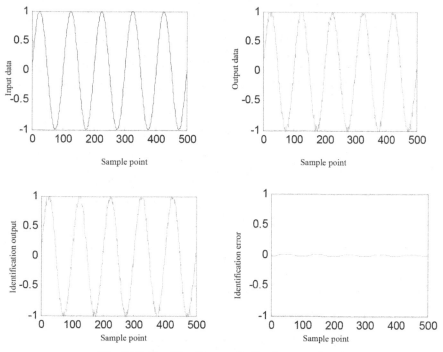

Fig. 7.17: Identification results for sine wave input

Figure 7.18: Aircraft used for flutter experiment

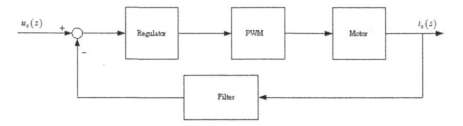

Figure 7.19: Closed loop structure for aircraft flutter experiment

Figure 7.20: Equivalent open loop structure for aircraft flutter experiment

Here, all the transfer functions and their corresponding parameters in Fig. 7.21 are given as follows:

$$G(q) = \frac{0.72(0.68q+1)}{(24q^2 + 13q + 1)(0.00054q + 1)(0.27q + 1)} \qquad (7.121)$$

The above rational transfer function model is actually unknown and needs to be identified. In the system identification theory, it is only used as a true value to measure the accuracy of identification. Assume the number of input-output data is 1000, two observed noises are all independent and identically distributed while noises with zero mean and unit variance, i.e. $\sigma_y^2 = \sigma_u^2 = 1$. Separable least squares strategy is used to identify the unknown model parameters existing in numerator and denominator polynomials respectively. To verify the identification accuracy, one triangle wave input is used to excite the system (7.121). After comparing the actual output and its identification output in Fig. 7.22, we see that the identification output is similar to its corresponding actual output, i.e. the identification error is around zero value.

Based on the obtained model parameters, identified by our proposed separable least squares strategy, we continue to study the optimal input signal for the whole identification experiment. After computing the trace operation of the asymptotic variance matrix, the optimal input signal is derived by minimizing one numerical optimization problem, whose decision variable is the power spectral of the optimal input signal. The optimal input signal is plotted in Fig. 7.23, where the optimal input signal is a binary, and its amplitude is ± 0.5, except for some switches at the total lengths, for example, $1s$, $2s$, $3s$ $\cdots$.

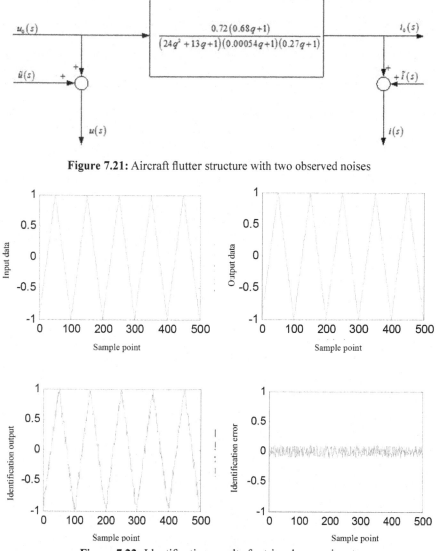

Figure 7.21: Aircraft flutter structure with two observed noises

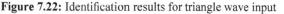

Figure 7.22: Identification results for triangle wave input

7.3.5 Conclusion

In this second application, aircraft flutter model parameters are identified based on one stochastic model of aircraft flutter, whose observed input and output are corrupted by noises respectively. In order to identify the aircraft flutter model parameters with statistical noise and unknown but bounded noise, separable least squares strategy and set membership strategy are proposed to achieve this goal.

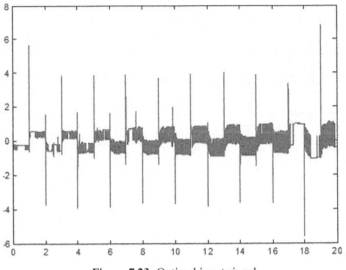

Figure 7.23: Optimal input signal

The parameter set used to include the unknown parameter is obtained through solving one minimization and maximization problem, and their simplified forms are also given by our own mathematical derivations. Trace operation with respect to the asymptotic variance matrix is minimized to solve the power spectral for the optimal input signal for statistical noise. And the radius of information, corresponding to our established parameter uncertainty interval, is minimized to give that optimal input signal for unknown but bounded noise.

7.4   Synthesis cascade estimation for aircraft system identification

One of the oldest and most fundamental of all human scientific pursuits is developing mathematical models for physical systems based on observations or measurements. This activity is known as system identification. Aircraft system identification describes the theory and practice of system identification applied to aircraft modeling. System identification is based on observations of input and output for the aircraft system under test. In practice, these observations are corrupted by measurement noise. This requires the introduction of statistical theory and methods, such as estimation strategy, validation method, etc. The most important requirement for a mathematical model is that it be useful in some way, sometimes, this means that the model can be used to predict some aspect of the behavior of a physical system, while at other times just the values of parameters in the model provide the desired insight. In any case, the synthesized model must be simple enough to be useful and at the same time complex enough to capture important dependencies and features embodied in the observations or measurements. System identification is one of three general problems in aircraft dynamics and control, i.e. simulation, control, and system identification. For many applications, an aircraft can be assumed to be a rigid body, whose motion is governed by the laws of Newton physics. System identification can be used to characterize applied forces and moments acting on the aircraft that arise from aerodynamics and propulsion. Typically, thrust forces and moments are obtained from ground tests, so aircraft system identification is applied to model the functional dependence of aerodynamic forces and moments on aircraft motion and control variables. Modern computational methods and wind tunnel testing can provide comprehensive data about the aerodynamic characteristics of an aircraft. However, there are still several motivations for identifying aircraft models from flight data, for example, being used to obtain more accurate and comprehensive mathematical models of aircraft dynamics for use in later designing stability augmentation and flight control systems.

In the early days of powered flight, only the most basic information about aircraft aerodynamics was obtained from measurements in steady flight. One of the first approaches for obtaining static and dynamic parameters from flight data was frequency response data and a simple semi-graphical method for the detailed analysis. Several years later, more general and rigorous ways were established to determine aerodynamic parameters from transient maneuvers. Parameter estimation methods were dependent on ordinary and nonlinear least squares. Dramatic improvement in aircraft system identification came in the later 1960s and early 1970s, due to the availability of digital computers and progress in the new technical discipline called system identification. The goal of system identification is similar to aircraft system identification, i.e. to use the observations to identify the unknown plant, such as an aircraft, robot, missile, or some other advanced plant. The continuous development for system identification includes four steps, i.e. experiment design, model structure choice, parameter estimation, and model validation, so these four related steps are beneficial for the special case-aircraft system identification. More specifically, the input excites the system and can usually be specified by the experimenter. It is therefore an extended disturbance that can be directly measured. The input must be distinguished from unmeasured disturbances, which are observed only through their influence on the system response. In some cases, the experimenter may not have the capability to specify the input, so that input for normal operation of the system must be used, i.e. satisfying the property of persistent excitation or rich enough. The output is an observable signal indicating the system response to the input and disturbances. It means the intrinsic characteristics, corresponding to the considered system, are all included in these inputs or output sequences implicitly. Our mission is to extract these unseen characteristics or physical principles from these outputs through some advanced technologies, such as statistical inference, machine learning, deep learning, and reinforcement learning. Typical the output at any time is a function of the current state and input. The state is a variable that completely specifies the status of the system at any given time. The state is not unique, for example, a change in a coordinate system would give a different but equivalent state. For deterministic physical systems, knowledge of the state at a time, combined with knowledge of the system dynamics and inputs from the initial time to the current time, is sufficient to compute the state at the current time. This state at the current time reflects what has happened to the system, and can be used in the later system theory. As a consequence, the process of deriving the state is named state estimation or filter theory.

The cascade control system was firstly studied in our previous contributions, but in this part, we turn to the problem of cascade system identification for aircraft systems, i.e. the cascade system exists for aircraft truly. After introducing our mentioned cascade system and its identification problem, the prediction error method is proposed to achieve the identification mission for all the parameterized models, exiting in the cascade system. Roughly, when all unknown plants in the cascade system are parameterized by their corresponding unknown parameter vectors, the problem of identifying the cascade system is transferred to estimating these unknown parameter vectors through the prediction error method. To increase the identification speed and reduce the computational complexity, the online subgradient descent optimization algorithm is yielded for estimation recursively. As the prediction error method is well suited for the parameterized model, so for the general case, we proposed other identification strategies to obtain the identified models for those all unknown plants in the cascade system by virtue of our derivations. As a consequence, the above identified process corresponds to the third step for system identification theory, and the choice of a cascade system is the second step, i.e. model structure choice. To give complete consideration to aircraft system identification, the last step-model structure validation is added to show our new contribution. From a practical point of view, a numerical example of cascade system identification for aircraft is given in detail.

7.4.1   Cascade system

Before the given consider cascade system, some basic knowledge of aircraft system identification is reviewed. An aircraft can be considered as an input-output system, as shown in Fig. 7.24. The aircraft dynamics are excited by the control inputs, which in the current case are the conventional aerodynamic surfaces: aileron for roll control, elevator for pitch control, rudder for yaw control, and throttle for speed control. We can record the aircraft's dynamic response to the control inputs in numerical form using an onboard measurement system. Typical flight measurements for flight dynamics consideration are shown in Fig. 7.24: translational velocities, angular velocities, attitudes, linear accelerometers, and aerodynamic angles. Additional measurements could include the engine response, wing strams and aircraft position.

A dynamic model relates the control inputs to the aircraft's dynamic response. This model can be as simple as a graph of the input-to-output response or as

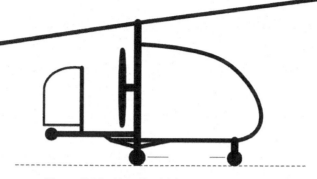

Figure 7.24: Aircraft with input-output

complex as a set of differential equations of motion. Dynamic models are needed for many applications, including analysis of aircraft stability and control, pilot simulations, flight control design, and analysis of aircraft handling characteristics. Aircraft system identification is a highly versatile procedure for rapidly and efficiently extracting accurate dynamic models of the aircraft from the measured response to specific control inputs. Models might be desired to characterize the aircraft system as a whole or to characterize an aircraft subsystem, such as an actuator or the engine, so at its simplest definition, system identification is a process that provides a model that best characterizes the measured responses to controls. Specialized flight test maneuvers are used to excite the dynamics of concern for a particular application, such as the study of flight dynamics and control. Considering the aspect of aircraft system identification, four important aspects of aircraft system identification have to be carefully treated, namely maneuver, measurements, methods, and model (shown in Fig. 7.25).

(1) Design of the control input share to excite different modes of the aircraft's dynamic motion;
(2) Selection of instrumentation and filters for high accuracy measurements;
(3) Quality of data analysis by selecting the most suitable time or frequency domain identification method;
(4) Type of flight aircraft under investigation to define the structure of a possible mathematical model.

Generally, each part in Fig. 7.25 is carefully investigated for each aircraft and is also the key to successful flight aircraft system identification. They lead to an interdisciplinary task, covering the fields of statistical theory, control theory, numerical techniques, sensor and instrumentations, signal processing, flight test techniques, and flight mechanics.

Consider the following system structure in Fig. 7.26, where the models are connected in a cascade manner without process noise, and the output of each model is either measured or there is an excitation signal added to it. In this case,

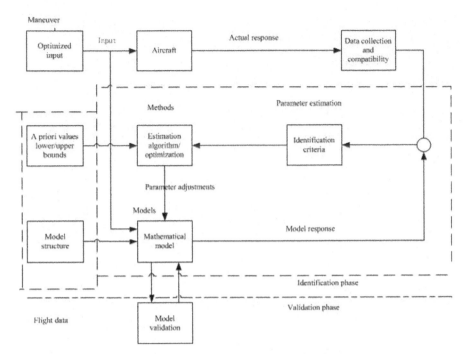

Figure 7.25: Aircraft system identification

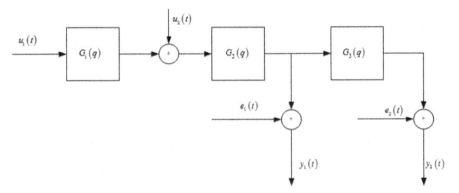

Figure 7.26: Cascade system identification

it is not practical to use specific input-output signals to identify particular models, as they lead to a non-standard model structure, it means the existing linear system identification results can be applied directly.

Where in Fig. 7.26, $\{G_1(q),\ G_2(q),\ G_3(q)\}$ are three unknown models, being identified in this chapter, $\{u_1(t),\ u_2(t)\}$ are two external excitation input, $\{y_1(t),\ y_2(t)\}$ correspond to two output responses, $\{e_1(t),\ e_2(t)\}$ are two additional measurement noise sequences given as the output of unknown linear stable filters driven by white noise, q is the delay operator.

The fact about this cascade system really exists in aircraft system, will be explained in section 5. Observing that cascade system in Fig. 7.26 again, two inputs $\{u_1(t), u_2(t)\}$ and two outputs $\{y_1(t), y_2(t)\}$ can be measured through some physical devices, such as sensors, so the input-output sequence $\{u_1(t), u_2(t), y_1(t), y_2(t)\}_{t=1}^{N}$ are known in priori, where number N is the total number of observations. The problem of cascade system identification is to use those input-output sequence $\{u_1(t), u_2(t), y_1(t), y_2(t)\}_{t=1}^{N}$ to extract the knowledge for those three unknown models $\{G_1(q), G_2(q), G_3(q)\}$, under the circumstances of two measurement noises $\{e_1(t), e_2(t)\}_{t=1}^{N}$. After the knowledge for the three unknown models $\{G_1(q), G_2(q), G_3(q)\}$ are formulated as their own mathematical models, such as transfer function or state space form, etc, then the later controller design depends on them.

7.4.2 Cascade system identification

Consider the cascade system in Fig. 7.26, where $\{G_1(q), G_2(q), G_3(q)\}$ are stable transfer functions, $\{u_1(t), u_2(t)\}$ are known input signals, $\{y_1(t), y_2(t)\}$ are measured outputs subject to two white noises $\{e_1(t), e_2(t)\}_{t=1}^{N}$. The relation among these signals can be formulated as follows:

$$\begin{cases} y_1(t) = e_1(t) + G_2(q)\left[G_1(q)u_1(t) + u_2(t)\right] \\ \qquad = e_1(t) + G_2(q)G_1(q)u_1(t) + G_2(q)u_2(t) \\ y_2(t) = e_2(t) + G_3(q)G_2(q)\left[u_2(t) + G_1(q)u_1(t)\right] \\ \qquad = e_2(t) + G_3(q)G_2(q)u_2(t) + G_3(q)G_2(q)G_1(q)u_1(t) \end{cases} \quad (7.122)$$

For convenience, define the following vectors:

$$y(t) = \begin{bmatrix} y_1(t) \\ y_2(t) \end{bmatrix}, u(t) = \begin{bmatrix} u_1(t) \\ u_2(t) \end{bmatrix}, e(t) = \begin{bmatrix} e_1(t) \\ e_2(t) \end{bmatrix} \quad (7.123)$$

Then equation (7.122) is rewritten as

$$\begin{bmatrix} y_1(t) \\ y_2(t) \end{bmatrix} = \begin{bmatrix} G_2(q)G_1(q) & G_2(q) \\ G_3(q)G_2(q)G_1(q) & G_3(q)G_2(q) \end{bmatrix} \begin{bmatrix} u_1(t) \\ u_2(t) \end{bmatrix} + \begin{bmatrix} e_1(t) \\ e_2(t) \end{bmatrix} \quad (7.124)$$

i.e.

$$y(t) = G(q)u(t) + e(t) \quad (7.125)$$

where

$$G(q) = \begin{bmatrix} G_2(q)G_1(q) & G_2(q) \\ G_3(q)G_2(q)G_1(q) & G_3(q)G_2(q) \end{bmatrix}$$

So the problem of aircraft system identification is to use input-output $\{u(t),$ $y(t)\}$ to identify that transfer function matrix $G(q)$.

For those three unknown models $\{G_1(q), G_2(q), G_3(q)\}$, we consider that each$\{G_1(q), G_2(q), G_3(q)\}$ is parameterized by a rational transfer function in the parameter vector $\{\theta_1, \theta_2, \theta_3\}$, i.e.

$$G_j\left(q,\theta_j\right) = \frac{B_j\left(q,\theta_j\right)}{A_j\left(q,\theta_j\right)} \tag{7.126}$$

where $B_j(q, \theta_j)$ and $A_j(q, \theta_j)$ are two polynomials, i.e.

$$
\begin{aligned}
B_j\left(q,\theta_j\right) &= b_1^j q^{-1} + \cdots + b_{m-1}^j q^{-(m-1)} \\
A_j\left(q,\theta_j\right) &= 1 + a_1^j q^{-1} + \cdots + a_m^j q^{-m} \\
\theta_j &= \begin{bmatrix} b_1^j & \cdots & b_{m-1}^j & a_1^j & \cdots & a_m^j \end{bmatrix}
\end{aligned}
\tag{7.127}
$$

Based on above defined rational transfer function forms, the cascade system is given as

$$
\begin{aligned}
y(t) &= G(q,\theta)u(t) + e(t) \\
\theta &= \begin{bmatrix} \theta_1 & \theta_2 & \theta_3 \end{bmatrix}^T
\end{aligned}
\tag{7.128}
$$

To apply the classical prediction error method, the one step ahead prediction $\hat{y}(t,\theta)$ is defined as follows:

$$\hat{y}(t,\theta) = G(q,\theta)u(t) \tag{7.129}$$

The prediction errors are given as

$$\varepsilon(t) = y(t) - \hat{y}(t,\theta) = y(t) - G(q,\theta)u(t) \tag{7.130}$$

As $y(t)$ is the observed output, $u(t)$ is the chosen input, combine them together to be input-output sequence $\{u(t), y(t)\}_{t=1}^{N}$ and construct a quadratic cost function to minimize, i.e.

$$V(\theta) = \frac{1}{N}\sum_{t=1}^{N}\varepsilon^2(t) = \frac{1}{N}\sum_{t=1}^{N}\left[y(t) - G(q,\theta)u(t)\right]^2 \tag{7.131}$$

where N is the sample size or the number of observed data.

Comment: Observing that the quadratic cost function $V(\theta)$, whose $\{u(t), y(t)\}_{t=1}^{N}$ are all known, as they can be collected in priori through some physical measurement. Only the transfer function $G(q)$ is unknown. On the other hand, as all transfer function $G_j(q)$ is parameterized by the unknown parameter vector θ_j, so the problem is turned to identify three unknown parameter vectors $\{\theta_1, \theta_2, \theta_3\}$, i.e.

$$\hat{\theta} = \arg\min_{\theta} V(\theta) \tag{7.132}$$

where $\hat{\theta}$ denotes the parameter estimation.

The global minimize $\hat{\theta}$ of this cost function is an asymptotically efficient estimate of θ.

To solve the quadratic cost function $V(\theta)$, one online subgradient descent algorithm is applied to generate the optimal parameter estimation. For convenience to show the online subgradient descent algorithm, the following recurrence is used.

$$\hat{\theta}_{t+1} = \Pi_v\left(\hat{\theta}_t - \gamma_t V'\left(\hat{\theta}_t\right)\right) \tag{7.133}$$

where $\gamma_t > 0$ are stepsize, $\Pi_v(\theta)$ is the standard projector on v, where v is one set, i.e. $\theta \in v$. $V'\left(\hat{\theta}_t\right)$ is a subgradient of V at $\hat{\theta}_t$, i.e.

$$V\left(\hat{\theta}_{t+1}\right) \geq V\left(\hat{\theta}_t\right) + V'\left(\hat{\theta}_t\right)\left(\hat{\theta}_{t+1} - \hat{\theta}_t\right) \tag{7.134}$$

To demonstrate the merit of online subgradient descent algorithm, the following two propositions are given to show it.

Proposition 7.4: After given the original value $\hat{\theta}(0)$, the vector $e = \theta - \Pi_v(\theta)$ forms an acute angle with every vector of the form $\eta - \Pi_v(\theta)$, $\eta \in v$, i.e.

$$\left(\theta - \Pi_v(\theta)\right)^T \left(\eta - \Pi_v(\theta)\right) \leq 0, \forall \eta \in v \tag{7.135}$$

Proof: Let $\eta \in v$, and $0 \leq t \leq 1$, we have

$$\phi(t) = \left\|\left[\Pi_v(\theta) + t\left(\eta - \Pi_v(\theta)\right)\right] - \theta\right\|_2^2 \geq \left\|\Pi_v(\theta) - \theta\right\|_2^2 = \phi(0) \tag{7.136}$$

Then

$$0 \leq \phi'(0) = 2\left[\Pi_v(\theta) - \theta\right]^T \left(\eta - \Pi_v(\theta)\right) \tag{7.137}$$

i.e.

$$\left(\theta - \Pi_v(\theta)\right)^T \left(\eta - \Pi_v(\theta)\right) \leq 0, \forall \eta \in v \tag{7.138}$$

In particular

$$\left\|\eta - \theta\right\|_2^2 = \left\|\eta - \Pi_v(\theta)\right\|_2^2 + \left\|\Pi_v(\theta) - \theta\right\|_2^2 + 2\left(\Pi_v(\theta) - \theta\right)^T \left(\eta - \Pi_v(\theta)\right)$$

$$\geq \left\|\eta - \Pi_v(\theta)\right\|_2^2 + \left\|\Pi_v(\theta) - \theta\right\|_2^2 \tag{7.139}$$

i.e.

$$\left\|\eta - \Pi_v(\theta)\right\|_2^2 \leq \left\|\eta - \theta\right\|_2^2 - \left\|\Pi_v(\theta) - \theta\right\|_2^2, \forall \eta \in v \tag{7.140}$$

Which completes the proof of the Proposition 1.

Proposition 7.5: For that online subgradient descent algorithm, then for every $\eta \in v$, we have

$$\gamma_t \left(\theta_t - \eta \right)^T V'\left(\theta_t \right) \leq \frac{1}{2}\left\| \theta_t - \eta \right\|_2^2 - \frac{1}{2}\left\| \theta_{t+1} - \eta \right\|_2^2 + \frac{1}{2}\gamma_t^2 \left\| V'\left(\theta_t \right) \right\|_2^2 \qquad (7.141)$$

Proof: By using above Proposition 1, we have*

$$\begin{cases} d_{t+1} = \dfrac{1}{2}\left\| \theta_{t+1} - \eta \right\|_2^2 \\[2mm] d_t = \dfrac{1}{2}\left\| \theta_t - \eta \right\|_2^2 \end{cases} \qquad (7.142)$$

Then

$$d_t \leq \frac{1}{2}\left\| \left[\theta_t - \eta \right] - \gamma_t V'\left(\theta_t \right) \right\|_2^2 = d_t - \gamma_t \left(\theta_t - \eta \right)^T V'\left(\theta_t \right) + \frac{1}{2}\left\| V'\left(\theta_t \right) \right\|_2^2 \qquad (7.143)$$

Summing up inequalities over $t = 1, 2 \ldots N$, we get

$$\sum_{t=1}^{N} \gamma_t \left(V\left(\theta_t \right) - V\left(\eta \right) \right) \leq d_1 - d_2 + \sum_{t=1}^{N} \frac{1}{2}\gamma_t^2 \left\| V'\left(\theta_t \right) \right\|_2^2 \qquad (7.144)$$

Furthermore, it holds that

$$d_t - d_N \leq \max_{\theta, \eta} \frac{1}{2}\left\| \theta - \eta \right\|_2^2 \qquad (7.145)$$

Then we have the following upper bound in equation (25)

$$\max_{t \in [1, N]} V\left(\theta_t \right) - V_* \leq \frac{\max\limits_{\theta, \eta} \dfrac{1}{2}\left\| \theta - \eta \right\|_2^2 + \sum\limits_{t=1}^{N} \dfrac{1}{2}\gamma_t^2 \left\| V'\left(\theta_t \right) \right\|_2^2}{\sum\limits_{t=1}^{N} \gamma_t} \qquad (7.146)$$

The above inequality shows the convenient results for the online subgradient descent algorithm.

In equation (7.146), the subgradient $V'(\theta_t)$ is written as

$$V'\left(\theta_t \right) = \frac{2}{N}\sum_{t=1}^{N}\left[y_t - G(q, \theta)u(t) \right]\left(-\frac{\partial G(q, \theta)}{\partial \theta} \right)u(t) \qquad (7.147)$$

All the above derived results hold for the special parameterized systems, $G(q, \theta) = \{G_1(q, \theta_1), G_2(q, \theta_2), G_3(q, \theta_3)\}$. For the general systems $G(q) = \{G_1(q), G_2(q), G_3(q)\}$, here we give our new way to identify them without any parameter vector.

Rewrite that equation (7.122) again.

$$y_1(t) = e_1(t) + G_2(q)\big[G_1(q)u_1(t) + u_2(t)\big]$$
$$= e_1(t) + G_2(q)G_1(q)u_1(t) + G_2(q)u_2(t) \tag{7.148}$$

During the whole identification process, two inputs $\{u_1(t), u_2(t)\}_t^N$ are all known in priori, so we multiply $u_1(t)$ on both sides of equation (7.148) to get

$$u_1^T y_1(t) = u_1^T e_1(t) + G_2(q)G_1(q)u_1^T u_1(t) + G_2(q)u_1^T u_2(t) \tag{7.149}$$

Taking expectation operation on both sides of equation (7.148), we have

$$E\big[u_1^T y_1(t)\big] = E\big[u_1^T e_1(t)\big] + G_2(q)G_1(q)E\big[u_1^T u_1(t)\big] + G_2(q)E\big[u_1^T u_2(t)\big] \tag{7.150}$$

i.e.

$$\phi_{u_1 y_1}(w) = G_2(q)G_1(q)\phi_{u_1}(w) + G_2(q)\phi_{u_1 u_2}(w) \tag{7.151}$$

where $\phi_{u_1 y_1}(w), \phi_{u_1}(w), \phi_{u_1 u_2}(w)$ are three power spectrums, and the property about input signal $u_1(t)$ be independent with external noise $e_1(t)$ is used on above equation (7.151).

Similarly, it holds that

$$\phi_{u_2 y_1}(w) = G_2(q)G_1(q)\phi_{u_1 u_2}(w) + G_2(q)\phi_{u_2}(w) \tag{7.152}$$

Taking subtract operation between equation (7.151) and (7.152), we have

$$G_1(q)\phi_{u_2 y_1}(w) - \phi_{u_2 y_1}(w) = G_2(q)G_1^2(q)\phi_{u_1}(w) - G_2(q)\phi_{u_2}(w)$$
$$= G_2(q)\big[G_1^2(q)\phi_{u_1}(w) - \phi_{u_2}(w)\big] \tag{7.153}$$

From equation (7.152), it holds that

$$G_2(q) = \frac{G_1(q)\phi_{u_1 y_1}(w) - \phi_{u_2 y_1}(w)}{\big[G_1^2(q)\phi_{u_1}(w) - \phi_{u_2}(w)\big]} \tag{7.154}$$

where in equation (7.153), $\phi_{u_1 y_1}(w), \phi_{u_2 y_1}(w), \phi_{u_1}(w), \phi_{u_2}(w)$ are all known, but $G_1(q)$ exists in the right side.

Consider the identification for system $G_1(q)$, we observe that the ideal case holds that

$$y_1(t) - G_2(q)u_2(t) = G_2(q)G_1(q)u_1(t) \tag{7.155}$$

It means $G_1(q)$ is estimated as follows:

$$G_1(q) = \arg\min_{G_1(q)} \sum_{t=1}^{N}\big[y_1(t) - G_2(q)u_2(t) - G_1(q)G_2(q)u_1(t)\big]^2 \tag{7.156}$$

Combining equation (7.154) and (7.156), the identification for systems $\{G_1(q), G_2(q)\}$ can be the following iterative steps.

Step 1: Given one initial value $G_1^0(q)$;

Step 2: Apply equation (7.154) to compute $G_2^0(q)$, i.e.

$$G_2^0(q) = \frac{G_1^0(q)\phi_{u_1 y_1}(w) - \phi_{u_2 y_1}(w)}{\left[\left(G_1^0(q)\right)^2 \phi_{u_1}(w) - \phi_{u_2}(w)\right]}$$

Step 3: Substituting $G_2^0(q)$ into equation (7.156) to optimize that quadratic cost function, then $G_1^1(q)$ is obtained, i.e.

$$G_1^1(q) = \arg\min_{G_1(q)} \sum_{t=1}^{N} \left[y_1(t) - G_2^0(q)u_2(t) - G_1(q)G_2^0(q)u_1(t)\right]^2$$

Step 4: Substituting $G_1^1(q)$ into equation (7.154) to get $G_2^1(q)$

$\vdots$

Step i: Iteratively generate the $(i-1)$ nonparametric estimations $\left\{G_1^{i-1}(q), G_2^{i-1}(q)\right\}$

$\vdots$

Step n: Until the following inequality is satisfied, then terminate the above iterative steps, i.e.

$$\left\|G_1^n(q) - G_1^{n-1}(q)\right\| + \left\|G_2^n(q) - G_2^{n-1}(q)\right\| \le \zeta \qquad (7.157)$$

where ζ is one small positive value, being one given the threshold, for example, $\zeta = 0.5$.

Finally, the estimation for the third system $G_3(q)$ is simple, only through some substitution operations.

7.4.3   Model structure validation process

Based on the equation (7.128), after using our proposed iterative identification steps to estimate that total unknown parameter vector θ, the next mission it to validate the identified model about the parameter estimation $\hat{\theta}$.

In the standard prediction error algorithm, when using the input-output data $Z^N = \{y(t), u(t)\}_{t=1}^{N}$ with the data number N, the parameter vector is identified by

$$y(t) = G(q,\theta)u(t) + e(t); \theta = \begin{bmatrix} \theta_1 & \theta_2 & \theta_3 \end{bmatrix}^T$$

$$\varepsilon(t) = y(t) - \hat{y}(t,\theta) = y(t) - G(q,\theta)u(t)$$

$$\hat{\theta} = \arg\min_{\theta} V\left(\theta, Z^N\right) = \arg\min_{\theta} \frac{1}{N}\sum_{t=1}^{N} \varepsilon^2(t,\theta) \qquad (7.158)$$

Defining the asymptotic limit parameter estimate θ^* as

$$\theta^* = \arg\min_{\theta} \lim_{N \to \infty} E\left\{V\left(\theta, Z^N\right)\right\}$$

where E denotes the expectation operator.

In the common identification process, assume that there always exists one true parameter vector θ_0 such that

$$G(q, \theta_0) = G_0(q)$$

This assumption shows that the identified model is contained in the considered model set. Based on some identification results, we get the asymptotic matrix of the parameter estimate.

$$\text{cov}\,\hat{\theta} = \lambda_0 \langle \varphi, \varphi \rangle^{-1} \tag{7.159}$$

where $\langle \varphi, \varphi \rangle$ denotes some inter product operator, φ is the negative gradient of the predictor error, i.e. it can be computed from

$$\varphi(t, \theta) = -\frac{\partial \varepsilon(t, \theta)}{\partial \theta} = \frac{\partial \hat{y}(t, \theta)}{\partial \theta}$$

Next, we give the calculation process of the negative gradient of the predictor error. As (7.159) is a basic formula in studying asymptotic analysis, we combine (7.128) and (7.129) to get

$$\hat{y}(t, \theta) = G(q, \theta) u(t) \tag{7.160}$$

substituting (7.129) into (7.160) and computing the partial derivative operations with respect to unknown parameter vector θ, and then we have

$$\frac{\partial \varepsilon(t, \theta)}{\partial \theta} = G'(\theta) u(t)$$

$$G'(\theta) = \frac{\partial G(\theta)}{\partial \theta} \tag{7.161}$$

where $G'(\theta)$ denotes the partial derivative operation with respect to θ, and the delay operator q are all ignored to simplify the derivations.

Using the uncorrelated assumption between white noise $e(t)$ and $u(t)$, i.e. it holds that $E(e(t)^T u(t)) = 0$.

According to (7.161), the asymptotic covariance matrix is

$$P_\theta = \text{cov}\,\hat{\theta} = \lambda_0 \left[E\varphi(t, \theta_0) \varphi^T(t, \theta_0) \right]^{-1} \tag{7.162}$$

On basis of (7.159), we have the asymptotic result

$$\hat{\theta} \xrightarrow{N \to \infty} \theta_0$$

It shows that the parameter estimator $\hat{\theta}$ will converge to its limit θ_0, and further, $\hat{\theta}$ will asymptotically converge ($N \rightarrow \infty$) to a normally distributed random variable with mean θ_0 and variance P_θ.

$$\sqrt{N}\left(\hat{\theta} - \theta_0\right) \rightarrow \mathbb{N}\left(0, P_\theta\right), \text{ as } N \rightarrow \infty$$

This asymptotic result can be rewritten in a quadratic form, and then we get one λ^2 distribution.

$$N\left(\hat{\theta}_N - \theta_0\right)^T P_\theta^{-1}\left(\hat{\theta}_N - \theta_0\right) \xrightarrow{N \rightarrow \infty} \lambda_n^2 \tag{7.163}$$

where n is the number of degrees of freedom in the λ^2 distribution, being equal to the dimension of the parameter vector. Equation (7.163) implies that the random variable $\hat{\theta}$ satisfies one uncertainty bound.

$$\hat{\theta}_N \in D\left(\alpha, \theta_0\right)$$
$$D\left(\alpha, \theta_0\right) = \left\{\theta_0 / N\left(\theta - \theta_0\right)^T P_\theta^{-1}\left(\theta - \theta_0\right) \le \lambda_{n,\alpha}^2\right\} \tag{7.164}$$

with $\lambda_{n,\alpha}^2$ corresponding to a probability level α in λ_n^2 distribution, But now in order to quantity the uncertainty on θ_0 rather than on $\hat{\theta}$. For every realization of $\hat{\theta}$, it holds that

$$\hat{\theta}_N \in D\left(\alpha, \theta_0\right) \Leftrightarrow \theta_0 \in D\left(\alpha, \hat{\theta}_N\right)$$

It signifies that

$$\theta_0 \in D\left(\alpha, \hat{\theta}_N\right) \quad \text{with probability } \alpha$$
$$D\left(\alpha, \hat{\theta}\right) = \left\{\theta / N\left(\hat{\theta} - \theta\right)^T P_\theta^{-1}\left(\hat{\theta} - \theta\right) \le \lambda_{n,\alpha}^2\right\} \tag{7.165}$$

Equations (7.164) and (7.165) give the confidence intervals of an unknown parameter estimator under the cascade condition. The probability level of the event $\hat{\theta} \in D\left(\alpha, \theta_0\right)$ holds is at least α.

7.4.4 Simulation examples

In this section, we give one simulation example to verify our identification strategies for aircraft system identification. This simulation example is about variable thrust axis aircraft flight control, being a multi-loop control system, which is divided according to its level. The main components are divided into rudder loop/thrust deflection loop, stabilization loop, and control (guidance) loop. Among them, the main difference between its stabilization loop and the conventional aircraft stabilization loop is the addition of an additional thrust deflection rotor system, whose system structure is shown in Fig. 7.27.

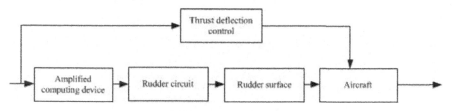

Figure 7.27: Thrust deflection rotor system

In Fig. 7.27, the rudder loop subsystem includes two negative feedback loops, one is the output angular position signal of the steering gear measured by the position sensor, which is fed back to the input end of the rudder loop, so that the control signal is proportional to the output signal of the steering gear or other functional relationship; The angular rate signal output by the steering gear is measured by the tachometer and fed back to the amplifier to increase the damping of the rudder loop and improve the dynamic performance of the rudder loop. The measurement component is used to measure the flight attitude information of the aircraft, and the amplification and calculation device is used to realize the stable control calculation of the aerodynamic part. The control and guidance loop of the aircraft with a variable thrust axis is constructed by the stability loop, the aircraft gravity center measurement component, and the kinematics link describing the geometric relationship of the aircraft's spatial position, which is used to stabilize and control the trajectory of the aircraft.

Observing Fig. 7.27 again, those three systems, i.e. rudder circuit, rudder surface and aircraft, are represented as follows.

$$G_1(q) = \frac{0.7q + 0.2}{q^2 - 0.7q + 0.5}, G_2(q) = \frac{0.6q^2 - 0.2q}{q^2 - 1.3q + 0.6}, G_3(q) = \frac{0.6q^2 + 0.8q + 1.2}{q^2 - 0.75q + 0.7}$$

$$(7.166)$$

To formulate the above three equations with our considered parameter vector, we define that

$$\theta_1 = \begin{bmatrix} 0.7 & 0.2 & -0.7 & 0.5 \end{bmatrix}^T$$

$$\theta_2 = \begin{bmatrix} 0.6 & -0.2 & -1.3 & 0.6 \end{bmatrix}^T$$

$$\theta_3 = \begin{bmatrix} 0.6 & 0.8 & 1.2 & -0.75 & 0.7 \end{bmatrix}^T$$

$$\theta = \begin{bmatrix} \theta_1 & \theta_2 & \theta_3 \end{bmatrix}^T \qquad (7.167)$$

One sensor (shown in Fig. 7.28) is used to collect the input-output sequence, is plotted in Fig. 7.29, then the cascade system is changed to identify the unknown parameter vector. During the identification process, all initial parameter values are set to be number 0, then our proposed prediction error method and online subgradient descent algorithm are used to identify those three unknown parameter

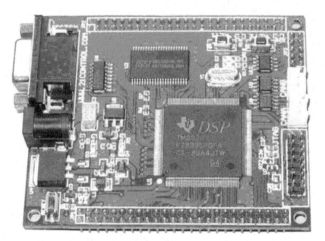

Figure 7.28: Sensor

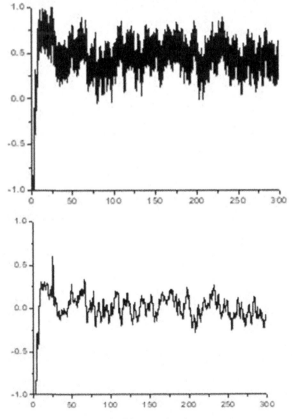

Figure 7.29: Observation input-output

vectors. Figure 7.30 gives the parameter estimation for the first unknown parameter vector θ_1. From Fig. 7.30, we see after the iterative steps are increased, the parameter estimations will approach their own true values.

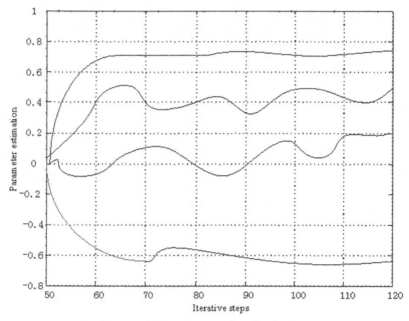

Figure 7.30: Parameter estimation for θ_1

In the design process of the aircraft, since the aerodynamic data of the aircraft is generally obtained through the wind tunnel blowing experiment, the aerodynamic parameters of the aircraft inevitably have errors. To show the asymptotic efficiency for the parameter vector, the initial state of the aircraft is considered as a straight and level flight along the track in the ground coordinate system. The altitude is 500 m, pitch attitude is –0.88 rad, attack angle and elevator are –0.88 rad and 0.5. Based on our three identified systems, the output trajectory of the considered aircraft is plotted in Fig. 7.31, where two curves correspond to the output trajectories for the true model and our identified model respectively. It means these two output trajectories are the same as each other, i.e. few errors exists.

7.4.5   Conclusion

In this section, one kind of special network system - cascade system is considered from the point of system identification, which extends the commonly used open loop system and closed loop system. Two different identification strategies are proposed for the parameterized system and non-parameterized system respectively, i.e. parameter estimation and non-parameter estimation. As a

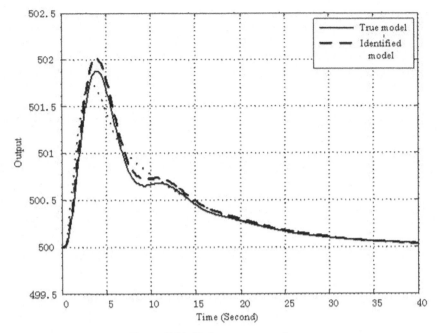

Figure 7.31: Trajectory comparisons

cascade system really exists in an aircraft system, so all identification results can be applied to aircraft system identification directly. This chapter paves a way for the next network system identification for aircraft system modeling.

7.5   Summary

The idea of our considered data driven is applied in the state of charge estimation and model parameter identification, which correspond to our research. Whatever state of charge estimation and model parameter identification, they all need the system identification theory. The parameter identification and optimal input design are included in the four processes of the system identification field, but the applied fields are different from each other.

References

[1] Mohamed, M.R. and Leung, P.K. 2015. Performance characterization of a vanadium redox of battery at different operating parameters under a standardized test-bed system, Apply Energy, 137(2): 402-412.
[2] Massimo, G. 2016. Vanadium redox of batteries: Potentials and challenges of an emerging storage technology, IEEE Ind. Electron. Mag., 10(3): 20-31.

[3] Hong, W.C., Li, B.Y. and Wang, B.G. 2015. Theoretical and technological aspects of batteries: Measurement, Energy Storage Sci. Technol., 56(4): 744-756.

[4] Petchsingh, C., Quill, N. and Joyce, J.T. et al. 2016. Spectroscopic measurement of state of charge in vanadium batteries with an analytical model of VIV-VV absorbance, J. Electrochem. Soc., 163(40): 5068-5083.

[5] Li, X., Xiong, J. and Tang, A. et al. 2018. Investigation of the use of electrolyte viscosity for online state-of-charge monitoring design in vanadium redox of battery, Appl. Energy, 211(6): 1050-1059.

[6] Ngamsai, K. and Arpornwichanop, A. 2015. Measuring the state of charge of the electrolyte solution in a vanadium redox flow battery using a four-pole cell device, J. Power Sources, 29(3): 150-157.

[7] Ressel, S., Bill, F. and Holtz, L. et al. 2018. State of charge monitoring of vanadium redox of batteries using half cell potentials and electrolyte density, J. Power Sources, 37(8): 776-783.

[8] Chou, Y.S., Hsu, N.Y. and Jeng, K.T. et al. 2016. A novel ultrasonic velocity sensing approach to monitoring state of charge of vanadium redox flow battery, Apply Energy, 182(3): 283-289.

[9] Zhong, Q., Zhong, F. and Cheng, J. et al. 2016. State of charge estimation of lithium-ion batteries using fractional order sliding mode observer, ISA Trans, 66(8): 448-459.

[10] Xiong, B., Zhao, J. and Su, Y. et al. 2017. State of charge estimation of vanadium redox of battery based on sliding mode observer and dynamic model including capacity fading factor, IEEE Trans. Sustainable Energy, 8(7): 1658-1667.

[11] Wei, Z., Tseng, K.J. and Wai, N. et al. 2017. An adaptive model for vanadium redox of battery and its application for online peak power estimation, J. Power Sources, 344(4): 195-207.

[12] Wei, Z., Tseng, K.J. and Wai, N. et al. 2016. Adaptive estimation of state of charge and capacity with online identified battery model for vanadium redox of battery, J. Power Sources, 332(6): 389-398.

[13] Wei, Z., Bhattaraia, A. and Zou, C. et al. 2018. Real-time monitoring of capacity loss for vanadium redox battery, J. Power Sources, 390(7): 261-269.

[14] Lin, C., Mu, H. and Xiong, R. 2017. A novel multi-model probability battery state of charge estimation approach for electric vehicles using H-infinity algorithm, Apply Energy, 344(6): 195-207.

[15] Jakob, K. 2011. A design algorithm using external perturbation to improve iterative feedback tuning convergence, Automatica, 47(2): 2665-2670.

[16] Rojas, C.R. 2010. The cost of complexity in system identification: Frequency function estimation of finite impulse response system, IEEE Transaction on Automatic Control, 55(10): 2298-2309.

[17] Ohlsson, H. and Ljung, L. 2010. Segmentation of ARX-models using sum-of-norms regularization, Automatica, 46(6): 1107-1111.

[18] Bravo, J.M., Alamo, T. and Vasallo, M. 2017. A general framework for predictions based on bounding techniques and local approximations, IEEE Transactions on Automatic Control, 62(7): 3430-3435.

[19] Tanaskovic, M., Fagiano, L. and Novara, C. 2017. Data driven control of nonlinear systems: An on line direct approach, Automatica, 75(1): 1-10.

[20] Gevers, M. 2009. Identification and information matrix: How to get just sufficiently rich, IEEE Transactions of Automatic Control, 54(12): 2828-2840.

[21] Wei, T., Shi, Z. and Li, H. 2006. Frequency generalization total least square algorithm of aircraft flutter model parameter identification, Control and Decision, 21(7): 726-729.

[22] Soderstorn, T. 2009. A covariance matching approach for identifying errors in variables systems, Automatica, 45(9): 2018-2031.

[23] Weyer, E. and Campi, M.C. 2017. Asymptotic properties of SPS confidence regions, Automatica, 82(1): 287-294.

[24] Ninness, B. and Hjalmarsson, H. 2005. Analysis of the variability of joint input signal-output estimation methods, Automatica, 41(6): 1123-1132.

[25] Hjalmarssion, H. 2011. A geometric approach to variance analysis in system identification, IEEE Transactions of Automatic Control, 56(5): 983-997.

[26] Wang Jianhong, Ricardo, A. and Ramirez-Mendiza. 2019. Combing instrumental variable and variance matching for aircraft flutter model parameters identification, Shock and Vibration, 28(6): 1-12.

[27] Wang, J. and Xu Y. 2015. A fast filtering algorithm in the aircraft flutter model parameter identification experiment, Computing Technology and Automation, 34(1): 6-10.

[28] Jianhong, W. and Zhu Yonghong. 2012. Instrumental variable covariance method and asymptotic analysis for aircraft flutter model parameter identification, Journal of Applied Sciences, 30(4): 433-440.

[29] Wang, J. and Wang D. 2009. Forgetting factor algorithm for aircraft flutter model parameter identification, Chinese Space Science and Technology, 38(6): 7-14.

[30] Wang Jianhong. 2019. Optimal input signal design for multi UAVs formation anomaly detection, ISA Transactions, 91(6): 157-165.

[31] Wang, J. and Wang, D. 2011. Application of subspace predictive control in active noise and vibration control, Journal of Vibration and Shock, 30(10): 129-135.

[32] Casini, M., Garulli, A. and Vicino, A. 2006. On input signal design in l2 conditional set membership identification, Automatica, 42(5): 815-823.

[33] Casini, M., Garulli, A. and Vicino, A. 2011. Input signal design in worst case system identification using binary sensors, IEEE Transactions of Automatic Control, 56(5): 1186-1191.

[34] Jauberthie, C. and Denis-Vidal, L. 2018. Optimal input signal design for parameter estimation in a bounded error context for nonlinear dynamical systems, Automatica, 92(6): 86-91.

CHAPTER 8

Data Driven Subspace Predictive Control

8.1 Introduction

In this chapter, the idea of data driven is combined with subspace predictive control, which is the extended form for classical subspace system identification. More specifically, the detailed contributions are formulated as follows.

(1) Subspace predictive control strategy is applied to design controller under fault conditions. The statistic description and explicit expression of the residual are analyzed after the output predictive estimations are derived. The problem of designing a predictive controller is formulated as one optimization problem with equality and inequality conditions, the dual decomposition is used to change the original constrained optimization problem into an unconstrained optimization problem. Further, the nearest neighbor gradient algorithm is proposed to solve the primal dual optimization problem.

(2) To pave a way for studying the nonlinear system based on the existed results of a linear system, in this paper, one novel subspace data driven control for linear parameter varying system with the scheduling parameters. The input-output observation data matrix is used to identify Markov parameters in the form of state space forms, and only the data matrix is used to represent the output prediction value at the future different time instants. Then the output prediction value is applied to the cost function in data driven control. For this quadratic cost function, a parallel distribution algorithm is used to solve its optimal control input value, and the iterative convergence of the parallel distribution algorithm is analyzed in detail Finally the proposed subspace data driven control strategy is applied to control the DC motor, whose mass distribution factor is deemed as one linear parameter varying system.

(3) We propose the local polynomial method to solve the problem of estimating the frequency response function in the linear system. Compared with other nonparametric identification methods based on the windowing strategies, this new identification method can be remarkably efficient in reducing the

effect caused by the leakage error when the discrete Fourier transform is used under non period input excited signal. Considering the constraints between the coefficients of the polynomials at neighbor frequencies, we modify the proposed local polynomial method to get one constrained local polynomial method. The modified local polynomial method reduces the mean square error about the frequency response function and the estimation of the frequency response function is identified by one multi-objective least squares criterion.

8.2   Nearest neighbor gradient algorithm in subspace predictive control under fault condition

Subspace predictive control is from the idea of subspace identification; whose goal is to construct one mathematical equation corresponding to the considered plant using only input-output measured sequences. During the subspace identification process, each matrix in a state space equation is identified by using the basic matrix singular value decomposition strategy, as this basic strategy is applied to decompose a matrix constituted by past and future input-output measured sequence. Instead, subspace predictive control can construct the future prediction of output directly from input-output measured sequences, and avoid constructing the state space equation, i.e. the state space equation is not needed in estimating the predictors from an estimated state sequence.

Subspace predictive control is one of the special data driven control methods because it combines system identification and predictive control and its ability to obtain output prediction directly by measured sequences, which is very important in predictive control theory. The basic output prediction in subspace predictive control is proposed in [1], and [2] compares this output prediction and another value coming from iterative correction tuning control strategy, and the equivalence between these two output predictions can be guaranteed through introducing one pre-filter. In [3] a novel subspace predictive control algorithm based on subspace identification was studied to solve actuator saturation limitations in a range of active vibration and noise control problems, and meanwhile, the subspace predictive control permits limitations on allowable actuator saturation. An upper bound of a maximal number of possible iteration steps is derived in subspace predictive control, and the proposed subspace predictive control is realized in an example of helicopter simulation [4]. When imposing the lower and upper bounds constraint conditions on the faults, a fast gradient method was applied to solve this problem, based on fault estimations the subspace predictive control was proposed to solve an optimization problem with linear matrix inequality constraints [5]. In recent years, there are no new references on this subspace predictive control strategy, other than our above mentioned references.

Here, we continue our work in studying subspace predictive control deeply. The essence of our chapter is that the problem of designing a dual decomposition controller is studied for subspace predictive control strategy under fault conditions, as the existence of fault conditions has not been considered in original research. For state space equation with fault condition, we establish one explicit function form between fault and residual using only input-output measured sequences and construct one least squares optimization problem to obtain fault estimation. The statistical property of the residual is analyzed based on our derived output prediction, then the Kronecker product is used to derive the detailed structure corresponding to the residual vector at every time instant. After substituting our output prediction into the objective function of predictive control, one quadratic programming problem with equality and inequality constraints is considered. When solving this constrained optimization problem, the fast gradient method is not suited for this complex optimization problem, as here one regularization term is added to our objective function. So to solve this complex quadratic optimization problem, a dual decomposition idea is proposed to convert the formerly constrained optimization into one unconstrained optimization, then one nearest neighbor gradient algorithm is given to solve its optimal value. Finally, the efficiency of the proposed control strategy and nearest neighbor gradient algorithm can be easily realized in one simple helicopter example.

8.2.1   Problem description

Consider the following stochastic discrete time state space model.

$$\begin{cases} x(k+1) = Ax(k) + Bu(k) + Ef(k) + Fw(k) \\ y(k) = Cx(k) + Du(k) + Gf(k) + v(k) \end{cases} \tag{8.1}$$

where equation (8.1) is a multi-input and multi-output state space model, and

$$x(k) \in R^n, y(k) \in R^l, u(k) \in R^m, f(k) \in R^{n_f} \tag{8.2}$$

denote state variable, output variable, control variable and fault respectively. Matrices A, B, C, D, E, F, G are real bounded matrices with approximate dimensions, and their dimensions are determined by the dimensions of matrices $x(k)$, $y(k)$, $u(k)$ and $w(k)$, $v(k)$. Disturbs include process noise $w(k) \in R^{n_w}$ and measurement noise $v(k) \in R^l$, further process noise $w(k) \in R^{n_w}$ and measurement noise $v(k) \in R^l$ are all assumed to be white Gauss noise with zero mean.

In the system identification field, input-output state space relation (8.1) is rewritten as following innovative form.

$$\begin{cases} \hat{x}(k+1) = A\hat{x}(k) + Bu(k) + Ef(k) + Ke(k) \\ y(k) = C\hat{x}(k) + Du(k) + Gf(k) + e(k) \end{cases} \tag{8.3}$$

where innovation $e(k)$ is defined as that

$$e(k) = y(k) - C\hat{x}(k) - Du(k) - Gf(k) \tag{8.4}$$

where K is the Kalman gain matrix, innovation $e(k)$ is determined by process noise $w(k) \in R^{n_w}$ and measurement noise $v(k) \in R^l$, here the variance of innovation $e(k)$ is σ_e. In the Kalman filter theory [8], $u(k)$ and $y(k)$ are two deterministic variables, then after substituting the definition of innovation $e(k)$ into the state space equation with innovation form, one closed loop input-output relation is obtained.

$$\begin{cases} \hat{x}(k+1) = (A-KC)\hat{x}(k) + (B-KD)u(k) + (E-KG)f(k) + Ky(k) \\ y(k) = C\hat{x}(k) + Du(k) + Gf(k) + e(k) \end{cases} \tag{8.5}$$

where $u(k)$ and $f(k)$ can be regarded as two external and deterministic input signals, and in equation (8.4), $f(k)$ denotes the given fault. When the output trajectory is given, the expected output trajectory is used to quality the output data at a future time instant. An approximate choice of the control input is obtained by minimizing the measurement error [6], i.e. it leads to one problem of designing predictive controller.

Define state, input and fault as that respectively

$$\phi = A - KC, \quad \tilde{B} = B - KD, \quad \tilde{E} = E - KG \tag{8.6}$$

When closed loop system is minimal realization, i.e. $\phi = A - KC$ is stable, $x(k)$, $u(k)$ and $y(k)$ are all bounded.

8.2.2 Fault estimation

Substituting our above defined state, input and fault into equation (8.4), we obtain that

$$\hat{x}(k+1) = \phi\hat{x}(k) + \tilde{B}u(k) + \tilde{E}f(k) + Ky(k) \tag{8.7}$$

Set time instant $k - p$ as initial time instant, and recursively iterate p sample instants to time instant k, the following relation is given

$$\hat{x}(k) = \phi^p\hat{x}(k-p) + \sum_{\tau=0}^{p-1}\phi^\tau\begin{bmatrix}\tilde{B} & K\end{bmatrix}\begin{bmatrix}u(k-\tau-1)\\y(k-\tau-1)\end{bmatrix} + \sum_{\tau=0}^{p-1}\phi^\tau\tilde{E}f(k) \tag{8.8}$$

Substituting equation (8.8) into (8.7), the iterative form corresponding to output equation is listed as that

$$\begin{aligned} y(k) = C\phi^p\hat{x}(k-p) + \sum_{\tau=0}^{p-1}C\phi^\tau\begin{bmatrix}\tilde{B} & K\end{bmatrix}\begin{bmatrix}u(k-\tau-1)\\y(k-\tau-1)\end{bmatrix} \\ + \sum_{\tau=0}^{p-1}C\phi^\tau\tilde{E}f(k) + \begin{bmatrix}D & 0\end{bmatrix}\begin{bmatrix}u(k)\\y(k)\end{bmatrix} + Gf(k) + e(k) \end{aligned} \tag{8.9}$$

Output equation (8.9) is one vector autoregressive model with external input;

this novel vector autoregressive model is seen in subspace identification theory. The output equation establishes one linear transformation from past input-output measurements to future outputs. The following unknown parameters are identified in equation (8.10).

$$\left\{ D, C\phi^\tau \tilde{B}, C\phi^\tau K, C\phi^\tau \tilde{B}, C\phi^\tau \tilde{E}, G \quad j = 0 \cdots p-1 \right\} \tag{8.10}$$

but not parameters about $\{A, B, C, D, E\}$ or $C\phi^p$. The time index k in equation (8.10) is replaced by time sequence $t, t + 1 \cdots t + N - 1$, then reformulate $y(t)$, $y(t + 1) \cdots y(t + N - 1)$ as one block row vector. This block row vector is denoted as Y_{id}, i.e.

$$Y_{id} = \left[y(t) \quad y(t+1) \quad \cdots \quad y(t+N-1) \right] \tag{8.11}$$

Reformulating all components in vector Y_{id} based on equation (8.11), and after some calculations, we have that

$$
\begin{aligned}
Y_{id} &= \left[y(t) \quad y(t+1) \quad \cdots \quad y(t+N-1) \right] \\
&= C\phi^p \underbrace{\left[\hat{x}(t-p) \quad \hat{x}(t-p+1) \quad \cdots \quad \hat{x}(t+N-p-1) \right]}_{x_{id}} \\
&+ \left[C\phi^{p-1}\tilde{B} \quad C\phi^{p-1}K \quad \cdots \quad C\tilde{B} \quad CK \quad D \right] \\
&\quad \underbrace{\begin{bmatrix} u(t-p) & u(t-p+1) & \cdots & u(t+N-p-1) \\ y(t-p) & y(t-p+1) & \cdots & y(t+N-p-1) \\ \vdots & \vdots & \vdots & \vdots \\ u(t-1) & u(t) & \cdots & u(t+N-2) \\ y(t-1) & y(t) & \cdots & y(t+N-2) \\ u(t) & u(t+1) & \cdots & u(t+N-1) \end{bmatrix}}_{z_{id}} \\
&+ \underbrace{\left[e(t) \quad e(t+1) \quad \cdots \quad e(t+N-1) \right]}_{E_{id}} \\
&+ \left[C\phi^{p-1}\tilde{E} \quad C\phi^{p-2}\tilde{E} \quad \cdots \quad C\tilde{E} \quad G \right] \\
&\quad \underbrace{\begin{bmatrix} f(t-p) & f(t-p+1) & \cdots & f(t+N-p-1) \\ f(t-p+1) & f(t-p+2) & \cdots & f(t+N-p) \\ \vdots & \vdots & \vdots & \vdots \\ f(t-1) & f(t) & \cdots & f(t+N-2) \\ f(t) & f(t+1) & \cdots & f(t+N-1) \end{bmatrix}}_{f_{id}}
\end{aligned} \tag{8.12}
$$

Define Markov parameter sequences in equation (8.12) as that

$$E_1 = \begin{bmatrix} C\phi^{p-1}\tilde{B} & C\phi^{p-1}K & \cdots & C\tilde{B} & CK & D \end{bmatrix};$$
$$E_2 = \begin{bmatrix} C\phi^{p-1}\tilde{E} & C\phi^{p-2}\tilde{E} & \cdots & C\tilde{E} & G \end{bmatrix} \tag{8.13}$$

To simplify notation, equation (8.12) can be simplified as that

$$Y_{id} = C\phi^p x_{id} + E_1 z_{id} + E_2 f_{id} + E_{id} \tag{8.14}$$

Using the priori assumptions that the closed loop system is minimal realization and matrix $\phi = A - KC$ is stable, then we see that if the past identification horizon p is chosen as large enough, then $\left\| \phi^p \right\| \ll 1$ holds, i.e. $\left\| \phi^p \right\| \approx 0$. So the first unknown term $C\phi^p x_{id}$ can be neglected, then only approximation relation holds.

$$Y_{id} \approx E_1 z_{id} + E_2 f_{id} + E_{id} \tag{8.15}$$

Reformulate the above equation again, and define the called residual vector as that

$$r_{id} = Y_{id} - E_1 z_{id} = E_2 f_{id} + E_{id} \tag{8.16}$$

where equation (8.16) is a closed relation from fault f_{id} to residual vector r_{id}. The estimation of fault f_{id} can be achieved in the sense of least squares, it means that one well pose least squares optimization problem is constructed to estimate the fault f_{id}.

$$\min_{f_{id}} \left\| r_{id} - E_2 f_{id} \right\|_2^2 \tag{8.17}$$

Further in equation (8.17) some constrain conditions on fault f_{id} can be imposed.

8.2.3   Output estimation in subspace predictive control

To study subspace predictive control, the first step is to give the future output estimation at a future time instant, and the output estimation can be computed by using equation (8.15). As the residual is generated in more than one sample sliding horizon, i.e. sliding horizon level is $[k - L + 1, k]$, L is the output horizon level. Similar to the derivation of equation (8.15), the time index k in equation (8.15) is replaced by time index $k - L + 1$, $k - L + 1$, $\cdots$ k, then we obtain one column vector as

$$y_{y,L} = \begin{bmatrix} y(k - L + 1) & y(k - L + 2) & \cdots & y(k) \end{bmatrix}^T \tag{8.18}$$

Similarly, define block vector $u_{k,L}$ and $e_{k,L}$, and let $[k - L - p + 1 \ \ k - L]$ be past sliding window, p is past horizon level, then output equation is formulated as follows:

$$
\begin{bmatrix} y(k-L+1) \\ y(k-L+2) \\ \vdots \\ y(k) \end{bmatrix} = \underbrace{\begin{bmatrix} C\phi^P \hat{x}(k-L-p+1) \\ C\phi^P \hat{x}(k-L-p+2) \\ \vdots \\ C\phi^P \hat{x}(k-p) \end{bmatrix}}_{b_{k,L}}
$$

$$
+ \begin{bmatrix} C\phi^{p-1}\tilde{B} & C\phi^{p-1}K & \cdots & C\phi\tilde{B} & C\phi K & C\tilde{B} \\ 0 & 0 & C\phi^{p-1}\tilde{B} & C\phi^{p-1}K & \cdots & C\phi\tilde{B} \\ \vdots & \vdots & \vdots & & & \\ 0 & \cdots & \cdots & 0 & C\phi^{p-1}\tilde{B} & \cdots \end{bmatrix}
$$

$$
\left. \begin{matrix} CK & D & 0 & 0 \\ C\phi K & C\tilde{B} & CK & D & 0 \\ \vdots & & & \vdots & \vdots \\ C\phi^{L-1}K & C\phi^{L-2}K & \cdots & C\tilde{B} & CK & D & 0 \end{matrix} \right]
$$

$$
\times \begin{bmatrix} u(k-L-p+1) \\ y(k-L-p+1) \\ \vdots \\ u(k-L) \\ y(k-L) \\ u(k-L+1) \\ y(k-L+1) \\ \vdots \\ u(k) \\ y(k) \end{bmatrix} + \begin{bmatrix} e(k-L+1) \\ e(k-L+2) \\ \vdots \\ e(k) \end{bmatrix}
$$

$$
+ \begin{bmatrix} C\phi^{p-1}\tilde{E} & C\phi^{p-2}\tilde{E} & \cdots & & C\tilde{E} & G & 0 \\ 0 & C\phi^{p-1}\tilde{E} & C\phi^{p-2}\tilde{E} & \cdots & & C\tilde{E} & G & 0 \\ \vdots & & & \vdots & & & & \vdots \\ 0 & \cdots & C\phi^{p-1}\tilde{E} & \cdots & C\phi^{L-1}\tilde{E} & C\phi^{L-2}\tilde{E} & \cdots & C\tilde{E} & G \end{bmatrix}
$$

$$
\begin{bmatrix} f(k-L-p+1) \\ \vdots \\ f(k-L) \\ f(k-L+1) \\ \vdots \\ f(k) \end{bmatrix} \tag{8.19}
$$

Let all block Hankel matrices be that respectively

$$H_z^{L,p} = \begin{bmatrix} C\phi^{p-1}\tilde{B} & C\phi^{p-1}K & \cdots & & C\tilde{B} & CK \\ 0 & 0 & C\phi^{p-1}\tilde{B} & C\phi^{p-1}K & \cdots & C\phi\tilde{B} & C\phi K \\ \vdots & & \vdots & & & \vdots \\ 0 & \cdots & 0 & C\phi^{p-1}\tilde{B} & \cdots & C\phi^{L-1}K \end{bmatrix},$$

$$H_f^{L,p} = \begin{bmatrix} C\phi^{p-1}\tilde{E} & C\phi^{p-2}\tilde{E} & \cdots & C\phi\tilde{E} & C\tilde{E} \\ 0 & C\phi^{p-1}\tilde{E} & \cdots & & C\phi\tilde{E} \\ \vdots & & & & \vdots \\ 0 & \cdots & C\phi^{p-1}\tilde{E} & \cdots & \end{bmatrix}$$

$$T_y^L = \begin{bmatrix} 0 & & & \\ CK & 0 & & \\ \vdots & & \vdots & \\ C\phi^{L-2}K & C\phi^{L-3}K & \cdots & 0 \end{bmatrix}, T_u^L = \begin{bmatrix} D & & & \\ C\tilde{B} & D & & \\ \vdots & & \vdots & \\ C\phi^{L-2}\tilde{B} & C\phi^{L-3}\tilde{B} & \cdots & D \end{bmatrix},$$

$$T_f^L = \begin{bmatrix} G & 0 & \cdots \\ C\tilde{E} & G & \cdots \\ \vdots & & \vdots \\ C\phi^{L-2}\tilde{E} & C\phi^{L-3}\tilde{E} & \cdots & G \end{bmatrix}$$

$$(8.20)$$

Combining input-output within past sliding window as

$$z(k) = \begin{bmatrix} u(k) & y(k) \end{bmatrix}^T \tag{8.21}$$

Collecting input-output measured sequence within past sliding window as that

$$z_{k-L,p} = \begin{bmatrix} z(k-L-p+1) & z(k-L-p+2) & \cdots & z(k-L) \end{bmatrix}^T \tag{8.22}$$

Using above defined notations, a simplified form corresponding to equation (8.22) is obtained.

$$y_{k,L} = b_{k,L} + \begin{bmatrix} H_z^{L,p} & T_u^L \end{bmatrix} \begin{bmatrix} z_{k-L,p} \\ z_{k,L} \end{bmatrix} + \begin{bmatrix} H_f^{L,p} & T_f^L \end{bmatrix} \begin{bmatrix} f_{k-L,p} \\ f_{k,L} \end{bmatrix} + e_{k,L} \tag{8.23}$$

Then equation (8.23) can be rewritten as

$$y_{k,L} = b_{k,L} + \underbrace{\begin{bmatrix} H_z^{l,,p} & T_u^{L} \end{bmatrix}}_{\varphi_z} z_{k,L+p} + \underbrace{\begin{bmatrix} H_f^{L,p} & T_f^L \end{bmatrix}}_{\varphi_f} f_{k,L+p} + e_{k,L} \tag{8.24}$$

Then

$$y_{k,L} = b_{k,L} + \varphi_z z_{k,L+p} + \varphi_f f_{k,L+p} + e_{k,L} \tag{8.25}$$

where equation (8.25) is our output estimation for subspace predictive control, but $z_{k,L+p}$ includes vector $y_{k,L}$, so one suitable matrix transformation is used to eliminate vector $y_{k,L}$ in $z_{k,L+p}$.

8.2.4 Residual analysis

Using equation (8.25) to construct residual vector.

$$r_{k,L} = y_{k,L} - \varphi_z z_{k,L+p} = b_{k,L} + \varphi_f f_{k,L+p} + e_{k,L} \tag{8.26}$$

As innovation $e(k)$ is assumed to be a white Gauss signal in standard Kalman filter theory, then the statistical distribution of residual $r_{k,L}$ is given as

$$r_{k,L} \sim \begin{cases} N\left(Eb_{k,L}, \sigma_e^L\right) & \text{fault free} \\ N\left(Eb_{k,L} + \varphi_f f_{k,L+p}, \sigma_e^L\right) & \text{fault} \end{cases} \tag{8.27}$$

To compute the detailed structure of a residual vector, we choose $i = 1, \cdots L$ iteratively in output equation (8.23).

$$y(k-L+i) = C\phi^P \hat{x}(k-p-L+i) + e(k-L+i) + \left(\hat{E}_1 + \Delta\hat{E}_1\right)$$
$$\begin{bmatrix} z_{k-L+i-1,p} \\ u(k-L+i) \end{bmatrix} + \left(\hat{E}_2 + \Delta\hat{E}_2\right) \begin{bmatrix} f_{k-L+i-1,p} \\ f(k-L+i) \end{bmatrix} \tag{8.28}$$

where in equation (8.28), E_1 and E_2 are replaced by $\hat{E}_1 + \Delta\hat{E}_1$ and $\hat{E}_2 + \Delta\hat{E}_2$. As vector

$$\Delta\hat{E}_1 \begin{bmatrix} z_{k-L+i-1,p} \\ u(k-L+i) \end{bmatrix}$$

is a column vector, so it holds that

$$\Delta\hat{E}_1 \begin{bmatrix} z_{k-L+i-1,p} \\ u(k-L+i) \end{bmatrix} = vec\left(\Delta\hat{E}_1 \begin{bmatrix} z_{k-L+i-1,p} \\ u(k-L+i) \end{bmatrix} \right) \tag{8.29}$$

Using the property of Kronecker product operation, we have that

$$vec\left(\Delta\hat{E}_1 \begin{bmatrix} z_{k-L+i-1,p} \\ u(k-L+i) \end{bmatrix} \right) = \left(\begin{bmatrix} z_{k-L+i-1,p} & u(k-L+i) \end{bmatrix} \otimes I \right) vec\left(\Delta\hat{E}_1\right) \tag{8.30}$$

Similarly, it holds that

$$vec\left(\Delta\hat{E}_2 \begin{bmatrix} f_{k-L+i-1,p} \\ f(k-L+i) \end{bmatrix} \right) = \left(\begin{bmatrix} f_{k-L+i-1,p} & f(k-L+i) \end{bmatrix} \otimes I \right) vec\left(\Delta\hat{E}_2\right) \tag{8.31}$$

After multiplying pseudo-inverse matrix z_{id}^+ on both sides of equation (9), we obtain that

$$Y_{id} z_{id}^+ = C\phi^P x_{id} z_{id}^+ + E_1 + E_2 f_{id} z_{id}^+ + E_{id} z_{id}^+ \tag{8.32}$$

Computing the error matrix of Markov parameter matrix E_1, then we have

$$\Delta \hat{E}_1 = E_1 - \hat{E}_1 = C\phi^P x_{id} z_{id}^+ + E_2 f_{id} z_{id}^+ + E_{id} z_{id}^+ \tag{8.33}$$

Taking vectored operation on above equation, then

$$vec\left(\Delta \hat{E}_1\right) = \left(z_{id}^+ \otimes I\right) vec\left(E_{id}\right) + \left[z_{id}^+ \otimes \left(C\phi^P\right)\right] vec\left(x_{id}\right) + \left[z_{id}^+ \otimes E_2\right] vec\left(f_{id}\right) \tag{8.34}$$

Similarly, we get

$$vec\left(\Delta \hat{E}_2\right) = \left(f_{id}^+ \otimes I\right) vec\left(z_{id}\right) + \left[f_{id}^+ \otimes \left(C\phi^P\right)\right] vec\left(x_{id}\right) + \left[f_{id}^+ \otimes E_1\right] vec\left(z_{id}\right) \tag{8.35}$$

Combining $y(k - L + i)$, $i = 1 \cdots L$ to be one vector $y_{k,L}$ and substituting all equities to give the residual structure.

$$r_{k,L} = \left(z_0^T \otimes I\right)\left(z_{id}^+ \otimes I\right) vec\left(E_{id}\right) + \left(z_0^T \otimes I\right)\left[z_{id}^+ \otimes \left(C\phi^P\right)\right] vec\left(x_{id}\right) + b_{k,L} + e_{k,L} \tag{8.36}$$

where data matrix z_0 is represented as follows:

$$z_0 = \begin{bmatrix} z_{k-L,p} & z_{k-L+1,p} & \cdots & z_{k-1,p} \\ u(k-L+1) & u(k-L+2) & \cdots & u(k) \end{bmatrix} \tag{8.37}$$

In the above mathematical derivation for a residual vector, equation (8.36) signifies the statistical distribution property of residual vector $r_{k,L}$, and the detailed structure of residual l vector $r_{k,L}$ is given in equation (8.37).

8.2.5 Nearest neighbor gradient algorithm for subspace predictive control

As subspace predictive control belongs to predictive control [7], so future control input is solved as an optimal solution. Assume the expected output trajectory is priori known as

$$\gamma_{k,L} = \left[\gamma_{k-L+1}^T \cdots \gamma_k^T\right]^T \tag{8.38}$$

The common used quadratic objective function is given in predictive control field.

$$J_1\left(u_{k,L}\right) = \left(\hat{y}_{k,L} - \gamma_{k,L}\right)^T Q_1 \left(\hat{y}_{k,L} - \gamma_{k,L}\right) + u_{k,L}^T R u_{k,L} \tag{8.39}$$

Two matrices Q_1 and R are positive definite weight matrices, and the decision variables are combined as follows:

$$u_{k,L} = \left[u^T \left(k - L + 1 \right) \quad \cdots \quad u^T \left(k \right) \right]^T \tag{8.40}$$

In objective function (8.39), only the second term is the explicit function of future control input, and the first term includes $u_{k,L}$ implicitly. To expand equation (8.39), we expand the first term as an explicit function about $u_{k,L}$. From equation (8.15), the predictions with respect to measured sequence at future time instant are derived.

$$\hat{y}_{k,L} = H_z^{L,p} z_{k-L,p} + \left[H_f^{L,p} \quad T_f^L \right] f_{k,L+p} + T_u^L z_{k,L} \tag{8.41}$$

Above $z_{k-L,p}$ includes past input-output measured data sequence, but $z_{k-L,p}$ also includes future control input $u_{k,L}$, so we rewrite the above output predictions as follows:

$$
\begin{bmatrix} \hat{y}(k-L+1) \\ \hat{y}(k-L+2) \\ \vdots \\ \hat{y}(k) \end{bmatrix}
= H_z^{L,p} z_{k-L,p} +
\begin{bmatrix} 0 & 0 & \cdots & 0 \\ K & 0 & \cdots & 0 \\ \vdots & & \vdots & \\ C\phi^{L-2}K & C\phi^{L-3}K & \cdots & 0 \end{bmatrix}
\begin{bmatrix} \hat{y}(k-L+1) \\ \hat{y}(k-L+2) \\ \vdots \\ \hat{y}(k) \end{bmatrix}
$$

$$
+ \begin{bmatrix} D & 0 & \cdots & 0 \\ C\tilde{B} & 0 & \cdots & 0 \\ \vdots & & \vdots & \\ C\phi^{L-2}\tilde{B} & C\phi^{L-3}\tilde{B} & \cdots & D \end{bmatrix}
\begin{bmatrix} u(k-L+1) \\ u(k-L+2) \\ \vdots \\ u(k) \end{bmatrix}
+ \left[H_f^{L,p} \quad T_f^L \right] f_{k,L+p} \tag{8.42}
$$

Transferring and reformulating terms to give

$$
\begin{bmatrix} 1 & 0 & \cdots & 0 \\ K & 1 & \cdots & 0 \\ \vdots & & \vdots & \\ -C\phi^{L-2}K & -C\phi^{L-3}K & \cdots & 1 \end{bmatrix}
\begin{bmatrix} \hat{y}(k-L+1) \\ \hat{y}(k-L+2) \\ \vdots \\ \hat{y}(k) \end{bmatrix}
$$

$$
= H_z^{L,p} z_{k-L,p} +
\begin{bmatrix} D & 0 & \cdots & 0 \\ C\tilde{B} & 0 & \cdots & 0 \\ \vdots & & \vdots & \\ C\phi^{L-2}\tilde{B} & C\phi^{L-3}K & \cdots & D \end{bmatrix}
\begin{bmatrix} u(k-L+1) \\ u(k-L+2) \\ \vdots \\ u(k) \end{bmatrix}
$$

$$
+ \left[H_f^{L,p} \quad T_f^L \right] f_{k,L+p} \tag{8.43}
$$

So the explicit relation between future output predictions and future control input $u_{k,L}$ is obtained.

$$
\hat{y}_{k,L} =
\begin{bmatrix}
1 & 0 & \cdots & 0 \\
K & 1 & \cdots & 0 \\
\vdots & & & \vdots \\
-C\phi^{L-2}K & -C\phi^{L-3}K & \cdots & 1
\end{bmatrix}^{-1}
$$

$$
\left(
H_z^{L,p} z_{k-L,p} +
\begin{bmatrix}
D & 0 & \cdots & 0 \\
C\tilde{B} & 0 & \cdots & 0 \\
\vdots & & & \vdots \\
C\phi^{L-2}\tilde{B} & C\phi^{L-3}K & \cdots & D
\end{bmatrix}
u_{k,L} +
\begin{bmatrix} H_f^{L,p} & T_f^L \end{bmatrix} f_{k,L+p}
\right)
$$

$$\tag{8.44}$$

Introducing three matrices to simplify the above equation.

$$
\hat{y}_{k,L} = \Lambda_1 z_{k-L,p} + \Lambda_2 u_{k,L} + \Lambda_3 f_{k,L+p} \tag{8.45}
$$

The construction of three matrices Λ_1, Λ_2, Λ_3 is to multiply some certain matrix on $z_{k-L,p}$, $u_{k,L}$, $f_{k,L+p}$ respectively. After substituting equation (8.45) into objective function, one following optimization problem is established.

$$
\min_{u_{k,L}} J_1\left(u_{k,L}\right) = \left(\Lambda_1 z_{k-L,p} + \Lambda_2 u_{k,L} + \Lambda_3 f_{k,L+p} - \gamma_{k,L}\right)^T
$$

$$
Q_1\left(\Lambda_1 z_{k-L,p} + \Lambda_2 u_{k,L} + \Lambda_3 f_{k,L+p} - \gamma_{k,L}\right) + u_{k,L}^T R u_{k,L} \tag{8.46}
$$

Reformulating above equation to get

$$
\min_{u_{k,L}} J_1\left(u_{k,L}\right) = u_{k,L}^T \left(\Lambda_2^T Q_1 \Lambda_2 + R\right) u_{k,L}
$$

$$
+ 2\left(z_{k-L,p}^T \Lambda_1 Q_1 \Lambda_2 + f_{k,L+p}^T \Lambda_3^T Q_1 \Lambda_2 - \gamma_{k,L}^T Q_1 \Lambda_2\right) u_{k,L} \tag{8.47}
$$

To solve an optimization problem about predictive controller with equality and inequality constrain conditions, the dual decomposition is used to convert the formerly constrained optimization into an unconstrained optimization.

To rewrite optimization problem (8.47) as its more general form, let

$$
\Lambda_2^T Q_1 \Lambda_2 + R = \frac{1}{2}H; \quad 2\left(z_{k-L,p}^T \Lambda_1 Q_1 \Lambda_2 + f_{k,L+p}^T \Lambda_3^T Q_1 \Lambda_2 - \gamma_{k,L}^T Q_1 \Lambda_2\right) = g^T \tag{8.48}
$$

One regularization term is added in equation (8.48), and its advantage is to guarantee the decision variables not abrupt the whole optimization process, i.e. one constrained optimization problem is given as follows:

$$
\min_{u_{k,L}} J_2\left(u_{k,L}\right) = \frac{1}{2}u_{k,L}^T H u_{k,L} + g^T u_{k,L} + \lambda \left\| M u_{k,L} - m \right\|_1
$$

$$
\text{subject to} \quad A_1 u_{k,L} = B_1, \quad A_2 u_{k,L} \leq B_2 \tag{8.49}
$$

Introducing Lagrange multiplier vector μ_1, μ_2, μ_3 to obtain the Lagrange function of the above dual problem.

$$\sup_{\mu_1,\mu_2 \geq 0,\mu_3} \inf_{u_{k,L},\alpha} \left\{ \begin{array}{l} \frac{1}{2}u_{k,L}^T H u_{k,L} + g^T u_{k,L} + \lambda\|\alpha\|_1 + \mu_1\left(A_1 u_{k,L} - B_1\right) \\ + \mu_2\left(A_2 u_{k,L} - B_2\right) + \mu_3\left(M u_{k,L} - m - \alpha\right) \end{array} \right\} \qquad (8.50)$$

Reformulating above Lagrange function to get

$$\sup_{\mu_1,\mu_2 \geq 0,\mu_3} \left\{ \begin{array}{l} \inf_{u_{k,L}}\left[\left(A_1\mu_1 + A_2\mu_2 + M\mu_3 + g\right)u_{k,L} + \frac{1}{2}u_{k,L}^T H u_{k,L} \right] \\ -\mu_1 B_1 - \mu_2 B_2 - \mu_3 m + \inf_{\alpha}\left[\lambda\|\alpha\|_1 - \mu_3\alpha \right] \end{array} \right\} \qquad (8.51)$$

Firstly taking infimum operation on α

$$\inf_{\alpha}\left[\lambda\|\alpha\|_1 - \mu_3\alpha \right] = \inf_{\alpha}\left\{ \sum_i\left(\lambda|[\alpha]_i| - [\alpha]_i[\mu_3]_i\right) \right\}$$

$$= \sum_i\left\{ \inf_{[\alpha]_i}\left(\lambda|[\alpha]_i| - [\alpha]_i[\mu_3]_i\right) \right\} = \begin{cases} 0 & \text{if } \|\mu_3\|_\infty \leq \lambda \\ -\infty & \text{else} \end{cases} \qquad (8.52)$$

where $[\ ]_i$ denotes the element of an vector, to take minimization operation on future control input $u_{k,L}$, set

$$A = \begin{bmatrix} A_1 & A_2 & M \end{bmatrix}^T ; B = \begin{bmatrix} B_1 & B_2 & m \end{bmatrix}^T ; \mu = \begin{bmatrix} \mu_1 & \mu_2 & \mu_3 \end{bmatrix}^T \qquad (8.53)$$

Then minimizing objective function (8.53) on future control input $u_{k,L}$, we have

$$\inf_{u_{k,L}}\left[\left(A^T\mu + g\right)^T u_{k,L} + \frac{1}{2}u_{k,L}^T H u_{k,L} \right] = -\frac{1}{2}\left(A^T\mu + g\right)^T H^{-1}\left(A^T\mu + g\right) \qquad (8.54)$$

The dual problem is that

$$\sup_{\mu}\left\{ -\frac{1}{2}\left(A^T\mu + g\right)^T H^{-1}\left(A^T\mu + g\right) - B^T\mu \right\} \qquad (8.55)$$

Define negative dual function as

$$f(\mu) = \frac{1}{2}\left(A^T\mu + g\right)^T H^{-1}\left(A^T\mu + g\right) + B^T\mu \qquad (8.56)$$

As $f(\mu)$ includes quadratic term and linear term coming from the semi-definite Hessian matrix, then the gradient of $f(\mu)$ is computed as

$$\nabla f(\mu) = AH^{-1}\left(A^T\mu + g\right) + B \qquad (8.57)$$

From min-max theorem in convex optimization theory [9], the minimum Lipschitz constant ζ of $\nabla f(\mu)$ is that

$$\zeta = \left\| AH^{-1}A \right\|_2 \tag{8.58}$$

Nearest neighbor gradient algorithm is used to solve the dual problem (8.56), and its iterative process is given as follows:

$$v^t = \mu^t + \frac{t-1}{t-2}\left(\mu^t - \mu^{t-1}\right); \mu^{t+1} = P_\mu\left(v^t - \frac{1}{\zeta}\nabla f\left(\mu^t\right)\right) \tag{8.59}$$

where t denotes the iteration step, P_μ is Euclidean projection operation of μ. New iteration value μ^{t+1} is a negative gradient projection on basis of the last iteration value. Control input of the optimization problem is set

$$u_{k,L}^t = H^{-1}\left(-A^T \mu^t - g\right) \tag{8.60}$$

substituting v^t into $\nabla f(\mu)$ to get

$$\nabla f\left(v^t\right) = -A\left(u_{k,L}^t + \frac{t-1}{t+2}\left(u_{k,L}^t - u_{k,L}^{t-1}\right)\right) + B \tag{8.61}$$

Set

$$\bar{u}_{k,L}^t = u_{k,L}^t + \frac{t-1}{t+2}\left(u_{k,L}^t - u_{k,L}^{t-1}\right), \mu = \begin{bmatrix} \mu_1 & \mu_2 & \mu_3 \end{bmatrix}^T \tag{8.62}$$

Parallel processing equation (8.62) to give

$$u_{k,L}^t = H^{-1}\left(-A^T \mu^t - g\right), \bar{u}_{k,L}^t = u_{k,L}^t + \frac{t-1}{t+2}\left(u_{k,L}^t - u_{k,L}^{t-1}\right)$$

$$\mu_1^{t+1} = \mu_1^t + \frac{t-1}{t+2}\left(\mu_1^t - \mu_1^{t-1}\right) + \frac{1}{\zeta}\left(A_1^T \bar{u}_{k,L}^t - B_1\right)$$

$$\mu_2^{t+1} = \max\left\{0, \mu_2^t + \frac{t-1}{t+2}\left(\mu_2^t - \mu_2^{t-1}\right) + \frac{1}{\zeta}\left(A_2^T \bar{u}_{k,L}^t - B_2\right)\right\}$$

$$\mu_3^{t+1} = \min\left\{\lambda, \max\left\{-\lambda, \mu_3^t + \frac{t-1}{t+2}\left(\mu_3^t - \mu_3^{t-1}\right) + \frac{1}{\zeta}\left(M^T \bar{u}_{k,L}^t - m\right)\right\}\right\} \tag{8.63}$$

8.2.6 Conclusion

In this section, the problem of subspace predictive control is considered under fault conditions, and the statistical distribution is derived for a residual vector. One nearest neighbor gradient algorithm is applied to obtain a predictive controller by solving one optimization problem with equality and inequality constraints. The asymptotic analysis of the nearest neighbor gradient algorithm in subspace predictive control is our next subject.

8.3 Subspace data driven control for linear parameter varying system

Subspace data driven control is a novel direct data driven control strategy from the subspace system identification of state space forms. Specifically, subspace system identification is to apply the past and future input-output data sequence to identify each system matrix existing in the original state space form, i.e. the singular value decomposition corresponding to the data matrix is implemented to obtain each system matrix. It is well known that subspace data driven control combines the common properties between direct data driven control strategy and subspace system identification, so that the one sequence of output predictions at different time instants are constructed based on the collective past and future input-output data sequence. After comparing these output predictions and their expected or desired output values, a cost function is established within the framework of the data driven idea. Consider this cost function with equality or inequality constraints as one constrain optimization problem, the future control input sequence is yielded through some minimization operations, and then the first element from this future control input sequence is chosen as the start point for the next minimization operation. This iterative process is called the roll horizon process, which is more widely studied and applied in some practices. As the subspace data driven data control strategy applies the past input-output observed data sequence to construct the future output predictions at some different future time instants directly without modeling all the system matrices for the original state space form, this nice subspace data driven control simplifies the three main steps for the classical linear Gaussian regulator, i.e. identification, filtering and control, and this is the main reason why in recent years, lots of research concern on this direct data driven control strategy.

Although lots of research on direct data driven control strategy exist in recent years, few references are seen about our considered subspace data driven control, which only is suited for state space forms. For example, reference [9] constructs the basic and important output prediction value from the point of subspace system identification, and the constructed output prediction value is compared with the other output prediction value within the iterative correlation control strategy in [10]. Reference [11] proposes the Bayesian framework for the subspace data driven control strategy, then the ellipsoidal optimization algorithm is proposed to solve that cost function for the subspace data driven control strategy, whose optimization problem corresponds to one uncertain optimization problem in [12]. Furthermore, the level set from convex analysis theory is applied to construct the original ellipsoid for the above mentioned ellipsoidal optimization algorithm. Considering the realization problem for this subspace data driven control strategy, reference [13] applies it in vibration suppression of active noise, which effectively suppresses the flutter of small helicopters in the hovering state. Reference [14] uses a fast gradient algorithm to estimate faults under the constraints of upper and lower fault limits. The overall framework of subspace data driven control is shown in Fig. 8.1, where the controller is designed using subspace data control.

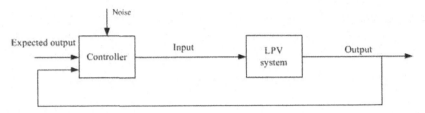

Figure 8.1: Structure of subspace data driven control

The models that appear in nature and industrial mass production are all nonlinear models. Due to the complexity of nonlinear models and their unknown forms, nonlinear system identification and nonlinear control are still under investigation. A common method for nonlinear model analysis is to use the Taylor series method to expand the original nonlinear system at a certain operating point or equilibrium stable point, and to approximately replace the original nonlinear system with a linear model that ignores high-order terms. The disadvantage of this linearization method is that when a certain operating point is changing, the linearization model is also changing all the time. To make up for the shortcomings of the linearization method, a compromise model is sought between the nonlinear model and the linear model. Although this compromise model is a nonlinear model, it also has the characteristics of a linear model. In the process of modeling a certain motor, it is found that a time-varying scheduling parameter sequence needs to be introduced to reflect the time-varying matrix of the linear state space system. The time-varying model is called a linear parameter changing system in controller design. Reference [15] proposes a variety of identification strategies for the identification of linear parameter change systems. The identification method is designed based on the given form of the linear parameter change system, such as state space equation form or transfer function form. However, the closed-loop controller design of this system is still in and it is still realized by the three major processes of first identification, then filter design, and then controller parameter selection.

In this section, subspace data driven control is applied to the controller design for a linear parameter varying system with the scheduling parameters. The input-output observation data matrix is used to identify Markov parameters in the form of state space forms, and only the data matrix is used to represent the output prediction value at the future different time instants. Then the output prediction value is applied to the cost function in data driven control. For this quadratic cost function, a parallel distribution algorithm is used to solve its optimal control input value, and the iterative convergence of the parallel distribution algorithm is analyzed in detail.

8.3.1 Linear parameter varying system

Consider the following linear parameter varying system with the scheduling parameter.

$$
\begin{cases}
x(k+1) = \sum_{i=1}^{n_\mu} \mu_k^{(i)} \left(A^{(i)} x(k) + B^{(i)} u(k) + K^{(i)} e(k) \right) \\
y(k) = Cx(k) + e(k)
\end{cases}
\tag{8.64}
$$

where in equation (8.64), $x(k) \in R^n$, $y(k) \in R^l$, $u(k) \in R^m$ are state variable, input variable and control input variable respectively. Noise signal or innovation signal $e(k)$ is one white noise with zero mean and variance σ_e^2. $\mu_k^{(i)} \in R$ corresponds to the scheduling parameter or weight value for each local model, $\{A^{(i)}, B^{(i)}, K^{(i)}\}$ is the ith local model. n_μ is the total number of all local models, k is time instant for this control problem. When all the scheduling parameters are all equal to be number 1, i.e. $\mu_k^{(i)} \equiv 1$, then we define

$$
A = \sum_{i=1}^{n_\mu} A^{(i)}, B = \sum_{i=1}^{n_\mu} B^{(i)}, K = \sum_{i=1}^{n_\mu} K^{(i)}
$$

Then equation (8.64) is reduced to the following linear time invariant system:

$$
\begin{cases}
x(k+1) = Ax(k) + Bu(k) + Ke(k) \\
y(k) = Cx(k) + e(k)
\end{cases}
\tag{8.65}
$$

According to some related definitions from subspace system identification theory, the following time varying matrices are set as

$$
A_k = \sum_{i=1}^{n_\mu} \mu_k^{(i)} A^{(i)}, B_k = \sum_{i=1}^{n_\mu} \mu_k^{(i)} B^{(i)}, K_k = \sum_{i=1}^{n_\mu} \mu_k^{(i)} K^{(i)}
\tag{8.66}
$$

Innovation signal $e(k)$ existing in output equation (8.64) is rewritten as

$$
e(k) = y(k) - Cx(k)
\tag{8.67}
$$

Substituting equation (8.67) into state equation (8.64), we have

$$
x(k+1) = A_k x(k) + B_k u(k) + K_k e(k) = [A_k - K_k C] x(k) + B_k u(k) + K_k y(k)
\tag{8.68}
$$

Based on the existed notation from subspace system identification, set the stable observation matrix ϕ_k as that

$$
\phi_k = A_k - K_k C = \sum_{i=1}^{n_\mu} \mu_k^{(i)} \phi^{(i)} = \sum_{i=1}^{n_\mu} \mu_k^{(i)} \left(A^{(i)} - K^{(i)} C \right)
$$

$$
\phi^{(i)} = A^{(i)} - K^{(i)} C
\tag{8.69}
$$

Combining equation (8.64), (8.68) and (8.69), the closed form for that formal linear parameter varying system is changed as follows:

$$\begin{cases} \hat{x}(k+1) = \phi_k \hat{x}(k) + B_k u(k) + K_k y(k) \\ y(k) = C\hat{x}(k) + e(k) \end{cases} \tag{8.70}$$

It is different with the mathematical derivations from subspace data driven control for the linear time invariant system, that here the state transition matrix is one time varying form, denoting as that

$$\varphi_{j,k} = \phi_{k+j-1}\phi_{k+j-2}\cdots\phi_k = \prod_{\tau=0}^{j-1}\phi_{k+\tau}, \forall j \geq 1, \varphi_{0,k} = I \tag{8.71}$$

8.3.2   Subspace data driven control strategy

During the following system identification and controller design process, s and f are the number of past data and future data, N is the number of all observed data, t is the time instant for the identification problem. In order to use equation (8.70), we combine some observed data as one vector form, i.e. define the following some vector forms as

$$y_{[k,k+f)} = \left[y(k)^T \ y(k+1)^T \ \cdots y(k+f-1)^T \right]^T,$$

$$u_{[k,k+f)} = \left[u(k)^T \ u(k+1)^T \ \cdots u(k+f-1)^T \right]^T$$

$$e_{[k,k+f)} = \left[e(k)^T \ e(k+1)^T \ \cdots e(k+f-1)^T \right]^T,$$

$$z_{[k-s,k)} = \left[u(k-s)^T \ y(k-s)^T \ \cdots u(k-1)^T \ y(k-1)^T \right]^T$$

where $y_{[k,k+f)}$ represent the output values at future time instants, and $u_{[k,k+f)}$ are the control input signals at future time instants, i.e. the future control inputs. $e_{[k,k+f)}$ is innovative sequence, $z_{[k-s,k)}$ is one vector, being full of past input-output observation data at the past time instants.

Applying the time varying Markov parameters, and iterating the above equation (8.70), the future output predictions at future time instants are described as follows:

$$y_{[k,k+f)} = b_x + H_{s,z} z_{[k-s,k)} + \tau_{f,u} u_{[k,k+f)} + \tau_{f,y} y_{[k,k+f)} + e_{[k,k+f)} \tag{8.72}$$

where all matrices in above equation (8.72) are formulated as follows:

$$b_x = \begin{bmatrix} C\varphi_{s,k-s}\hat{x}(k-s) \\ C\varphi_{s,k-s+1}\hat{x}(k-s+1) \\ \vdots \\ C\varphi_{s,k-s+f-1}\hat{x}(k-s+f-1) \end{bmatrix}, H_{s,z} = \begin{bmatrix} E_0 \\ E_1 \\ \vdots \\ E_{f-1} \end{bmatrix}$$

$$\tau_{f,u} = \begin{bmatrix} 0 & & & & \\ CB_k & 0 & & & \\ C\varphi_{1,k+1}B_k & CB_{k+1} & 0 & & \\ \vdots & & & \ddots & \\ C\varphi_{f-2,k+1}B_k & C\varphi_{f-3,k+2}B_{k+1} & \cdots & CB_{k+f-2} & 0 \end{bmatrix},$$

$$\tau_{f,y} = \begin{bmatrix} 0 & & & & \\ CK_k & 0 & & & \\ C\varphi_{1,k+1}K_k & CK_{k+1} & 0 & & \\ \vdots & & & \ddots & \\ C\varphi_{f-2,k+1}K_k & C\varphi_{f-3,k+2}K_{k+1} & \cdots & CK_{k+f-2} & 0 \end{bmatrix}$$

where Markov parameter is defined as follows, for $i = 0 \cdots f-1$, it holds that

$$E_i = \begin{bmatrix} 0 \cdots 0, C\varphi_{s-1,k-s+i+1}B_{k-s+i}, C\varphi_{s-1,k-s+i+1}K_{k-s+i} \cdots C\varphi_{i,k}B_{k-1}, C\varphi_{i,k}K_{k-1} \end{bmatrix}$$

In case of the internal stability for equation (8.64), then

$$\left\| \hat{x}(k-s+i) \right\|_2 < \infty, i = 0 \cdots f-1$$

If number or time horizon s is sufficiently large, based on the stability of that time varying transition matrix, it guarantees that

$$\left\| \varphi_{s,\tau} \right\|_2 < \varepsilon \ll 1, \forall \tau \geq 0$$

where ε is a small positive number, and biased term b_x is neglected.

Set the scheduling parameter vector be that

$$\mu_k = \begin{bmatrix} \mu_k^{(1)} & \cdots & \mu_k^{(n_\mu)} \end{bmatrix}^T$$

For $i = 1 \cdots f-1, j = 1 \cdots s$, define the following matrices.

$$P_{j/k+i-j} = \mu_{k+i-1} \otimes \mu_{k+i-2} \otimes \quad \otimes \mu_{k+i-j} \otimes I_{m+l} = \left(\overset{k+i-1}{\underset{=k+i-j}{\otimes}} \mu \right) \otimes I_{m+l}$$

$$P_{j\ k+i-j} = \left(\overset{k+i-1}{\underset{\tau=k+i-j}{\otimes}} \mu_\tau \right) \otimes I_m \quad P_{j\ k+i-j} = \left(\overset{k+i-1}{\underset{\tau=k+i-j}{\otimes}} \mu_\tau \right) \otimes I_l$$

where in above equation $\otimes$ is matrix direct product operation.

Collecting the n_j possible multiply operations of the local parameter matrices, we get

$$\left(\prod_{\tau=1}^{j-1} \phi^{(n_\tau)} \right) \gamma^{(n_j)} = \left(\prod_{\tau=1}^{j-1} \phi^{(n_\tau)} \right) \left[B^{(n_j)} \quad K^{(n_j)} \right] \tag{8.73}$$

where $\phi^{(n_\tau)}, \gamma^{(n_j)}$ and index n_j are defined as follows respectively:

$$\phi^{(n_\tau)} = A^{(n_\tau)} - K^{(n_\tau)}C, \quad \gamma^{(n_j)} = \left[B^{(n_j)} \quad K^{(n_j)} \right]$$

$$n_1 \cdots n_{j-1} \, n_j \in \left\{ 1 \cdots n_\mu \right\}$$

Then recursively compute to yield that

$$L_j = \left[\phi^{(1)} L_{j-1} \quad \phi^{(2)} L_{j-1} \quad \cdots \quad \phi^{(n_\mu)} L_{j-1} \right], L_1 = \left[\gamma^{(1)} \quad \gamma^{(2)} \quad \cdots \quad \gamma^{(n_\mu)} \right] \tag{8.74}$$

Using our constructed all vectors and matrices, the time varying Markov parameter in equation (8.72) is rewritten as the following explicit form with the scheduling parameter vector:

$$C\varphi_{j-1,k+i-j+1}\gamma_{k+i-j} = CL_j P_{j/k+i-j}^z, \gamma_{k+i-j} = \left[B_{k+i-j} \quad K_{k+i-j} \right] \tag{8.75}$$

The purpose of subspace system identification for linear parameter varying time with scheduling parameters is to estimate the Markov parameters in (8.72), so we collect the past input-output observation data sequence and scheduling parameter to establish the following information matrix

$$z_{k-s}^{2k+f-2} = \left[N_{k-s}^s z_{[k-s,k)}, N_{k-s+1}^s z_{[k-s+1,k+1)}, \cdots N_{2k+f-s-1}^s z_{[2k+f-s-1,2k+f-1)} \right] \tag{8.76}$$

where matrix N_{k-s+i}^s is defined as that

$$N_{k-s+i}^s = \begin{bmatrix} P_{s/k+i-s}^z & & & \\ & P_{s-1/k+i-s+1}^z & & \\ & & \ddots & \\ & & & P_{1/k+i-1}^z \end{bmatrix}$$

Based on equation (8.72), the future output prediction data sequence at future time instant is represented as that

$$y_{[k,k+f)} = E_0 z_{k-s}^{2k+f-2} + e_{[k,k+f)} \tag{8.77}$$

when matrix z_{k-s}^{2k+f-2} is a full row rank matrix, the least squares solution of Markov parameters is that

$$\hat{E}_0 = y_{[k,k+f)} \left(z_{k-s}^{2k+f-2} \right)^+ = \left[CL_s \quad \cdots \quad CL_1 \right] \tag{8.78}$$

where $\left(z_{k-s}^{2k+f-2} \right)^+$ in the above equation is the pseudo-inverse operation.

On the basis of the above least squares estimation $\hat{E}_0$, we can construct the output prediction value at a future time instant for our considered subspace data driven control strategy. After substituting the estimation in equation (8.78) into equation (8.72) and neglecting the biased term b_x or unknown innovative sequence $e_{[k,k+f)}$, we determine the deterministic future output prediction value at future time instant, i.e.

$$\hat{y}_{[k,k+f)} \approx H_{s,z}z_{[k-s,k)} + \tau_{f,u}u_{[k,k+f)} + \tau_{f,y}\hat{y}_{[k,k+f)} \tag{8.79}$$

Equation (8.72) is one closed observation form. As the unknown term $\hat{y}_{[k,k+f)}$ is included in both sides of equation (8.79), so we need rewrite the output prediction $\hat{y}_{[k,k+f)}$ as the other predictive form, being related with $u_{[k,k+f)}$ and $z_{[k-s,k)}$. Then the output prediction value for that linear parameter varying system with scheduling parameter is given as follows.

$$\hat{y}_{[k,k+f)} = \begin{bmatrix} \hat{y}(k) \\ \hat{y}(k+1) \\ \vdots \\ \hat{y}(k+f-1) \end{bmatrix} = \begin{bmatrix} \Gamma_0 \\ \Gamma_1 \\ \vdots \\ \Gamma_{f-1} \end{bmatrix} z_{[k-s,k)}$$

$$+ \begin{bmatrix} \Lambda_{1,1} & 0 & \cdots & 0 \\ \Lambda_{2,1} & \Lambda_{2,2} & \cdots & 0 \\ \vdots & \vdots & \vdots & \vdots \\ \Lambda_{f-1,1} & \Lambda_{f-1,2} & \cdots & \Lambda_{f-1,f-1} \end{bmatrix} u_{[k,k+f-1)}$$

$$= \Gamma z_{[k-s,k)} + \Lambda u_{[k,k+f-1)} \tag{8.80}$$

where parameters $\{\Gamma_i, \Lambda_i / i = 1 \cdots f-1, 1 \leq j < i\}$ can be generated recursively to be

$$\begin{cases} \Gamma_i = E_i + \sum_{\tau=0}^{i-1}\left(CL_{i-\tau}^y P_{i-\tau/k+\tau}^y\right)\Gamma_\tau, \Gamma_0 = E_0 \\ \Lambda_{i,j} = CL_{i-j+1}^u P_{i-j+1/k+j-1}^u + \sum_{\bar{\tau}=1}^{i-j}\left(CL_{i-j+1-\tau}^y P_{i-j+1-\tau/k+\tau+j-1}^y\right)\Lambda_{\tau+j-1,j} \\ \Lambda_{i,i} = CL_1^u P_{1/k+i-1}^u \end{cases} \tag{8.81}$$

By comparison, it can be seen that the estimated value of the output prediction of the linear parameter varying system is more complicated than the estimated value of the output prediction of the linear time-invariant system. Equation (8.80) is abbreviated as

$$\hat{y}_{[k,k+f)} = \Gamma z_{[k-s,k)} + \Lambda u_{[k,k+f-1)} \tag{8.82}$$

To establish the idea of subspace data driven control strategy, we assume the desired or expected output trajectory r is known, i.e.

$$r_{[k,k+f)} = \left[r^T(k) \quad r^T(k+1) \quad \cdots \quad r^T(k+f-1)\right]^T \tag{8.83}$$

The future control input sequence at a future time instant is designed through minimizing the following quadratic cost function or performance function.

$$J_k\left(u_{[k,k+f-1)}, \mu_{[k,k+f-1)}\right) = \left\|r_{[k,k+f)} - \hat{y}_{[k,k+f)}\right\|_Q^2 + \left\|\Delta u_{[k,k+f-1)}\right\|_{R_1}^2 + \left\|u_{[k,k+f-1)}\right\|_{R_2}^2$$

$$\tag{8.84}$$

where weighted matrices Q, R_1, R_2 are all positive matrices, and the second term is added in equation (8.84) to show the control input rate of change. The purpose is not to make large jumps in the control input sequence, and always maintain a smooth control input.

After expanding the second term to be that

$$\Delta u_{[k,k+f-1)} = S_\Delta u_{[k,k+f-1)} - S_{uz} z_{[k-s,k)}$$

$$S_\Delta = \begin{bmatrix} I_m & & & \\ -I_m & I_m & & \\ & & \ddots & \\ & & -I_m & I_m \end{bmatrix}, \quad S_{uz} = \begin{bmatrix} 0 & I_m & 0 \\ & \ddots & \\ 0 & 0 & 0 \end{bmatrix} \tag{8.85}$$

That cost function (8.84) satisfies the following hard constrain condition.

$$\begin{cases} \hat{y}_{[k,k+f)} = \Gamma z_{[k-s,k)} + \Lambda u_{[k,k+f-1)} \\ A\left[u_{[k,k+f-1)} \quad \hat{y}_{[k,k+f)}\right] \leq b \end{cases} \tag{8.86}$$

Consider the optimization problem with the cost function (8.84) and the equality or inequality constrain condition (8.86), the following parallel distribution algorithm is applied to solve its optimal solution.

8.3.3   Parallel distribution algorithm

Expanding the cost function (8.84) and neglecting those constant terms without control input, we have

$$J_k(u,\mu) = u^T\left[\Lambda^T Q\Lambda + S_\Delta R_1 S_\Delta + R_2\right]u + 2\left[(r - \Gamma z)^T Q\Lambda - (S_{uz}z)^T\right]u$$

$$= \frac{1}{2}u^T \Lambda_1 u - u^T b_1 \tag{8.87}$$

where during the above mathematical derivations, the first equality condition in equation (8.86) is used, then we rewrite the second inequality condition in equation (8.86) as that

$$[A_1 \quad A_2] \begin{bmatrix} u \\ \Gamma z + \Lambda u \end{bmatrix} \le b$$

i.e.

$$(A_1 + A_2\Lambda)u \le b - A_2\Gamma z$$

Simplifying it to be that

$$A_3 u \le c \tag{8.88}$$

Combining equation (8.87) and (8.88) to be one quadratic programming problem, which is formulated to be that

$$\min_{u} \frac{1}{2} u^T \Lambda_1 u - u^T b_1 \tag{8.89}$$

$$s.t.\ A_3 u \le c$$

The dual function of equation (8.89) is that

$$q(\lambda) = \inf \left\{ -u^T \Lambda_1 u - u^T b_1 - \lambda^T \left(A_3 u - c \right) \right\}$$

That decision variable λ in the above dual function is named the Lagrange multiply, and the dual function is minimized in the case of $u = \Lambda_1^{-1}(b_1 - A_3\lambda)$. After substituting this optimal value u into that dual function, we have

$$q(\lambda) = -\frac{1}{2} \lambda^T A_3 \Lambda_1^{-1} A_3 \lambda - \lambda^T c - A_3 \Lambda_1^{-1} b_1 - \frac{1}{2} b_1 \Lambda_1^{-1} b_1 \tag{8.90}$$

Through some basic variation transformations, the dual function for that quadratic programming problem is deem as that

$$\min_{\lambda} \frac{1}{2} \lambda^T M \lambda + \lambda^T d \tag{8.91}$$

$$s.t. \quad \lambda \ge 0, M = A_3 \Lambda_1^{-1} A_3, d = c - A_3 \Lambda_1^{-1} b_1$$

We see that if the optimal solution λ^* for that dual function is obtained, then the optimal solution for that original quadratic programming problem is that

$$u^* = \Lambda_1^{-1} \left(b_1 - A_3 \lambda^* \right) \tag{8.92}$$

The parallel distribution algorithm is proposed to solve that dual problem here. Set a_j as the jth column for matrix A_3, and all elements of a_j is not zero. If $a_j = 0$, it is not senseless due to constrain condition $a_j u \le c$. As weighted matrix Q is positive and definite, so the jth diagonal element of matrix M is that $M_{jj} = a'_j \Lambda \ a_j$ and it is positive too. It means that for each j, the dual cost function is strictly convex along the jth coordinate.

Taking the first-order partial derivative with respect to λ_j for that dual cost function, it holds that.

$$d_j + \sum_{k=1}^{m} M_{jk} \lambda_k \tag{8.93}$$

where M_{jk} and d_j are elements coming from matrix M and vector d respectively.

Set that above partial derivative be zero, the process of unconstrained minimization of the dual cost function from the initial point λ to reach $\tilde{\lambda}_j$ along the jth coordinate is that

$$\tilde{\lambda}_j = -\frac{1}{M_{jj}}\left(d_j + \sum_{k \neq j} M_{jk} \lambda_k\right) = \lambda_j - \frac{1}{M_{jj}}\left(d_j + \sum_{k=1}^{m} M_{jk} \lambda_k\right) \tag{8.94}$$

Due to nonnegative constraint $\lambda_j \geq 0$, the iterative form with the jth coordinate is updated as follows.

$$\lambda_j = \max\left\{0, \tilde{\lambda}_j\right\} = \max\left\{0, \lambda_j - \frac{1}{M_{jj}}\left(d_j + \sum_{k=1}^{m} M_{jk} \lambda_k\right)\right\} \tag{8.95}$$

Based on the first-order partial derivative (8.93) of the dual cost function with respect to λ_j, equation (8.95) can be adjusted as

$$\lambda_j(t+1) = \max\left\{0, \lambda_j(t) - \frac{\eta}{M_{jj}}\left(d_j + \sum_{k=1}^{m} M_{jk} \lambda_k(t)\right)\right\} \tag{8.96}$$

where $\eta > 0$ is the step size, and this recursive form is suited for parallel distribution algorithm.

From the numerical optimization theory, we see that in order to ensure that the iterative process of (8.96) can converge to its global minimum, the step size parameter η selected should be sufficiently small, that is, the convergence can be achieved for the special case, i.e. $\eta = \frac{1}{m}$.

8.3.4   Simulation

Our considered subspace data driven control of linear parameter varying system with scheduling parameter vector is applied to the modeling and controller design of DC motor. The DC motor can be approximated by a linear model while ignoring a variety of external factors. However, when the consideration of the mass distribution is added to the rotating disk, a nonlinear model is required to characterize it. The reconsideration graph of the DC motor mass is shown in Fig. 8.2.

The mass distribution on a homogeneous disk will become inhomogeneous. The mathematical description of the DC motor can be divided into two parts: the motor part and the mechanical part. Some calibration parameters in the DC motor are shown in Table 8.1. Firstly, we use Kirchhoff's voltage law to get

$$L_m \dot{I}(t) = v(t) - K_i \omega(t) - R_m I(t) \tag{8.97}$$

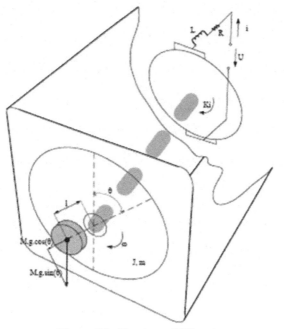

Figure 8.2: Structure of DC motor

where $I(t)$ is current intensity, $v(t)$ is control input voltage, and $\omega(t)$ is rotation velocity.

The variable relationship between the motor part and the mechanical part is that

$$\begin{cases} J\dot{\omega}(t) = K_i I(t) - b\omega(t) + Mgl\sin\big(\theta(t)\big) \\ \dot{\theta}(t) = \dot{\omega}(t) \end{cases} \tag{8.98}$$

where $\theta(t)$ is rotation angle.

When an appropriate observation parameter is chosen to replace the nonlinear term, equation (8.98) will become a linear parameter varying system.

Table 8.1: Nominal parameters in DC motor

Parameters	Number value
Motor torque constant	$K_i = 53.6 \times 10^{-3}$ Nm/A
Resistance	$R_m = 9.5\ \Omega$
Damping	$L_m = 0.84 \times 10^{-3}$ H
Disc inertia	$J = 2.2 \times 10^{-4}$ Nm^2
Friction coefficient	$b = 6.6 \times 10^{-5}$ Nms/rad
Extra quality	$M = 0.07$ kg
Mass distribution at the center of the disk	$l = 0.042$ m

Set the scheduling parameter as that

$$\mu(t) = \frac{\sin(\theta(t))}{\theta(t)}$$

When the rotation angle is measurable, the scheduling parameters can be estimated. Select the following state variables as that

$$x(t) = [\theta(t) \quad \omega(t) \quad I(t)]^T$$

The state space form of continuous time linear parameter change can be obtained, i.e.

$$\begin{bmatrix} \dot{\theta}(t) \\ \dot{\omega}(t) \\ \dot{I}(t) \end{bmatrix} = \left(\begin{bmatrix} 0 & 0 & 0 \\ 0 & -\dfrac{b}{J} & \dfrac{K_i}{J} \\ 0 & -\dfrac{K_i}{L_m} & -\dfrac{R_m}{L_m} \end{bmatrix} + \begin{bmatrix} 1 & 0 & 0 \\ \dfrac{Mgl}{J} & 0 & 0 \\ 0 & 0 & 0 \end{bmatrix} \mu(t) \right) \times \begin{bmatrix} \theta(t) \\ \omega(t) \\ I(t) \end{bmatrix} + \begin{bmatrix} 0 \\ 0 \\ \dfrac{1}{L_m} \end{bmatrix} v(t)$$

$$y(t) = \begin{bmatrix} 1 & 0 & 0 \\ 0 & 0 & 1 \end{bmatrix} \begin{bmatrix} \theta(t) \\ \omega(t) \\ I(t) \end{bmatrix} \tag{8.99}$$

Neglecting the fast motor system, then equation (8.99) is reduced to be a second order system.

$$\begin{bmatrix} \dot{\theta}(t) \\ \dot{\omega}(t) \end{bmatrix} = \left(\begin{bmatrix} 0 & 1 \\ 0 & -\dfrac{1}{\tau} \end{bmatrix} + \begin{bmatrix} 0 & 0 \\ \dfrac{Mgl}{J} & 0 \end{bmatrix} \mu(t) \right) \begin{bmatrix} \theta(t) \\ \omega(t) \end{bmatrix} + \begin{bmatrix} 0 \\ \dfrac{K_m}{\tau} \end{bmatrix} v(t) \tag{8.100}$$

For formula (8.100), the discretization method in linear system theory is used. According to the discretization process of the continuous linear time-invariant system, h is taken as the period of discrete use. The corresponding relationship between each system matrix between the continuous system and the discrete system is as follows:

$$\begin{bmatrix} \theta(k+1) \\ \omega(k+1) \end{bmatrix} = e^{h\left(\begin{bmatrix} 0 & 1 \\ 0 & -\frac{1}{\tau} \end{bmatrix} + \begin{bmatrix} 0 & 0 \\ \frac{Mgl}{J} & 0 \end{bmatrix} \mu(k) \right)} \begin{bmatrix} \theta(k) \\ \omega(k) \end{bmatrix} + \int_0^h e^{\tau\left(\begin{bmatrix} 0 & 1 \\ 0 & -\frac{1}{\tau} \end{bmatrix} + \begin{bmatrix} 0 & 0 \\ \frac{Mgl}{J} & 0 \end{bmatrix} \mu(t) \right)} d\tau \begin{bmatrix} 0 \\ \frac{K_m}{\tau} \end{bmatrix} v(k)$$

$$y(k) = \begin{bmatrix} 1 & 0 \\ 0 & 0 \end{bmatrix} \begin{bmatrix} \theta(k) \\ \omega(k) \end{bmatrix} \tag{8.101}$$

The exponential matrix in the above formula is expanded according to the existing calculation method in the linear system theory, and the linear parameter variation system such as equation (8.64) can be obtained. This complex expansion process can directly call the matrix calculation program in MATLAB. A pseudo-real-time simulation environment is established for the DC motor. The software driver of the motor is established under the real-time Microsoft Windows XP using the real-time MATLAB/Simulink environment. The communication protocol is realized through the USB interface, and the sampling frequency in the pseudo-real-time condition is set to 20 HZ. A series of observation data sets are collected for the DC motor module, and a reference trajectory is selected based on the given scheduling parameters to construct a calibration parameter sequence. The subspace data driven control method is used to realize the tracking of the position and speed of the DC motor. The entire closed-loop control experiment is initiated using data acquisition and offline physical parameter identification.

The least squares algorithm is applied to identify the Markov parameter, and at this time, some physical parameters are chosen as $N = 2000$, $T_s = 0.05$ s. When exciting the scheduling parameter, it must satisfy that

$$\lim_{\theta \to 0} \left\| \mu(t) A_1 \right\|_2 = 0$$

If the above equation is not satisfied, the original system will lose nonlinear performance. The reference selection of the rotational speed should ensure that the scheduling parameters $\mu(t)$ are excited sufficiently continuously. Based on the idea of data driven for future output prediction estimates (8.82) and possible hard constraints (8.86), the cost function is minimized with respect to the future control input sequence. The rolling time domain strategy is used for the obtained control input sequence, i.e., only the first element is selected each time as the construction basis of the predicted value in the next optimization process. For the optimization problem at each sampling instant, the future time-domain level is chosen as $f = 3$, and each weighted matrices are $Q = 100$ I, $R_1 = 0.01$ I, $R_2 = 0.1$ I. That hard constraint about the control input is -7 $V \le v \le 7$ V.

The simulation results using subspace data driven control for this DC motor are shown in Figs. 8.3-8.5. Figure 8.3 shows the position tracking curve of the DC motor under the action of the subspace data driven controller, where the blue curve is the given reference trajectory, and the red is the actual controller output curve. It can be seen from Fig. 8.3 that there is a large deviation between the two curves at the initial moment, and as time goes on, the deviation between the two curves will gradually decrease, and an approximate effect can be achieved. Figure 8.4 shows the corresponding DC motor position tracking error curve, which can also better explain the fitting effect of Fig. 8.3. Figure 8.5 shows the rotation speed response curve of the DC motor, which is a continuous smooth curve, then it indicates that the speed of the DC motor is completely determined by the designed subspace data driven controller.

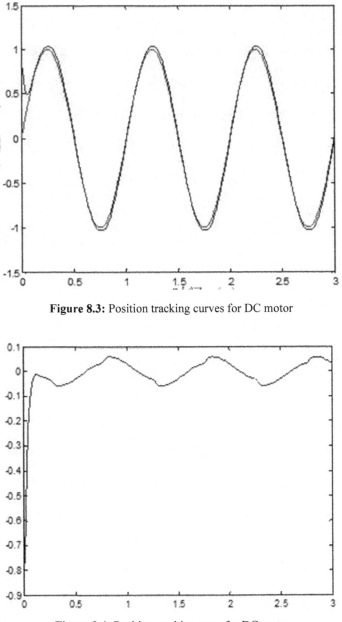

Figure 8.3: Position tracking curves for DC motor

Figure 8.4: Position tracking error for DC motor

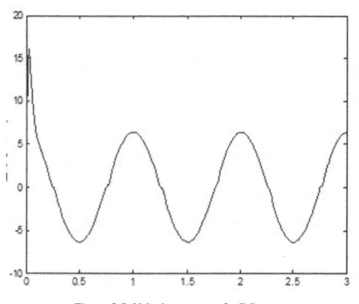

Figure 8.5: Velocity response for DC motor

The DC motor shown in Fig. 8.2 is placed in the servo closed-loop control loop of the flight simulation turntable. Flight simulation is a follow-up servo mechanism, which can simulate various flight attitudes of the aircraft in the air. The servo control loop of the flight simulation turntable, as a high-precision servo device, puts forward high requirements for the position follow-up control: on the one hand, it is expected that the position tracking should not have overshot, and the dynamic response process should be fast and smooth; on the other hand, to ensure the tracking accuracy requires position tracking with a small steady-state error. In the servo control loop used to control the rotation of the flight simulation turntable, the DC motor uses the subspace predictive control strategy to control the rotation of the flight simulation turntable through the position signal fed back by the photoelectric encoder or the code disc.

To order to test the subspace predictive controller of the DC motor using the linear parameter change system, the unit step signal is selected as the input signal of the entire flight simulation turntable system, and the output of the system is simulated. The physical parameters of the DC motor still use the values in Table 8.1. We select the unit step signal with the input signal $r(t) = 1(t)$ in the upper computer, and simulate the system output and error. The simulation results are displayed on the lower computer. The simulation graph is shown in Fig. 8.6. It can be seen from the input signal response curve that when the system input is a unit step signal, the adjustment time of the system is less than 0.2 s, the output has no overshoot, and the steady-state error is less than 0.05, which is in line with the time domain index of the positioning system.

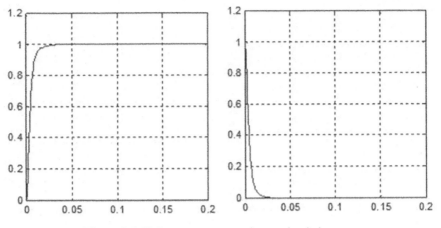

Figure 8.6: Unit step response and error simulation curve

8.3.5   Conclusion

Because the linear parameter varying system is a transition model of the nonlinear system and the linear system, it has both common properties and can approximately describe the real system. The subspace data driven control is applied to a linear parameter varying system with a scheduling parameter vector. According to the structural characteristics of the linear parameter varying system, the whole derivation process of the subspace data driven control is given, while introducing the vector product operator to simplify the derivation. A parallel distribution algorithm is used to solve the optimal control input value for the established cost function under the idea of data driven control. A linear parameter varying system considering the mass distribution factor is added to the DC motor, and the subspace prediction is applied to the control example of the DC motor.

8.4   Local polynomial method for frequency response function identification

The nonparametric frequency response function can not only provide an initial knowledge for more a complex system or plant, but also guide the selection of a model structure in estimating the model parameter for the researcher. The nonparametric estimation of the noise spectrum obtained in identifying the nonparametric frequency response function can be used to parameterize the frequency weighting of the considered input-output model, thereby the prediction identification criterion is simplified and the risk of local minimization is also reduced. Although frequency response estimation for the input-output transfer function can be obtained by taking the continuous or discrete Fourier transform of the input-output observed data with a finite number of the data points. However, the leakage error is inevitably introduced from taking that Fourier transform, as the leakage error in the frequency domain is equivalent to the transient error in the time domain. Leakage error is from the unknown past input signal and missed future output signal, i.e. the effects of these two signals will cause the leakage error. This leakage error largely hinders the accuracy estimation of the nonparametric frequency response function estimation with the condition of the random input signal. Leakage error can be alleviated by using a periodic input signal, but the selection of a periodic input signal is not advisable. Leakage error in frequency domain can be expressed as one additional rational term at the output. To reduce this great influence from the rational function, i.e. the leakage error, on the accuracy estimation of the frequency response function, one window function method can be used to achieve our goal. The use of the window function is based on the difference form of the input-output signal at some special frequency points. Although this window function method can reduce the influence of the leakage error, it also increases the interpolation error.

At present, few references exist about the frequency response function estimation, but they often exist in the field of mechanical vibration engineering. If the frequency response function is known in priori, then the model parameters of the mechanical vibration can be obtained through decomposing the frequency response function estimation. Reference [15] uses the subspace method to identify the output predictions of one state space model, which is replaced by our considered frequency response function. Then the model parameters of a beam under free vibration are obtained by using singular value decomposition. In [16], a periodic sinusoidal wave is applied as the input signal during the whole model parameter identification test. From the experiment results, we see this periodic sinusoidal signal can improve the accuracy estimation for the frequency response function and reduce the leakage error. In [17], the frequency response function is applied in the detection of structural damage, further, a new weighted digital filter is constructed to perform the Fourier transform on the input-output data sequence [18]. This new weighted digital filter can smooth the effect of colored noise. Generally, the frequency response function estimation can be divided into parameter estimation and nonparameter estimation. As the additional unknown parameters in parameter estimation, the solution of the unknown parameters will not be unique. On the contrary, as nonparameter estimation directly obtains the estimated value for the frequency response function, then it greatly simplifies the whole identification process. In [19], a transient impulse response modeling method is proposed to identify the frequency response function. This transient impulse response modeling method combines the approximation of the frequency response function and the leakage term as a function of the global parameters. Then these global parameters can be estimated by solving a global least squares problem.

Based on the transient impulse response method proposed in [20], in this chapter, we continue to study the identification problem of one nonparametric frequency response function. Here one new local polynomial method is proposed to estimate the nonparametric frequency function and the power spectrum of the external noise simultaneously. The Taylor series expansion is used to expand the frequency response function and the leakage term around a narrow window at a certain center frequency, then two local polynomial models are get. The least squares method is applied to solve the unknown coefficients of these two local polynomials based on the input-output data sequence. The least squares estimations corresponding to the local polynomial coefficients can give an estimate of the considered frequency response function at a specified frequency, while also reducing the leakage error and transient error. By comparing our local polynomial method and transient impulse response method in [20], we observe that although the two methods are based on the approximation of the considered frequency response function and the leakage term, the leakage term here is around the local bandwidth at each frequency point. The estimations of these local parameters are identified by solving the N local least squares problems, where N is the number of total data points. The advantage of our proposed local polynomial method over the transient impulse response method is that the leakage error depends on the system and the record length closely, and its negative effect is reduced by applying multi-level amplitude. Since more information will be a benefit for the estimation of each polynomial coefficient, so to reduce the variance error of the estimated coefficient, some constraint conditions about the local parameter variables are introduced in our proposed local polynomial method. Then a simple optimization method is used to solve this optimization problem, such as the first order gradient method, Newton method, etc.

8.4.1　Problem description

Consider one linear time invariant system $G(q)$, and this system is excited by one known random input signal $u(t)$. The output of this system is expressed as one sum form of the input signal $u(t)$ and one disturb term $v(t)$. Set $u(t)$ and $v(t)$ be two quasi-stationary signals, it means that disturb term $v(t)$ can be modeled as one output by driving one white noise $e(t)$ through a shaping filter. So the relation between input-output is written as

$$y(t) = G(q)u(t) + v(t) = G(q)u(t) + H(q)e(t) \qquad (8.102)$$

where q is one time shift operator, $G(q)$ and $H(q)$ are two causal rational functions about the time shift operator q. $e(t)$ is an white noise with zero mean value and variance σ_e^2.

The transfer function form in equation (8.102) can be rewritten as the following state space form from the minimum state space theory of linear system:

$$\begin{cases} x(t+1) = Ax(t) + Bu(t) + Ke(t) \\ y(t) = Cx(t) + e(t) \end{cases} \qquad (8.103)$$

From the basic modern control theory, we see that the rational function $G(q)$ is equivalent to its state space realization (A, B, C), i.e. the relation between them exists.

$$G(q) = C(qI - A)^{-1} B$$

The problem of finding matrices (A, B, C) for the rational function $G(q)$ corresponds to the realization problem for one rational function. When the system is controllable and observable, it is equivalence and uniqueness between the transfer function form (8.102) and the state space form (8.103). Many vibration response equations in the field of vibration engineering can be transformed into equations (8.102) or (8.103), such as spring-mass-damping systems, etc. As equation (8.13) is one linear model, so after linearizing the nonlinear system, then the nonlinear system can be reduced to this linear model through the Taylor series expansion method.

Here we give a partial fraction expansion method to realize the given rational transfer function, for example, set that rational transfer function as $G(q) = \dfrac{n(q)}{d(q)}$, where $n(s)$ and $d(q)$ are two polynomials. Furthermore, we factorize the polynomial $d(q)$ as that

$$d(q) = (q - \lambda_1)(q - \lambda_2) \cdots (q - \lambda_r)$$

where $\lambda_1, \lambda_2 \cdots \lambda_r$ are roots for polynomial $d(q)$.

Then $G(q) = \dfrac{n(q)}{d(q)}$ has one partial fraction expansion as follows:

$$G(q) = \frac{n(q)}{d(q)} = \sum_{i=1}^{r} \frac{w_i}{q - \lambda_i}$$

Set B, C as two constant matrices and they satisfy that

$$w_i = c_i b_i$$

Then one state space realization (A, B, C) for that rational transfer function $G(q)$ is given as follows:

$$A = \begin{bmatrix} \lambda_1 & & \\ & \ddots & \\ & & \lambda_r \end{bmatrix}, B = \begin{bmatrix} b_1 \\ \vdots \\ b_r \end{bmatrix}, C = \begin{bmatrix} c_1 & \cdots & c_r \end{bmatrix}$$

After collecting lots of observed data for the input-output model (8.102), for $t = 1, 2 \cdots N$, consider the existence of initial condition terms and transient terms in equation (8.102), then denote them as τ_G and τ_H respectively.

$$y(t) = G(q)u(t) + \tau_G(t) + H(q)e(t) + \tau_H(t) \tag{8.104}$$

Taking discrete Fourier transform on both sides of equation (8.104) to obtain

$$X(k) = \frac{1}{\sqrt{N}} \sum_{t=0}^{N-1} x(t) e^{-j\frac{2\pi kt}{N}}$$

The exact frequency domain expression for equation (8.104) is that

$$Y(k) = G(\Omega_k)U(k) + T_G(\Omega_k) + H(\Omega_k)E(k) + T_H(\Omega_k) \qquad (8.105)$$

where index k is the frequency point $\dfrac{kf_s}{N}$, f_s is the sample frequency, $\Omega_k = e^{-j\frac{2\pi kt}{N}}$.

Set $\omega_k = \dfrac{2\pi k}{N}$, then Ω_k is simplified as $\Omega_k = e^{-j\omega_k}$.

Two leakage error terms T_G and T_H are two rational forms about z^{-1}, and they are also two smooth functions of frequency.

Define

$$\tau(t) = \tau_G(t) + \tau_H(t)$$

Similarly, its discrete Fourier transform is given as

$$T(\Omega_k) = T_G(\Omega_k) + T_H(\Omega_k) \qquad (8.106)$$

Substituting equation (8.106) into (8.105) to get

$$\begin{cases} Y(k) = G(\Omega_k)U(k) + T(\Omega_k) + H(\Omega_k)E(k) \\ V(k) = H(k)E(k), \qquad k = 1, 2 \cdots N \end{cases} \qquad (8.107)$$

Formulating about equation (8.107) as the following proposition form.

Proposition 8.1: In the input-output equation (8.102), discrete Fourier transform relationship in the presence of initial condition term and transient term is obtained as

$$\begin{cases} Y_k = G_k U_k + T_k + H_k E_k \\ V_k = H_k E_k, \qquad k = 1, 2 \cdots N \end{cases}$$

If considering the initial condition and terminal condition $x(0)$ and $x(N)$ in equation (8.103), then the transient term $\tau(t)$ can be modified as one function form about one state space model. Its relations in time domain and frequency domain are given as

$$\tau(t) = CA^t(x_0 - x_N)$$

$$T_k = \frac{1}{\sqrt{N}} \sum_{t=0}^{N-1} \tau(t) e^{-j\omega_k t} = \frac{1}{\sqrt{N}} C\left(I - e^{-j\omega_k} A\right)^{-1}(x_0 - x_N) \qquad (8.108)$$

In equation (8.107) T_k and G_k are all smooth functions about frequency point, and equation (8.107) is one basic relation to identify the nonparametric functions for the frequency response function G_k and transient term T_k.

8.4.2 Local polynomial method

There are $2N$ unknown parameters $\{G_k, T_k, k = 1, 2 \cdots N\}$ in equation (8.107) but we have only N equations. It is impossible to identify $2N$ unknown parameters by using only N equations, we need to generate more equations to approximate T_k and G_k. This requirement can be achieved by applying the smooth function at each window of frequency Ω_k. These additional equations may include the unknown parameter G_k and other additional parameters in approximating T_k and G_k to construct one least squares problem. It means that our proposed local polynomial method establishes N local least squares problems for the local parameter vectors in each frequency window.

Assume we want to estimate T_k and G_k at frequency point ω_k, then consider data $\{Y_{k+r}, U_{k+r}, r = -L, \cdots 0 \cdots L\}$ around frequency point ω_k to construct

$$
\begin{aligned}
Y_{k+r} &= G_{k+r}U_{k+r} + T_{k+r} + V_{k+r} \\
&= G_k U_{k+r} + [G_{k+r} - G_k]U_{k+r} + T_{k+r} + V_{k+r}, r = -L, \cdots 0 \cdots L
\end{aligned}
\tag{8.109}
$$

Applying Taylor series expansion on difference $G_{k+r} - G_k$ and transient term T_{k+r} to obtain

$$
\begin{cases}
G_{k+r} = G_k + \sum_{s=1}^{\infty} g_s(k)r^s = G_k + \sum_{s=1}^{R} g_s(k)r^s + O\left(\left(\frac{r}{N}\right)^{R+1}\right) \\
T_{k+r} = T_k + \frac{1}{\sqrt{N}}\sum_{s=1}^{\infty} t_s(k)r^s = T_k + \frac{1}{\sqrt{N}}\sum_{s=1}^{R} t_s(k)r^s + \frac{1}{\sqrt{N}}O\left(\left(\frac{r}{N}\right)^{R+1}\right)
\end{cases}
\tag{8.110}
$$

where the following derivations are used in equation (8.110):

$$
G_{k+r} = G_k + \sum_{s=1}^{\infty} g_s(k)(\Omega_{k+r} - \Omega_k)^s = G(\Omega_{k+r}) - G(\Omega_k)
$$

$$
\Omega_{k+r} - \Omega_k = e^{-j\frac{2\pi(k+r)t}{N}} - e^{-j\frac{2\pi kt}{N}} = e^{-j\frac{2\pi kt}{N}}\left(e^{-j\frac{2\pi rt}{N}} - 1\right) = e^{-j\frac{2\pi kt}{N}}\left(e^{-j\frac{2\pi rt}{N}} - e^{0}\right)
$$

For that subtraction term $e^{-j\frac{2\pi rt}{N}} - e^{0}$, we apply the Taylor series expansion at $r = 0$ to get one form about r.

Observing that Taylor series expansion in equation (8.110) again, we use one R-order polynomial term to do the truncation. Value R determines the accuracy for the Taylor series expansion. It is common to choose R as $R = 3r$, then the obtained approximation error is tolerable, as the approximation error is determined by the infinite amount of the optimal one.

Furthermore, $R = 3r$ can guarantee that the infinite amount of the optimal one will approach to zero value.

Substituting equation (8.110) into (8.109) to get

$$Y_{k+r} = \left[G_k + \sum_{s=1}^{R} g_s(k) r^s \right] U_{k+r} + T_k + \frac{1}{\sqrt{N}} \sum_{s=1}^{R} t_s(k) r^s$$

$$+ V_{k+r} + \left[\sum_{s=R+1}^{\infty} g_s(k) r^s \right] U_{k+r} + \frac{1}{\sqrt{N}} \sum_{s=R+1}^{\infty} t_s(k) r^s \qquad (8.111)$$

Neglecting the high-order term, then we have

$$Y_{k+r} \approx \left[G_k + \sum_{s=1}^{R} g_s(k) r^s \right] U_{k+r} + T_k + \frac{1}{\sqrt{N}} \sum_{s=1}^{R} t_s(k) r^s + V_{k+r} \quad r = -L, \cdots 0 \cdots L$$

$$(8.112)$$

Combining $2(R + 1)$ unknown parameters to construct one local parameter vector

$$\theta_k = \left[G_k, g_1(k) \cdots g_R(k), T_k, t_1(k) \cdots t_R(k) \right]^T \qquad (8.113)$$

Define one vector with $2(L + 1)$ dimension as

$$\overline{Y}_{k,L} = \left[Y_{k-L}, Y_{k-L+1} \cdots Y_k \cdots Y_{k+L-1}, Y_{k+L} \right]^T,$$

$$\overline{U}_{k,L} = \left[U_{k-L}, U_{k-L+1} \cdots U_k \cdots U_{k+L-1}, U_{k+L} \right]^T \qquad (8.114)$$

Rewriting equation (8.112) as other vector form

$$\overline{Y}_{k,L} = K_{k,L}\left(R, \overline{U}_{k,L} \right) \theta_k + \overline{V}_{k,L} \qquad (8.115)$$

where in equation (8.115), $2(R + 1)$ unknown parameters θ_k are in local data $\{U_{k+r}, Y_{k+r}, r = -L, \cdots 0 \cdots L\}$, and there are $2(L + 1)$ equations. $K_{k,L}\left(R, \overline{U}_{k,L} \right)$ is a known matrix, which is related with vector $\overline{U}_{k,L}$.

Theorem 8.1: In our proposed local polynomial method, the unknown parameter vector $\hat{\theta}_k$ is identified through a least squares problem, i.e.

$$\min_{\theta_k} \left[\overline{Y}_{k,L} - K_{k,L}\left(R, \overline{U}_{k,L} \right) \theta_k \right]^T \left[\overline{Y}_{k,L} - K_{k,L}\left(R, \overline{U}_{k,L} \right) \theta_k \right] \qquad (8.116)$$

In the case of no constraint condition, by differentiating the cost function concerning the unknown parameter vector and by setting the derivative equal to zero, the parameter vector estimation $\hat{\theta}_k$ is obtained through some basic inverse matrix operation.

$$\hat{\theta}_k = \left[K_{k,L}\left(R, \overline{U}_{k,L} \right) K_{k,L}\left(R, \overline{U}_{k,L} \right)^T \right]^* \left[K_{k,L}\left(R, \overline{U}_{k,L} \right) \overline{Y}_{k,L} \right] \qquad (8.117)$$

where in equation (8.117) the pseudo-inverse operation is applied.

Comparing the above equation (8.113) and (8.117) in reference [20], we see the estimations of $[T_k, t_1(k) \cdots t_R(k)]^T$ are added in the defined unknown parameter vector, and these added parameter estimations can be used on the latter optimal signal design, so that the variance error may also be reduced.

8.4.3 Constrained local polynomial method

In the above proposed local polynomial method, each parameter vector θ_k is identified based on local data set $\{U_{k+r}, Y_{k+r}, r = -L, \cdots 0 \cdots L\}$. But when $r \leq L$, estimations θ_k and θ_{k+r} are solved by solving two separable least squares problems. Therefore, partial data overlap exists between the data set used for the two separable least squares problems. This overlap effect means that there is a correlation between these estimations because of correlation among the data.

Proposition 8.2: If $|r| \leq L$, parameters θ_k and θ_{k+r} are not independent, because they are all related through the polynomial constraints (8.110). For the remained terms, some relations exist in θ_k and θ_{k+r}.

$$\begin{cases} G_{k+r} = \theta_{k+r}(1) = \theta_k(1) + \sum_{s=1}^{R} \theta_k(s+1)r^s \\ T_{k+r} = \theta_{k+r}(R+2) = \theta_k(R+2) + \sum_{s=1}^{R} \theta_k(s+R+2)r^s \end{cases}$$

(8.118)

The above equation (8.118) is not considered in the equation (8.116), and the equation (8.118) is used as a constraint to reduce the mean square error of the parameter estimations θ_k, $k = 1 \cdots N$.

Adding a penalty function to combine constrain condition (8.118) and (8.116), then one multi objective least squares criterion is constructed as one constrain optimization problem.

To rewrite equation (8.118) as the matrices, we introduce two matrices for two positive integers R and L.

$$M(R, -L) = \begin{bmatrix} 1 & -L & (-L)^2 & \cdots & (-L)^R \\ \vdots & & \vdots & & \vdots \\ 1 & -2 & (-2)^2 & \cdots & (-2)^R \\ 1 & -1 & (-1)^2 & \cdots & (-1)^R \end{bmatrix}, \quad M(R, L) = \begin{bmatrix} 1 & 1 & 1 & \cdots & 1 \\ 1 & 2 & 2^2 & \vdots & 2^R \\ \vdots & \vdots & \vdots & \cdots & \vdots \\ 1 & L & L^2 & \cdots & L^R \end{bmatrix}$$

(8.119)

From the structures of the above introduced matrices, two matrices $M(R, -L)$ and $M(R, L)$ are all two Vandenmonde matrices. Thus the constrain condition (8.118) can be rewritten into a matrix form.

$$\begin{bmatrix} M(R,L) & 0 \\ 0 & M(R,L) \\ M(R,-L) & 0 \\ 0 & M(R,-L) \end{bmatrix} \theta_k = \begin{bmatrix} \varphi_k \\ \vdots \\ \phi_k \end{bmatrix} \qquad (8.120)$$

Simplifying the above matrix form as

$$\bar{M}\theta_k = E_k, \quad E_k = [\varphi_k, \phi_k]^T$$

$$\varphi_k = [\theta_{k+1}(1)\cdots\theta_{k+L}(1), \theta_{k+1}(R+2)\cdots\theta_{k+L}(R+2)]^T$$

$$\phi_k = [\theta_{k-L}(1)\cdots\theta_{k-1}(1), \theta_{k-L}(R+2)\cdots\theta_{k-1}(R+2)]^T \qquad (8.121)$$

The constraint form (8.120) divides the constraint (8.118) into two subsets. The upper constraint corresponds to the parameter vector θ_k at the high frequency in the entire frequency domain window, and the lower constraint corresponds to the parameter vector θ_k at the low frequency.

Combining two constrain conditions (8.121) and (8.116) to construct one multi objective least squares problem as

$$\min_{\theta_k} \left[\bar{Y}_{k,L} - K_{k,L}(R, \bar{U}_{k,L})\theta_k \right]^T \left[\bar{Y}_{k,L} - K_{k,L}(R, \bar{U}_{k,L})\theta_k \right]$$

$$+ \lambda\phi_u(\Omega_k) \left[\bar{M}\theta_k - E_k \right]^T \left[\bar{M}\theta_k - E_k \right] \qquad (8.122)$$

where $\phi_u(\Omega_k)$ is the power spectral corresponding to the input signal, weighting factor λ is used to correct the relative importance of the constraint matching and the error of the observed data. When weighting factor λ is increased, it will apply stronger smoothness in the estimated value of the frequency response function, and also reduce the variance error and increase the biased error.

Expanding the cost function in equation (8.122) and neglecting the constant term to obtain

$$\min_{\theta_k} \left\{ \theta_k^T \left[K_{k,L}(R, \bar{U}_{k,L})^T K_{k,L}(R, \bar{U}_{k,L}) + \lambda\phi_u(\Omega_k)\bar{M}^T\bar{M} \right] \right.$$

$$\left. \theta_k - 2\theta_k \left[K_{k,L}(R, \bar{U}_{k,L})^T \bar{Y}_{k,L} + \lambda\phi_u(\Omega_k)\bar{M}^T E_k \right] \right\}$$

The above equation is one quadratic program problem.

By differentiating with respect to parameter θ_k and by setting the derivative equal to zero, then we have

$$\left[K_{k,L}^T K_{k,L} + \lambda\phi_u\bar{M}^T\bar{M} \right]\theta_k = \left[K_{k,L}^T \bar{Y}_{k,L} + \lambda\phi_u\bar{M}^T E_k \right]$$

Formulating the above equity to obtain the parameter vector estimations.

$$\hat{\theta}_k = \left[K_{k,L}{}^T K_{k,L} + \lambda \phi_u \overline{M}^T \overline{M} \right]^* \left[K_{k,L}{}^T \overline{Y}_{k,L} + \lambda \phi_u \overline{M}^T E_k \right] \tag{8.123}$$

where pseudo-inverse operation is used here, and the above whole process shows that an optimization problem with constraints is transformed into an unconstrained optimization problem by introducing a weighting factor. Then the optimality necessary condition is used to solve the optimal value.

8.4.4 Simulation example

To verify the feasibility and effectiveness of the local polynomial method proposed in this chapter, we compare this local polynomial method and the method from reference [5]. Firstly, the flutter test flight model of the aircraft in reference [20] is selected as a simulation example to compare the deviation or fitting degree of the frequency response function estimation obtained by the local polynomial method and the method from reference [20]. Secondly, the proposed local polynomial method and the method from reference [20] are all applied to the Box-Jenkins structure, then the frequency response function estimation obtained by the Monte Carlo test is compared.

(1) Firstly, the aircraft flutter model is used here, i.e. in this simulation part, we only use wind tunnel test to construct the flutter mathematical model of the two-dimensional wing. In the whole wind tunnel test for a two-dimensional wing, the input is chosen as an artificially applied excitation signal, and the output is an accelerometer measurement, collected from many sampling data points. The simulation is based on 100 independent experiments and $N = 4096$ data points. Set these $N = 4096$ data points into four groups of data blocks, i.e. each data block includes 1024 sample data.

The true state space equation is described as follows:

$$A = \begin{bmatrix} 0.6541 & 0.6500 & -0.0009 & -0.0042 & 0.0003 \\ 0.6977 & 0.6512 & 0.0004 & 0.0031 & -0.0001 \\ -0.0007 & 0.0050 & -0.0211 & -1.0106 & 0.0131 \\ 1 & 0 & 0 & 0 & 0 \\ 0 & 1 & 0 & 0 & 0 \end{bmatrix}$$

$$B = \begin{bmatrix} -0.0018 & 0.001 & -0.0014 & 0 & 0 \end{bmatrix}^T$$

$$C = \begin{bmatrix} 0.1735 & -0.0301 & -0.2569 & 0.0922 & 0.1920 \end{bmatrix} \tag{8.124}$$

Our proposed local polynomial method and the method from reference [20] are applied to identify the above state space matrices respectively, and the simulation results are shown in Fig. 8.7. In Fig. 8.7, the red curve is the true or ideal frequency response function, the blue curve is the frequency response function, obtained by method from reference [20], and the black curve denotes the frequency response function, obtained by our proposed local polynomial method. From Fig. 8.1, we observe that the three curves are very close, which indicates that

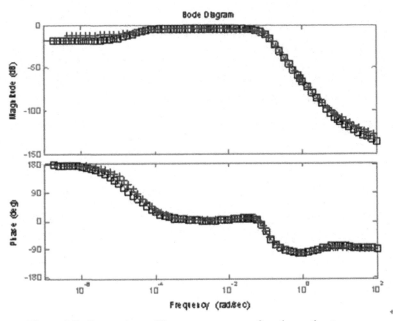

Figure 8.7: Comparison of frequency response function estimates

the local polynomial method from this chapter can achieve the same performance as the method in reference [20], and the local polynomial method can better fit the original actual curve at the low frequency of the amplitude, but there is a small deviation for method from reference [20].

(2) Secondly, consider the following Box-Jenkins model structure:

$$y(t) = 0.1943 \frac{q^2 + 2q + 1}{q^2 + 0.7125q + 0.7449} u(t) + 0.1084 \frac{q^2 + 2q + 1}{q^2 - 0.8773q + 0.3111} e(t)$$

$$(8.125)$$

where $e(t)$ is a white noise with zero mean value, and its standard variance is σ_e. Input signal $u(t)$ is a colored noise, and it can be obtained by passing one standard noise $w(t)$ through one shaped filter.

$$u(t) = 0.5276 \frac{(q+1)^3}{q^3 + 1.76q^2 + 1.1829q + 0.2781} w(t) \qquad (8.126)$$

This Box-Jenkins model is subjected to 200 Monte Carlo tests to obtain 200 clusters of input-output data sets, and the number of data contained in each cluster is 8196. We remove 1024 data from the first column in 8196 data per cluster to eliminate transient effects in the simulation experiments.

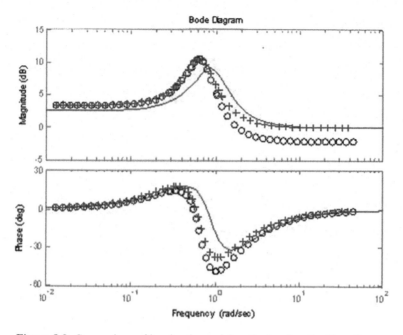

Figure 8.8: Comparison of local polynomial method and method in reference

In the simulation process, the frequency response function estimation curves obtained by the local polynomial method and the method in reference [20] are shown in Fig. 8.8. From Fig. 8.8, it can be seen that when the considered noise is a colored noise, the local polynomial method can better approximate the original true frequency response function with the frequency increase. When considering the existence of data coincidence at adjacent frequency points, the constrained local polynomial method is used to identify the frequency response function. The simulation results for the constrained local polynomial method and the local polynomial method are shown in Fig 8.9, then we observe the constrained local polynomial method can better fit the original real frequency response function curve.

8.4.5 Conclusion

Based on the transient impulse response method for identifying the nonparametric frequency response function, a new local polynomial method is proposed to identify the nonparametric frequency response function and transient response function. To avoid the correlations among the estimations caused by the overlap of the observed data, a further constrained local polynomial method is also considered. For the optimization problem with the constrain condition, the weighting factor is used to obtain the penalty criterion function, then the optimal necessary condition is applied to solve its optimal value. For the in-depth study of the approximation

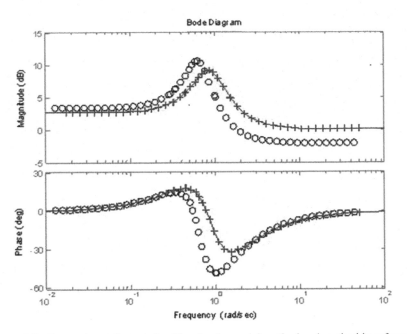

Figure 8.9: Comparison of constrained local polynomial method and method in reference

error in this chapter, the exact analytical formula of the approximation error will be deduced as our next research focus.

8.5   Conclusion

In this chapter, the problem of subspace predictive control is considered under fault conditions, and the statistical distribution is derived for the residual vector. One nearest neighbor gradient algorithm is applied to obtain a predictive controller by solving one optimization problem with equality and inequality constraints. Because the linear parameter varying system is a transition model of nonlinear system and linear system, it has both common properties and can approximately describe the real system. The subspace data driven control is applied to a linear parameter varying system with a scheduling parameter vector. According to the structural characteristics of the linear parameter varying system, the whole derivation process of the subspace data driven control is given, while introducing the vector product operator to simplify the derivation. A parallel distribution algorithm is used to solve the optimal control input value for the established cost function under the idea of data driven control. A linear parameter varying system considering the mass distribution factor is added to the DC motor, and the subspace prediction is applied to the control example of the DC motor.

References

[1] Chiuso, A. 2007. The role of vector autoregressive modeling in predictor based subspace identification, Automatica, 43(6): 1034-1048.

[2] Chiuso, A. 2008. On the relation between CCA and predictor based subspace identification, IEEE Transactions of Automatic Control, 52(10): 1795-1811.

[3] Wang, J. 2011. Application of subspace predictive control in active noise and vibration control, Journal of Vibration and Shock, 30(10): 129-135.

[4] Wang, J. 2010. Application of ellipsoid optimization in subspace predictive control, Journal of Applied Science, 28(4): 424-429.

[5] Wang, J. 2013. Fast gradient algorithm in subspace predictive control under fault estimation, Journal of Shanghai Jiaotong University, 47(7): 1015-1021.

[6] Ljung, L. 1999. System Identification: Theory for the User: Prentice Hall.

[7] Boyd, S. and Vandenberghe, L. 2008. Convex Optimization. UK: Cambridge University Press.

[8] Zeilinger, M.N. 2011. Real time suboptimal model predictive control using a combination of explicit MPC and online optimization, IEEE Transactions of Automatic Control, 56(7): 1524-1534.

[9] Jakob, K. 2011. A design algorithm using external perturbation to improve iterative feedback tuning convergence, Automatica, 47(2): 2665-2670.

[10] Feller, C. 2013. An improved algorithm for combinatorial multi-parameteric quadratic programming, Automatica, 49(5): 1370-1376.

[11] Casini, M., Garulli, A. and Vicino, A. 2017. A linear programming approach to online set membership parameter estimation for linear regression models, International Journal of Adaptive Control and Signal Processing, 31(3): 360-378.

[12] Zhang, X., Kamgarpour, M. and Georghiou, A. 2017. Robust optimal control with adjustable uncertainty sets, Automatica, 75(1): 249-259.

[13] Tanaskovic, M., Fagiano, L. and Novara, C. 2017. Data driven control of nonlinear systems: An on line direct approach, Automatica, 75(1): 1-10.

[14] Novara, C., Formentin, S. and Savaresi, S.M. 2016. Data driven design of two degrees of freedom nonlinear controllers, Automatica, 72(10): 19-27.

[15] Blackman, M. and Sharma, R. 2017. Model based control of utility scale wind turbines, Control and Intelligent Systems, 45(1): 1-9.

[16] Mascod, R.J. 2017. Adaptive VRFT based on MFAC for the speed control of PMDC motor, Control and Intelligent Systems, 45(2): 30-40.

[17] Wang, J. and Wang, D. 2011. Global nonlinear separable least square algorithm for the aircraft flutter model parameter identification, Journal of Vibration and Shock, 30(2): 210-213.

[18] Tang, W., Shi, Z. and Li, H. 2006. Frequency generalization total least square algorithm of aircraft flutter model parameter identification, Control and Decision, 21(7): 726-729.

[19] Wang, J. 2011. Research on control strategies based on advanced identification and their application. Nanjing: College of Automation, Nanjing University of Aeronautics and Astronautics.

[20] Wang, J. and Wang, D. 2011. Bias compensated instrumental variable algorithm for the aircraft flutter model parameter identification, Electronic Optics & Control, 18(12): 70-74.

CHAPTER 9

Conclusions and Outlook

9.1 Introduction

The main focus of this book is the combination of system identification theory and advanced control theory, then one novel control method-data driven control is proposed to solve our present urgent problems about control and identification. All the contents, existing in this book, are our contributions in these recent years, our goal is to pave a way for further research on control theory, i.e. achieve the mission of identification for control. More specifically, the basic theory for our study is system identification, which means our previous studies on system identification are extended to the advanced control theory, then the idea of data driven control appears. The entire content of this book is divided into two kinds, i.e. one is the data driven estimation or identification, the other is the data driven control. In the first case, closed loop system identification is considered from our points, such as stealth identification, performance analysis, model structure validation, etc. Then the obtained results about closed loop identification are extended to the iterative correlation tuning control and minimum variance control strategy. Furthermore, set membership identification is reviewed in case of unknown but bounded noise, then we combine it with model predictive control to get the novel data driven model predictive control strategy. Based on the classical linear matrix inequality proposed in convex optimization theory, stability analysis is considered for cooperative distributed model predictive control.

In this book, our mission is to introduce a direct data driven idea into model predictive control, i.e. to combine them with our proposed new concept-direct data driven model predictive control. Our combination with these two control strategies can not only remedy their shortcomings, but also be extended to other forms, such as set membership uncertainty. It is well known that there are two kinds of uncertainties for describing noise, whether process noise and measurement noise, i.e. probabilistic uncertainty and deterministic uncertainty. Probabilistic uncertainty, whose probability density function is known, is always that zero mean white Gaussian and mutually uncorrelated. In this case of probabilistic uncertainty, all classical control methods and identification algorithms hold, for example, the linear Gaussian regulator, the linear least squares method, and the Kalman filter theory. To date, theory research on probabilistic uncertainty is very mature. But the requirement for probabilistic uncertainty is that the probability density function of the noise must be known in priori, so before moving to the next identification problem or control design problem, time is needed to testify what is the probability density function for the noise. To avoid this unnecessary work on noise, deterministic uncertainty is now used to limit the considered noise within one interval or set, and this second description is more realistic than the probabilistic description in practice. As deterministic uncertainty always requires the considered noise in one interval or set, so deterministic uncertainty is called set membership uncertainty too. In recent years, lots of research within this case of set membership uncertainty have been studied, for example, set membership identification and set membership control of filtering. Based on our above detailed formulations, we propose our direct data driven model predictive control with set membership uncertainty. More specifically, consider one simple linear system with a set membership uncertainty on process noise and measurement noise. The final goal is to apply model predictive control to guarantee the output predictive to track one given desired output value at each time instant, so the most important factor is to obtain or construct the output prediction at each time instant. But when constructing these output predictions, no information on the state is given in case of set membership uncertainty. To get information about the state estimations at all-time instants, classical Kalman filtering is useless, due to this set membership uncertainty, not the probabilistic uncertainty. Therefore the direct data driven idea is to solve this state estimation problem with set membership uncertainty, i.e. the recursive state estimation is given. Based on our recursive state estimation, the output predictive at each time instant is easily obtained only by some simple mathematical operations. After comparing the output predictive with its own desired output value at each time instant, one cost function for model predictive control is yielded, while also considering some constrain conditions on the input signal, such as persistent excitation or limited amplitude. Then the one novel ellipsoid optimization algorithm is proposed to solve this above constrain optimization problem, whose decision variables are a sequence of ellipsoids with decreasing volumes. Then the central point of that final ellipsoid is deemed as the acceptable decision variable, i.e. the optimal input signal. Generally, direct data driven is used to achieve the goal of state estimation in case of set membership uncertainty, then the state estimation is extended to yield the output prediction at each time instant. Combining the output prediction and constrain condition to form one constrain optimization problem for model predictive control, then one novel ellipsoid optimization algorithm is proposed to solve this constrain optimization problem, as the nice property for model prediction control is to consider the constrain condition completely, which cannot deal with other control strategies.

More specifically, the main contributions of this book are generalized as follows.

(1) Recursive state estimation is derived to estimate the system state, which is beneficial for the output prediction. Some analyses based on linear matrix inequality are given to complete the state estimation with set membership uncertainty.
(2) Given a sequence of desired output values, the output predictions based on state estimation are substituted to be one constrain optimization problem, whose constrain conditions are with respect to the input signal. This constrain optimization problem corresponds to the main essence of model predictive control.
(3) To achieve the goal of identification for control, one interval predictor model is used in one min-max optimization problem, where the max operation is a piecewise affine function form. This goal is similar to the robust model predictive control.
(4) Contemplate the considered system with many variables, i.e. multi-inputs and multi-outputs. Further, two error functions are defined based on some prior knowledge, such as measurements or linear relations, etc. We derive two conditions for two control matrices while guaranteeing the considered system satisfies the non-interaction property.
(5) The equivalence between the nonlinear feedback system, block structure nonlinear system and the polynomial nonlinear state space system are established. The cost function of the model error is constructed to solve the unknown parameter vector.
(6) The problem of identifying a piecewise affine system, which combines the linear and nonlinear properties. The separate regions are determined as a multi class classification problem, which is solved by the classical first order algorithm of the convex optimization theory. In the presence of unknown but bounded noise, a zonotope parameter identification algorithm is proposed to identify the unknown parameter vector in each separated region.

9.2 Outlook

The work in this book represents only a few problems in our study path for identification for control. It means all the theories are based on system identification, i.e. the existing research for system identification is extended to advanced control theory. The next research ideas and directions are summarized as follows:

(1) Apply the functional analysis and geometry to system identification and advanced control theory.
(2) Apply game theory and dynamic programming to the distributed model predictive control.
(3) Apply our existing results on system identification and data driven control in other more general engineering problems, such as UAVs formation flights.

Index